RENEWABLE RESOURCES: OBTAINING, PROCESSING AND APPLYING

Renewable Resources: Obtaining, Processing and Applying

Ryszard Kozlowski,
Gennady Zaikov
and
Frank Pudel
Editors

Nova Science Publishers, Inc.
New York

For permission to use material from this book please contact us:
Telephone 631-231-7269; Fax 631-231-8175
Web Site: http://www.novapublishers.com

LIBRARY OF CONGRESS CATALOGING-IN-PUBLICATION DATA

Renewable resources : obtaining, processing, and applying / editors, Ryszard Kozlowski, Gennady Zaikov, Frank Pudel.
p. cm.
Includes index.
ISBN 978-1-60692-534-8 (hardcover)
1. Biomass energy. 2. Plant biotechnology. 3. Renewable natural resources. I. Kozlowski, R. (Ryszard) II. Zaikov, Gennadii Efremovich. III. Pudel, Frank.
TP339.R497 2009
660.6--dc22

2008050309

Published by Nova Science Publishers, Inc. ✈ New York

CONTENTS

PREFACE

Every year in the German city of Magdeburg, ÖHMI Consulting GmbH from Germany and the Institute of Natural Fibres from Poland, arrange the International Conference for Renewable Resources and Plant Biotechnology NAROSSA®.

The event takes place one year in Magdeburg, and next year in Poznan. This book contains selected topics presented in years 2007 and 2008, which were chosen by the conference Scientific Committee.

The specialists from whole world presented their achievements in the filed of renewable resources.

Generally the topics presented in the book cover wide aspects connected with renewable resources, how they can be obtained, assessed, processed and finally applied.

The new crops, very often with the improved genetics, become the key to the successful plant biotechnology, bio-fuels, biocomposites and agro-fine chemicals.

New agriculture resources like Jatropha as well as older ones like flax and hemp, kenaf, jute, sisal were discussed, too.

The special focus is on the diversification of renewable resources for non-food applications, including fibres, pulp and paper, agro-fine chemicals and energy.

Agro-fine chemicals like essential oils and some phyto-hormones from oilseed become valuable additives for nutrition and medical products.

Changing of climate across the world connected with green-house effect also turns scientists' attention to the problem of biotic and abiotic stress of cultivars.

One of key issues is how production of renewable resources for energy influences the increase of the costs of food and feed and how to make the energy production from plant sources cost-efficient.

Biofuels of first generation produced from oil plants like rape (Canola) or African palm contribute to the disturbance of the environmental balance. In scope of the biofuel second generation like bio-ethanol and bio-n-buthanol, we observe significant progress.

By using such processing as plasma or ammonia treatments, we can more easily improve the accession for decomposition of lignocellulosic raw materials and even production of carbon dioxide and hydrogen, which are next processed by microorganisms to bio-ethanol and bio-n-butanol.

In this book, we present to readers valuable new results of research from different part of globe in 23 chapters.

BREEDING FOR NON-FOOD APPLICATIONS AND BIOTECHNOLOGY

In: Renewable Resources: Obtaining, Processing... ISBN 978-1-6060692-534-8
Editors: R. Kozlowski, G. Zaikov and F. Pudel

Chapter 1

THE EFFICIENT *IN VITRO* MICROPROPAGATION OF WHITE MULBERRY (*MORUS ALBA* L.)

A. Luwanska, A. Talarek, G. Mankowska and K. Wielgus
Department of Biotechnology and Molecular Biology,
Institute of Natural Fibres, Poznan, Poland

ABSTRACT

White mulberry (Morus alba L.) is an important plant not only for breeders of silk worms, for which the mulberry leaves are the basic food, but also because of the healing properties of the plant. Due to many applications of this species an efficient method of reproduction and cultivation of white mulberry is the object of wide interest. The aim of this chapter was development of a methodology for micropropagation of 'Zolwinska Wielkolistna' cultivar of white mulberry. Plant regeneration from apical and axillary buds was carried out on MS basal solid and semi liquid media, both supplemented with 2 mg/l BAP. Explants development was also checked on the DT medium containing 2mg/l BAP and 0.1 mg/l NAA. Moreover the influence of the different source of carbon (glucose, fructose, sucrose) on mulberry cultures in vitro was examined. This chapter comprised also explants rooting and acclimatization in ex vitro conditions. The best result was obtained on MS solid medium containing fructose. MS semi liquid media gave similar results in numbers of buds or shoots per explant but contaminations causes elimination of the infected culture. MS medium with sucrose as sources of carbon and the DT medium gave worse results. Rooting on ½ MS medium containing 0.05 mg/l IBA as well as 0.3 mg/l IAA was successful. The explants from apical buds rooted better than from axillary buds. The acclimatization of healthy and vital seedlings in ex vitro conditions was at the level of 86%. The study proved that the in vitro micropropagation method with the use of the suitable medium allowed for highly efficient propagation of white mulberry.

Keywords: White mulberry, *Morus alba, in vitro* culture, micropropagation.

INTRODUCTION

The white mulberry belongs to the *Moraceae* family and for years has been associated with sericulture. Leaves of this plant constitute basic and the best feed source for caterpillars of silk worms (*Bombyx mori* L.) and ensure their proper growth and reproduction [1]. Countries of Central Asia as India or China, where the production of the natural silk is widely developed, are focused on enhancing foliage production in mulberry breeding [2]. However the cultivation of the mulberry results in a lot of different benefits. Mulberry (*Morus alba* L.) leaves, bark and branches have long been used in Chinese medicine to treat fever, protect the liver, improve eyesight, strengthen joints, facilitate discharge of urine and lower blood pressure [3]. Recently mulberry fruits have received much attention as potential sources of functional foods due to several biological and pharmacological effects of anthocyanins and flavonoids.

Mulberry fruits have been reported to have several anti-diabetic, anti-inflammtory and anti-hyperlipidemic activities [4]. In experiments on atherosclerosis in rabbits mulberry extract (fruit macerated with water, MWE) was shown to decrease cholesterol and triglyceride and attenuate the development of atherosclerosis in HCD-fed rabbits. Additionally, MWE had no toxic effects on the liver or renal functions during the experimental period. These data suggests a promising application of MWE in cardioprotection [5].

Recent research has revealed that the extract of mulberry may have neuroprotective effects and inhibit cancer cell migration and invasion [6]. Due to many applications of this species an efficient method of reproduction and cultivation of mulberry is the object of wide interest.

Mulberry is a wind pollinated, dioecious plant therefore constitutes aggregation huge heterozygous plants. This has adverse effects because daughter plants differ in force of growth, size and quality of leaves. Moreover the heterozygous nature of mulberry is a major barrier for genetic improvement by conventional hybridization and using traditional methods [7].

Unsatisfactory results in conventional method of mulberry breeding by the means of nursery cause searching for most modern methods related to plant biotechnology. Micropropagation has become an alternative for traditional methods of vegetative propagation.

In vitro culture after obtaining stability and in optimal conditions is one of the methods which allow for intensive plant propagation. Thus mulberry micropropagation gains more significance in reproduction of this plant. *In vitro* techniques such as tissue and organ culture offer new opening not only for the production of homozygous lines but also for genetic manipulation.

One of the main objectives of mulberry research has been to shorten the time needed for the improvement of existing varieties. In this regard genetic engineering could assist the genetic improvement of *Morus alba* [8].

The aims of mulberry improvement are thus directed towards development of high productivity genotypes with quality leaves, breeding for enhanced yield and quality for production areas [9], breeding for stress tolerant varieties [10].

Application of biotechnological tools for genetic improvement of mulberry gains greater significance, which in turn depends upon the availability of an efficient regeneration system [7]. Therefore aim of our study was to improve the methodology for micropropagation of 'Zolwinska Wielkolistna' cultivar of white mulberry which will allow for development in mulberry breeding in Poland.

Material and Methods

Plant Material

The study was carried out on 'Zolwinska Wielkolistna' cultivar of white mulberry. This genotype was cultivated before the outbreak of World War II, and it is the most adjusted to the climate conditions in Poland.

As preliminary plant material seeds were used. Seeds were sterilized in the 70% ethanol for 1 minute and next were incubated in solution of the sodium hypochlorite (1:2 $NaOCl:H_2O$) for 15 minutes. After sterilization the seeds were rinsed three times in sterile, deionized water.

Sterile seminal material of *Morus alba* was lined on MS medium, which was modified with two plant growth regulators: kinetin (0.04 mg/l) and gibberellic acid (1 mg/l) and cultivated in growth chamber at 26°C in dark. Seedlings served as the source of apical and axillary bud explants for establishing cultures.

Media and Plant Growth Regulator

In vitro cultures were grown on media with applying different variants of factors influencing the cultivating effectiveness. Plant regeneration from apical and axillary buds was carried out on MS basal solid (7g agar) and semi liquid (3.5g agar) media, both supplemented with 2 mg/l BAP.

Explant development was also checked on the DT medium containing *2mg/l BAP and 0.1 mg/l NAA.* Moreover, the influence of the different source of carbon on mulberry cultures *in vitro* was examined. Sucrose and fructose were used for the MS medium and the glucose for the DT medium. pH of each medium was adjusted to 5.8 before twice autoclaving at 117°C for 15 minutes.

Micropropagation

Apical and axillary buds were implanted vertically on the culture media. After 10-th and 20-th days of cultivation number of each explant, percentage of buds and *in vitro* behavior of explants were recorded. Subsequently, apical and axillary buds forming multiple shoots were cut and transferred to the same culture medium. After the transfer observation was repeated in the same way.

Rooting and Acclimatization

For root induction shoots were transferred to two type of half-strength MS medium containing 0.05 mg/l IBA or 0.3 mg/l IAA. After two weeks the total number of rooting explants and *in vitro* behavior were determined. Experiments on the induction of roots were carried out in two repetitions for each medium used at propagations stage. Plantlets with well developed roots were removed from culture vessel and thoroughly washed of medium remains.

If it was necessary the root length was shortened. Plantlets were planted to the sterile garden soil. The observation was carried out after two-week acclimatization.

Results

Initiation of the Sterile Culture

In the initial phase of the experiment aseptic mulberry seeds were lined on MS medium, which was modified with two plant growth regulators: kinetin (0.04 mg/l) and gibberellic acid (1 mg/l).

Such combination of phytohormones stopped the seeds rest and it has ensured high power of germination. The experiment was carried out in three repetitions (table 1).

Table 1. Power of seeds germination

Repetitions	Power of germination [%]	Purity [%]	Contamination [%]
I	84.2	93.33	6.67
II	90.0	93.33	6.67
III	67.1	93.33	6.67
Average	80.4	93.33	6.67

Micropropagation

The seedlings were taken as a source of explants for establishing aseptic *in vitro* culture. Apical and axillary buds were implanted on different variants of media. The best result was obtained on MS solid medium containing *2mg/l BAP and* fructose (Table 2). The possibility of obtaining large amounts of vital plantlets was confirmed through using many passages.

MS semi liquid media gave similar results in numbers of buds or shoots per explant to MS solid medium (Table 3). However in this case pollutants of the culture were the problem.

In experiments on applying the suitable carbon sources the MS medium with sucrose (Table 4) and the DT medium with glucose (Table 5) gave even less satisfactory results.

Table 2. The result obtained on MS solid medium containing *2mg/l BAP and* fructose

Observation No./ Passage	Type of explants	Explants with buds or shoots [%]	No. of buds or shoots per one explant	Growing explants [%]	Vittificated explants [%]	Necrosed explants [%]	Contaminated explants [%]	Callus		
								small	medium	big
1st observation	apical buds	100	1.92	100	0	0	0	0	0	0
	axillary buds	88.06	1.82	88.06	0	3.13	0	1.5	1.83	0
2nd observation	apical buds	96.88	3.27	96.88	0	3.13	15.63	23.44	20.31	10.94
	axillary buds	97.32	4.08	97.2	0	2.68	0.89	24.11	20.54	8.93
Observation after 1st passage	apical buds	91.43	3.23	91.43	0	1.9	36.67	5.71	3.81	0
	axillary buds	53.93	2.06	53.93	0	46.07	43.3	7.68	8.21	4.64
Observation after 2nd passage	apical buds	96.36	4.98	96.36	0	3.64	1.82	9.09	34.55	5.45
	axillary buds	83.93	3.91	83.93	0	16.07	29.46	0	0	0
Observation after 3rd passage	apical buds	94.29	3.92	94.29	0	2.86	22.86	5.71	2.86	0
	axillary buds	90	3.1	90	10	0	16.67	6.67	3.33	0
Observation after 4th passage	apical buds	98.3	6.13	98.3	0.57	1.14	0	13.07	23.3	13.64
	axillary buds	100	5	100	0	0	0	0	0	0
Observation after 5th passage	apical buds	100	3.31	100	0	0	0	0	0	0
	axillary buds	100	6.97	100	0	0	0	0	0	0

Table 3. The result obtained on MS semi liquid medium containing *2mg/l BAP and* fructose

Observation No./ Passage	Type of explants	Explants with buds or shoots [%]	No. of buds or shoots per one explant	Growing explants [%]	Vittificated explants [%]	Necrosed explants [%]	Contaminated explants [%]	Callus		
								small	medium	big
1st observation	apical buds	88.89	1.99	88.89	0	11.11	25	5.83	6.94	0
	axillary buds	0	0	0	0	100	100	0	0	0
2nd observation	apical buds	100	2.83	100	0	0	25	0	0	0
Observation after 1st passage	apical buds	92	4.2	92	0	8	40	4	8	0
	axillary buds	100	4.6	100	0	0	50	0	0	10
Observation after 2nd passage	apical buds	100	4.08	100	0	0	100	0	0	0
Observation after 3rd passage	apical buds	100	5.8	100	0	0	0	60	0	20
	axillary buds	100	6.57	100	0	0	0	56.67	26.67	0

Table 4. The result obtained on MS medium containing sucrose

Observation No./ Passage	Type of explants	Explants with buds or shoots [%]	No. of buds or shoots per one explant	Growing explants [%]	Vittificated explants [%]	Necrosed explants [%]	Contaminated explants [%]	Callus		
								small	medium	big
1st observation	apical buds	100	2.23	100	0	0	0	46.88	0	0
	axillary buds	93.75	1.81	93.75	0	6.25	0	0	33.33	18.75
2nd observation	apical buds	100	2.27	100	0	0	34.17	22.5	45.83	10.83
	axillary buds	89.44	2.02	89.44	0	10.56	4.44		34.03	18.89
1st observation after 1st passage	axillary buds	84.71	2.66	84.71	0	15.29	8.24	50.59	18.82	1.18
2nd observation after 1st passage	axillary buds	82.5	3.84	60	0	40	5	17.5	35	15
Observation after 1st passage	axillary buds	100	2.03	100	0	0	22.86	8.57	2.86	0

Table 5. The result obtained on DT medium containing glucose

Observation No./ Passage	Type of explants	Explants with buds or shoots [%]	No. of buds or shoots per one explant	Growing explants [%]	Vittificated explants [%]	Necrosed explants [%]	Contaminated explants [%]	Callus		
								small	medium	big
1st observation	apical buds	90.63	1.09	90.63	0	9.37	0	0	0	0
	axillary buds	100	1.34	100	0	0	0	1.25	0	0
2nd observation	apical buds	78.13	1.42	78.13	4.38	21.87	0	10.63	5.63	0
	axillary buds	86.25	2.07	86.25	2.5	13.75	0	17.5	7.5	1.25
Observation after passage	apical buds	41.88	1.55	41.88	0	58.12	0	26.88	33.75	13.13
	axillary buds	64.17	2.3	64.17	0	35.83	0	25.83	27.92	22.08

Rooting and Acclimatization

Explants obtained at the propagation stage were later transferred for taking root. Because of the good condition only explants from the solid and semi liquid MS medium containing *2mg/l BAP and* fructose brought good effects in rooting. The explants from apical buds rooted better than from axillary buds. Rooting ability on half-strength MS medium containing 0.05 mg/l IBA is presented in Table 6. Very satisfying result was also obtained for the half-strength MS medium containing 0.3 mg/l IAA (table 7).

After acclimatization in *ex vitro* conditions 86.16 % growing plantlets and 13.84 necrosed plantlets were obtained. Plantlets in average contained 7 leaves and were 5.54 cm high.

Table 6. Rooting ability on half-strength MS medium containing 0.05 mg/l IBA

Medium	Repetitions	Type of explants	Growing explants [%]	Necrosed explants [%]	Contaminated explants [%]	Rooted explants [%]
MS solid medium	I	apical buds	87.27	12.73	44.85	52.12
		axillary buds	82.5	17.5	17.5	18.75
	II	apical buds	99.04	0.96	4.33	65.58
		axillary buds	98.66	1.34	1.04	17.73
	III	apical buds	97.4	2.6	3.39	43.54
		axillary buds	100	0	0	57.5
	Average	apical buds	94.57	5.43	17.52	53.75
		axillary buds	93.72	6.28	6.18	31.37
MS semi liquid medium	I	apical buds	100	0	0	68.75
		axillary buds	100	0	10	51.25
	II	apical buds	100	0	7.69	76.92
		axillary buds	100	0	11.61	41.07
	Average	apical buds	100	0	3.85	72.84
		axillary buds	100	0	10.81	46.16

Table 7. Rooting ability on half-strength MS medium containing 0.3 mg/l IAA

Type of explants	Growing explants [%]	Necrosed explants [%]	Contaminated explants [%]	Rooted explants [%]
apical buds	100	0	0	100
axillary buds	100	0	0	72,57

Conclusion

Despite many advantages of traditional methods of mulberry breeding, techniques aiming at shortening the time of breeding simultaneously with increasing its efficiency and quality of obtained plants are being intensively examined. Furthermore, our study confirmed difficulties in vegetative propagation of white mulberry [unpublished] and indicated the necessity to use more modern propagation methods. Possibility of mulberry propagation (*Morus* sp.) with the application of *in vitro* culture has been reported earlier [11, 12, 13]. The best results were achieved in India, where mulberry is an important plant in agriculture. The key factor for *in vitro* culture was the medium composition and content of plant growth regulators. MS medium supplemented with plant growth regulator 2,4-D was reported as giving the best result [11, 12] However, later on medium containing BAP was proved to be more suitable [7, 14]. In the study the best results in the shoot proliferation were obtained at concentrations of BAP lower than those applied in our research. There are also articles reporting combination of auxin with cytokines to be the best for shoot proliferation [13].

Carbon source is also of great importance. There are reports indicating that the highest number of shoots per one explant was formed when sucrose was used [14]. However, in this study fructose was proven to be the best carbon source.

Chemical factors are closely connected to physical factors. For that reason the parameters such as temperature, humidity and light should be appropriately selected in plant *in vitro* culture.

A big number of publications concerning *in vitro* cultivating conditions show that the mulberry is the object of unfading interest. Our study confirmed possibility of using *in vitro* cultures for the propagation of the mulberry and has proven the described technique for micropropagation plants to be useful for commercial applications.

References

[1] Pieprzyk-Kokocha D. and Burczyk H. (2006). Uprawa morwy jako pokarmu dla jedwabników. *Zagadnienia doradztwa rolniczego, 4*, 105-112.

[2] Ercisli S. and Orhan E. (2007). Chemical composition of white (*Morus alba*), red (*Morus rubra*) and black (*Morus nigra*) mulberry fruits. *Food Chem., 103*, Issue 4, 1380-1384.

[3] Katsube T., Imawaka N., Kawano Y., Yamazakib Y., Shiwakuc K., Yamane Y. (2006). Antioxidant flavonol glycosides in mulberry (*Morus alba* L.) leaves isolated based on LDL antioxidant activity, *Food Chem., 97*, 25-31.

[4] Bae S. and Suh H. (2007). Antioxidant activities of five different mulberry cultivars in Korea. *Lebenson Wiss Technol., 40*, 955-962.

[5] Chen C-C, Liu L-K, Hsu J-D, Huang H-P, Yang M-Y, Wang C-J. (2005). Mulberry extract inhibits the development of atherosclerosis in cholesterol-fed rabbits. *Food Chem., 91*, 601-607.

[6] Pan G. and Lou C. (2008) Isolation of an 1-aminocyclopropane-1-carboxylate oxidase gene from mulberry (*Morus alba* L.) and analysis of the function of this gene in plant development and stresses response. *J. Plant Physiol.*, *165,* 1204-13.

[7] Vijaya Chitra D.S. and Padmaja G., (2005). Shoot regeneration via direct organogenesis from in vitro derived leaves of mulberry using thidiazuron and 6-benzylaminopurine. Scientia Horticulturae 106 (2005) 593-602.

[8] Agarwal S., Kanwar K., Saini NS., Jain RK. (2003). *Agrobacterium tumefaciens* mediated genetic transformation and regeneration of *Morus alba* L. *Scientia Horticulturae, 100*, 183-191.

[9] Madan M. and Sharma S. (1999). Biomass yield of hybrid varieties of mulberry in a non-moriculture area. Biomass and Bioenergy 17 (1999) 427-433.

[10] Reddy A. R., Chaitanya K. V., Jutur P. P. (2004). Differential antioxidative responses to water stress among five mulberry (*Morus alba* L.) cultivars. *Environ. Exp. Bot*, *52* 33–42.

[11] Vijaya Chitra D. S. and Padmaja G. (1999). Clonal propagation of mulberry (*Morus indica* L. cultivar M-5) through *in vitro* culture of nodal explants. *Sci Hortic, 80*, 289-298.

[12] Vijaya Chitra D. S. and Padmaja G. (2002). Seasonal influence on axillary bud sprouting and micropropagation of elite cultivars of mulberry. *Sci. Hortic, 92*, 55-68.

[13] Anis M., Faisal M. and Singh S. K. (2003). Micropropagation of Mulberry (*Morus alba* L.) through *in vitro* culture of shoot tip and nodal explants. *Plant Tissue Cult.*, *13*, 47-51.

[14] Bhau B S. and Wakhlu A K. (2003). Rapid micropropagation of five cultivars of mulberry. *Biol. Plant., 46*, 349- 355.

In: Renewable Resources: Obtaining, Processing... ISBN 978-1-6060692-534-8
Editors: R. Kozlowski, G. Zaikov and F. Pudel

Chapter 2

THE EFFECT OF MEDIUM COMPOSITION AND EXPLANT SOURCE ON PLANT REGENERATION OF *SALVIA OFFICINALIS* L.

A. Luwańska[1], K. Wielgus[1], G. Mankowska[1], M. Szalata[1], M. Szalata[2] and R. Słomski[2]

[1]Department of Biotechnology and Molecular Biology, Institute of Natural Fibres, Poznan, Poland

[2]Department of Biochemistry and Biotechnology, Agricultural University, Poznan, Poland

ABSTRACT

In the study the influence of different medium composition and explants source on plant regeneration was investigated in Bona - a Polish sage cultivar. The material used in research included explants from different sources: apical and axillary buds, cotyledons, young leaves, fragments of stem and roots. Plant regeneration from apical and axillary buds was carried on on four types of mediums. The best effects obtained on MS basal medium supplemented with plant growth regulators (0,3 mg/l BAP). Half-strength MS basal medium (with 0,3 mg/l BAP), MS basal medium containing two fitohormon (BAP, IAA) and BM basic culture medium, containing the macronutrients according to Murashige and Skoog, the micronutrients of Nitsch and Nitsch, supplemented with a plant growth regulator (BAP) did not give satisfactory results. Callus as a phase of plant regeneration was induced from cotyledons, young leaves, fragments of stem and roots. The experiment was carried out on three types of Daria medium (Daria ind +, Daria pro + and Daria root +) containing three kinds of plants growth regulators used in variable concentrations (BAP, NAA, IAA). The effect of medium composition and explants source on plant regeneration was determined by recording the percentage of necrosed and contaminated explants and explants developing shoots.

Keywords: *Salvia officinalis*, callus induction, medium composition, explants source, plant regeneration.

INTRODUCTION

Salvia officinalis L. (sage) is an aromatic herb with woody stems, grayish leaves and purplish flowers. Sage is the most important species within *Lamiaceae*. This species is native of southern Europe and the Mediterranean region. It is cultivated in several countries mainly to obtain dried leaves to be used as raw material in medicine, perfumery and food industry [Santos-Gomes P. C *et al.*, 2002]. Sage is grown especially for distillation of essential oil. Compounds from *Salvia* essential oil have been shown to exhibit high antibacterial activity against *Strephyloccocus aureus*, *Escherichia coli*, *Pseudomonas aeruginosa* and *Bacillus subtilis* [Radulescu V. *et al.*, 2004]. The biological properties of *Salvia* essential oil were attributed mainly to camphor, 1,8-cineole and α-thujone and β-thujone [Radulescu V. *et al.*, 2004]. Because of wide antibacterial activity sage is one of favorite species, which gives the opportunity to produce natural products in health care. *S. officinalis* essential oil is applied in the treatment of wide range of diseases of nervous system, heart and blood circulation, respiratory, digestive, metabolic and endocrine diseases, etc. *S. officinalis* infusion is easily obtained and commonly used for haemostatic, estrogenic, anti-perspiration, anti-neuralgic, antiseptic, hypoglycemic and many other therapeutic applications [Radulescu V. *et al.*, 2004]. Modern day research has shown that sage essential oil can improve the memory and has shown promising prospects for the treatment of Alzheimer's disease [Durling N. E *et al.*, 2006].

S. officinalis is the focus of research because this species has antioxidant properties. Many components have been isolated from *S. officinalis* such as diterpenoids, triterpenoids, flavonoids, phenolic glycosides and phenolic acid derivatives [Miura K. *et al.*, 2001]. Some of the phenolic compounds have revealed excellent scavenging activity free oxygen radicals such as superoxide anion radicals, hydroxyl radicals and singlet oxygen, inhibiting lipid peroxidation and consequently, the extracts have been widely used to stabilize fat and fat-containing foods [Santos-Gomes P. C *et al.*, 2002].

Due to the essential oil and antioxidant compounds *S. officinalis* becomes medically and economically interesting plant to study.

The aim of our study was to develop an effective method of micropropagation of *S. officinalis*. In the chapter the influence of different medium composition and explant sources on plant regeneration was investigated in Bona - a Polish sage cultivar. Initially, the technique of propagation by shoot tips was used, however the research also included explants from different parts of the plant. The experiments comprised determination of the effect of media composition and selection of the best plant growth regulator. Efficient micropropagation of *S. officinalis* in culture *in vitro* allows to obtain high numbers of regenerated plants, which can be used for extracting essential oil.

MATERIAL AND METHODS

Plant Material

Seeds of *S. officinalis* were the preliminary plant material. The sterilized seeds were germinated on aseptic medium.

For inoculating seeds MS (Murashige Skoog) medium supplemented with 1 mg/l giberellic acid (GA) and 0,04 mg/l kinetin was used. The seed cultures were kept in growth chamber at 26°C in dark. After germination seedlings were cultivated with a continuous photoperiod light. Different parts of three weeks old seedlings were used as a sources of explants. There were apical and axillary buds, cotyledons, young leaves, fragments of stem and roots used.

Sterilization

Seed sterilization in commercial sodium hypochlorite solution (ACE in proportion 1:4 with sterilized distilled water) for 15 min was done to obtain aseptic plant material After sterilization it was necessary to rinse the seeds thoroughly (three times) with sterile water each for 15 min. The effect of different method of explant sterilization on result on micropropagation efficiency was determine. Axillary and apical buds obtained from adult plants grown in soil were sterilized (dipped in 70% ethanol solution for 5 min.) before using 15% commercial sodium hypochlorite solution for 20 min. Plant explants were rinsed with sterile water.

Media and Plant Growth Regulators

The culture media used for the present study were: MS medium, half-strength MS basal medium, BM culture medium and Daria medium. Each type of medium was supplemented with different combinations of growth regulator and supplementary substances. pH of each medium was adjusted to 5.8 before twice autoclaving at 121°C for 15 min.

Plant regeneration from apical and axillary buds was carried out on four types of media: (1) MS medium supplemented with 30g/l sucrose, 10g agar (Merck), 100 mgl^{-1} myo-inositol and vitamins B_6:B_3:B_1 (0.5:0.5:1). This medium was further supplemented with cytokines - 0.3 mg/l BAP (6-Benzylaminopurine). (2) Half-strength MS basal medium supplemented with 20 g/l sucrose and the same concentration of vitamins as in MS medium. (3) MS medium supplemented with 20g sucrose, containing 0.5 mg/l BAP and 0.1 mg/l IAA (indole-3-butyric acid) and vitamins B_6:B_3:B_1 (0.5:0.5:0.1). (4) BM basic culture medium, containing the macronutrients according to MS, the micronutrients of Nitsch and Nitsch, supplemented with 1mg/l BAP.

On experiment which was carried out on Daria medium cotyledons, young leaves, fragments of stem and roots were used as explants. Three types of Daria medium were used: Daria ind+, Daria pro+ and Daria root+. The aim of cultivating on Daria ind+ (containing 1mg/l kinetin and 0.05mg/l NAA) was induction of callus cellular creation, which grew intensively after transferring to Daria pro+ medium (supplemented with 0.2 mg/l BAP and 0.03 mg/l NAA). In the last phase of the experiment the explants were transferred to Daria root+ (containing 2 mg/l IAA) to grow roots. pH of each medium was adjusted to 5.8 before twice autoclaving at 121°C for 15 min.

Micropropagation

Apical and axillary buds were implanted vertically on the culture media. Two weeks later shoots number on each explant, percentage of buds and *in vitro* behavior of explants were recorded.

Apical and axillary buds forming multiple shoots were transferred to the same culture medium. After transfer second phase of observation was conducted.

Rooting of Shoots

For root induction shoots were transferred to half-strength MS medium with auxin - IAA (0.3 mg/l). After 7, 10 and 20 days the total number of rooting explants was counted.

Results

Multiple shoot induction from apical and axillary buds was carried out on four types of medium. The best effects were obtained on MS basal medium supplemented with plant growth regulator - 0.3 mg/l BAP. Under these conditions 81% - 100% of explants formed buds or shoots (table 1). It is important to note that almost four shoots per one explant were obtained.

Furthermore *in vitro* behavior of shoots on this medium was very satisfying because 81% - 100% explants were growing. After each observation the explants were transferred on the same medium.

Table 1. *In Vitro* behavior of *S. officinalis* shoots on MS medium containing 0.3 mg/l BAP

	Explants with buds or shoots (%)	No. of buds or shoots per explant	Growing explants (%)	Necrosed explants (%)	Vittificated explants (%)	Contaminated explants (%)
1st obs.	100	3.08	100	0	0	0
2nd obs.	100	3.82	100	0	0	0
3rd obs.	81.04	2.41	81.04	4.9	0	12.5
4th obs.	79.48	3.07	79.48	5.68	2.34	12.5
5th obs.	88.12	2.68	88.12	1.87	0	10
6th obs.	94.41	2.78	86.84	0.85	4.77	5.59
7th obs.	94.2	3.11	77.89	5.8	15.93	5.88
8th obs.	100	3.44	100	0	0	6.25
9th obs.	93.75	3.52	93.75	4.69	1.56	0

The results obtained on this medium were satisfactory in comparison with the other types of medium.

MS medium containing 0.5 mg/l BAP and 0.1 mg/l IAA caused appearing significantly more buds per explant after 4 weeks of culture (Table 2). However an increased number of vittificated and necrosed explants were noticed. It might have resulted from higher content of cytokinin.

First explants transfer on fresh medium was done after second observation. Following transfers were done always after observations.

Table 2. *In Vitro* behavior of *S. officinalis* shoots on MS medium containing 0.5 mg/l BAP and 0.1 mg/l IAA

Obs.	Explants with buds or shoots (%)	No. of buds or shoots per explant	Growing explants (%)	Necrosed explants (%)	Vittificated explants (%)	Contaminated explants (%)
1st	97.22	2.22	100	0	0	0
2nd	94.84	5.29	41.37	4.17	58.63	0
3rd	95.28	1.89	19.61	5.17	75.23	0
4th	97.5	2.46	97.5	0	2.5	0
5th	88.46	2.98	88.46	1.59	1.25	10
6th	98.61	2.28	98.61	0.69	0	0.69
7th	100	1.87	92.19	0	7.81	8.33
8th	95.45	2.51	76.7	4.55	18.75	0
9th	100	2.48	100	0	0	14.06
10th	94.44	2.83	87.5	4.86	7.67	0
11th	100	2.58	100	0	33.75	23.75

Half-strength MS medium containing 0.3 mg/l BAP allowed to obtain high efficiency of multiple shoots (Table 3) what demonstrates optimal level of BAP as growth substance. However, this kind of medium did not give very satisfying results. *In vitro* shoots behavior showed different symptoms of stress with high percentage of vitrificated and necrosed explants.

Explants grown on BM medium with high concentration of BAP (1 mg/l) showed comparatively poor multiple shoot formation (Table 4).

The growing ability of explants depended considerably on culture purity. The method of sterilization is of fundamental significance. The method of seed sterilization used in this chapter brought satisfying results.

There were no contaminated explants of *in vitro* germinated seedlings.

Table 3. ***In Vitro*** **behavior of** ***S. officinalis*** **shoots on ½ MS medium after one week (1st observation) and two weeks (2nd observation) of cultivaton**

Obs.	Explants with buds or shoots (%)	No. of buds or shoots per explant	Growing explants (%)	Vittificated and necrosed explants (%)	Contaminated explants (%)
1st obs.	100	3.41	89.86	10.14	0
2nd obs.	100	1.94	16.67	83.33	0

Table 4. ***In Vitro*** **behavior of** ***S. officinalis*** **shoots on BM medium after two weeks (1st observation) and four weeks of cultivation (2nd observation)**

Obs.	Explants with buds or shoots (%)	No. of buds/shoots per explant	Length of shoots [mm]	Growing explants (%)	Necrosed explants (%)	Vittificate d explants (%)	Contaminat ed explants (%)
1st	93.56	2.3	54	45.10	3.85	51. 05	0
2nd	81.71	1.47	36	5.71	14.29	80	0

In case of sterilization of explants obtained from adult plants grown in soil the result was not as good (Table 5). After 4 weeks of cultivation 28% of explants were contaminated. This combined with older explants had negative influence on micropropagation efficiency.

Table 5. ***In Vitro*** **behavior of** ***S. officinalis*** **shoots collected from plant grown on soil on BM medium after two weeks (1st observation) and four weeks of cultivation (2nd observation)**

Obs.	Explants with buds or shoots (%)	No. of buds/shoots per explant	Length of shoots [mm]	Growing explants (%)	Necrosed explants (%)	Vittificated explants (%)	Contaminated explants (%)
1st	0	0	85	80	20	0	16
2nd	33.33	1.33	75	5.56	66.66	27.78	28

The percentage of rooting shoots on half-strength MS medium with auxin - IAA (0,3 mg/l) after 7 and 14 days was rising. On 14th day the total number of rooting shoot was 87.5% (Table 6).

The experiment on Daria medium was an attempt to multiply *S. officinalis* L. in *in vitro* environment and plant regeneration through organogenesis of callus cell, using cotyledons, leaves, fragments of stems and roots.

Table 6. Rooting of shoots on half-strength MS medium with IAA (0,3 mg/l)

	After 7 days	After 14 days
Rooting of shoots (%)	75	87.5

Callus was present on all types of explants besides leaves. Fragments of stems showed the highest ability for callus induction. However callus appeared to be unable to morphogenesis and was characterized with low potential of plant regeneration. During the later phase of growth explants gradually changed the color to brown and wilted (Table 7). Rooting of arrival-shoots was also not satisfying what in consequence resulted in low efficiency of plant regeneration (Table 8).

Table 7. Callus induction and proliferation from different part of *S. officinalis* plants on two types of Daria medium containing different concentrations of plant growth regulators

		Daria ind +		Daria pro +	
		1 obs.	2 obs.	1 obs.	2 obs.
Cotyledon	green	58.33	3.33	0	The end of experiment
	brown - green	36.67	53.33	1.67	
	brown	5	43.33	98.33	
	contamination	0	0	0	
	green callus	8.33	16.67	16.67	
	arrival-shoots	0	0	0	
	arrival-roots	0	0	0	
	regenerated plants	0	0	0	
Leaf	green	85.56	25.56	0	The end of experiment
	brown - green	12.22	20.11	16.67	
	brown	2.22	33.33	83.33	
	contamination	0	0	0	
	green callus	0	0	0	
	arrival-shoots	0	0	0	
	arrival-roots	0	0	0	
	regenerated plants	0	0	0	
Stem	green	86.49	49.04	12.02	0
	brown	13.51	50.96	87.98	100
	contamination	0	0	0	0
	creation of callus	60.70	80.88	100	100
	yellow callus	32.10	35.35	76.32	100
	green callus	28.60	45.53	23.68	0
	arrival-shoots	10.35	12.11	34.65	62.63
	arrival-roots	5	5	5	5
	regenerated plants	0	0	0	0
Root	white	100	14.44	0	The end of experiment
	brown	0	85.56	100	
	contamination	0	0	0	
	creation of callus	43.89	67.78	75.56	
	white callus	0	8.89	0	
	yellow callus	43.89	58.89	75.56	
	arrival-shoots	0	0	0	
	arrival-roots	22.22	37.78	37.78	
	regenerated plants	0	0	0	

Table 8. Rooting of arrival-shoots of *S. officinalis* on Daria root + medium

	1 obs.	2 obs.
Average No. of arrival-shoots from steam	19.33	19.33
green	54.17	54.17
brown	45.83	45.83
contamination	0	0
creation of callus	100	100
yellow callus	75	75
green callus	25	25
arrival-roots	5.07	5.07
regenerated plants	0.72	0.72

CONCLUSION

There are only a few reports on the *in vitro* propagation of genus *Salvia*. In the chapter the influence of different medium composition and explant sources on plant regeneration of *S. officinalis* L. has been reported. The best effects were obtained on MS basal medium supplemented with 0.3 mg/l BAP. Explants developed multiple shoots with 81%-100% efficiency, and after two weeks cultivated shoots formation was almost four shoots per explant. It must be added that explants were growing and in good condition. This result is the best in comparison with those reported in the available literature. The stimulating effect of BAP on multiple shoots formation has been reported earlier e.g. in a study of the multiplication rate of 8 sage species in the presence of 0.3 mg/l BAP [Marcarello C. *et al.*, 2006]. In our study the use of higher concentration of BAP (0.5 mg/l) combined with 0.1 mg/l IAA has resulted in initial multiplication rate by 2.7 shoots per explants (in average). However high percentage of vitrificated and necrosed explants was obtained. On the same type of medium production of up to 3 shoots per explant was previously described [Grzegorczyk I. and Wysokińska H., 2004]. Similar results were obtained in research with media supplemented with 1.5 mg/l BAP and 0.05 mg/l 2,4-D (dichlorophenoxyacetic acid) [Santos-Gomes P. C *et al.*, 2002]. Application of other growth regulators for multiple shoot formation as kinetin [Grzegorczyk I. and Wysokińska H., 2004; Santos-Gomes P. C *et al.*, 2002], TDZ (tidiazuron) and zeatin [Grzegorczyk I. and Wysokińska H., 2004] or GA and IAA [Avato P *et al.,* 2005] did not give desirable results.

In current research MS medium proved to be superior to ½ MS and BM medium. Low percentage of growing explants (45 – 47%) was described on BM medium containing 1 and 0.5 mg/l BAP [Avato P *et al.*, 2005].

Since MS medium is considered as a source of high salt content, superiority of this medium may reflect the requirement of a relatively high concentration of salts for growth of genus *Salvia*. The research of *S. fruticosa* and *S. miltiorrhiza* indicated that low-ion strength MS medium resulted in poor growth compared to full-strength medium [Arikat N. A *et al.*, 2004]. In our study cultivation on ½ MS medium resulted in low number of shoots per explant after two weeks because of necrosis.

Observed percentage of rooting shoots on half-strength MS medium with auxin - IAA (0,3 mg/l) was high - 87.5%.

The experiment on Daria medium did not bring the satisfying results in plant regeneration and calls for improvement of the method.

The results described in this chapter suggest that *in vitro* protocols for micropropagation of *S. officinalis* L. by apical and axillary buds is highly efficient and gives fast growth. Therefore *in vitro* sage multishoots propagation could be used in an improved program including *in vitro* selection and genetic manipulation, as well as it may be a good alternative for production of essential oil.

REFERENCES

[1] Arikat, N. A; Jawad, F. M; Karama, N. S; Shibli, R. A. (2003). Micropropagation and accumulation of essential oils in wild sage (*Salvia fruticosa* Mill.). *Sci. Hortic, 100,* 193-202.

[2] Avato, P; Fortunato, I. M; Ruta, C; D'Elia R. (2005). Glandular hairs and essential oils in micropropagated plants of *Salvia officinalis* L. *Plant Sci, 169,* 29-36.

[3] Durling, N. E; Catchpole, O. J; Grey, J. B; Webby, R. F; Mitchell, K. A; Yeap Foo, L; Perry, N. B. (2007). Extraction of phenolics and essential oil from dried sage (*Salvia officinalis*) using ethanol-water mixtures. *Food Chem., 101,* 1417- 1424.

[4] Grzegorczyk , I. and, Wysokińska H. (2004): Micropropagation of Salvia officinalis L. by shoot tips. *Biotechnologia, 2*, 212-218.

[5] Mascarello C.; Mantovani E.; Ruffoni B. (2006). In vitro culture of several ornamental and medical Salvia species.*Acta. Hort. (ISHS), 723,* 375-380.

[6] Miura, K; Kikuzaki, H; Nakatani, N. (2001) Apianane terpenoids from *Salvia officinalis*. *Phytochemistry*, *58,* 1171- 1175.

[7] Radulescu ,V; Silvia Chiliment, S; Oprea, E. (2004). Capillary gas chromatography–mass spectrometry of volatile and semi-volatile compounds of *Salvia officinalis*. *J. Chromatogr A, 1027*, 121-126.

[8] Santos-Gomes, P. C; Seabra, R. M; Andrade, P. B; Fernandes-Ferreira, M; (2002). Phenolic antioxidant compounds produced by in vitro shoots of sage (*Salvia officinalis* L.). *Plant Sci., 162*, 981- 987.

In: Renewable Resources: Obtaining, Processing... ISBN 978-1-6060692-534-8
Editors: R. Kozlowski, G. Zaikov and F. Pudel

Chapter 3

METHODS TO INCREASE FIBER FLAX RESISTANCE TO BIOTIC AND ABIOTIC STRESS

K. Wielgusz*[*]*, K. Heller, A. Andruszewska and M. Byczyńska

Institute of Natural Fibers, ul. Wojska Polskiego 71 B,
60-630 Poznań, Poland

ABSTRACT

Presented are the results of research carried out at the Institute of Natural Fibres in Poznan on protection of flax against agro phages and abiotic stress of the environment.

Fiber flax protection increasing plants resistance to biotic stress concerns diseases management, mainly fusarium wilt caused by Fusarium fungi, septoriosis (*Septoria lini*) and anthracnose (*Colletotrichum lini*). It also covers methods of weed and pest control (*flea beetle Longitarsus parvulus* Payk., thrips e.g. Trips lini Uz.).

The main goal of research on reduction of losses caused by abiotic stress is to increase resistance of flax to water deficiency in the environment and problem of heavy metal fixation by this plant.

In basic research the genotypes of *Linum* are evaluated for the resistance to stress of drought and diseases as a baseline for selection of parental components in flax breeding. In breeding for resistance, besides the priority of developing cultivars resistant to diseases, also developing genotypes resistant to water deficiency in environment becomes more and more important.

Application studies focus on evaluation of plant protection chemicals for reduction of losses caused by diseases, pests and weeds. Recently, also the effect of drought has been taken into account. The main goal of this research is to develop methods of pesticides application in flax that are environmental friendly.

The core of this research is the study on ontogenesis (plant development) of flax. The condition for developing integrated, methods of plant protection compatible with sustainable agriculture is adjusting agronomic treatments (including plant protection ones) to morphogenesis and functiogenesis of a crop.

---

[*] E-mail: langner@inf.poznan.pl

Keywords: fiber flax, growing technologies, ontogenesis, fiber yield, fiber quality.

Introduction

Reserach on flax protection against abiotic and biotic elements of the environment are carried out at the Institute of Natural Fibres in Poznan for more than 40 years [4]. The initial period of this research (1960-1990) was predominated by protecting flax against weeds, diseases, and pests. Than, the scope of work was broaden by problems connected with protection of plantation against unfavorable abiotic factors in the environment (heavy metals, drought). In last two decades (1990-2007) one of the objectives of research is to develop pro-ecological technologies of flax protection against agro phages and abiotic factors.

The studies discussed here combine approach of the basic and applied research. The basic research concern the agro phages biology, resistance of flax to abiotic and biotic stress and the effect of these stress factors on fenological development and yielding ability of fibrous flax.

The goal of the research conducted at the Institute of Natural Fibres in Poznan is to develop methods that allow for increasing the resistance to the above mentioned stresses.

Basic research focus on evaluation of *Linum* genotypes collected at INF Gene Bank for resistance to drought and disease stress. It is a baseline for selection of parental form in flax breeding.

Application research focus on testing the selected biological substances and microelement-containing fertilizers that enhance resistance of plants to diseases and plant protection chemicals reducing losses caused by pests.

Materials and Methods

The research on protecting flax against weeds conducted at inf cover evaluation of weed communities, degree of weed infestation in crops, weed onthogenesis and development of herbicides and adjuvants application methods. The research aiming at increasing the resistance of flax to drought cover determination of flax water requirements, testing the resistance to drought of particular flax cultivars and developing the antiperspirants application methods improving the plant water management.

Studies on increasing the flax resistance to *Fusarium* wilt included testing of selected biological measures and fertilizers containing the optimal for flax composition of microelements that were used as seed treatments or as foliage treatment. Besides that the cultivars of flax and breeding material from the INF *Linum* collection have form many years been selected for resistance to *Fusarium* wilt.

The baseline for the study were the field trials set in a complete randomized block system with four replications. The experiments were carried out in 3 Experimental Farms of INF : EF Białobrzezie and EF Pętkowo, EF Sielec Stary.

RESULTS

Flax Protection Against Weeds

Basic Research

Filed Weeds Communities in Flax in Poland

The results of 40-year observations (means from 315 experiments) showed that the most frequent weeds in flax are: *Chenopodium album* L., *Viola arvensis* Murr., *Thlaspi arvense* L., *Stellaria media* Vill., *Lamium amplexicaule* L., *Fallopia convolvulus* (L.) Á. Löve., *Capsella bursa-pastoris* (L.) Med. and *Poa annua* L. [4].

The second, in terms of occurrence frequency weeds are: *Anthemis arvensis* L., *Polygonum lapathifolium* L., *Galium aparine* L., *Sinapis arvensis* L. and *Polygonum aviculare* L.

The average frequency of weed occurrence in evaluated period was 139,2 weeds per 1 m^{-2} of which 125,2 weeds m^{-2} were dicotyledonous ones, while 14,0 weeds m^{-2} – grasses.

Observations carried out within past 40 years showed no occurrence of so called "flax specialists" such as: *Lolium remotum* Sark., *Spergula arvensis* L. spp. *maxima*, *Camelina alysum* (Mill.) Tell. and *Cuscuta epilinum* (Weihe).

Changes in Weed Occurrence in Flax During Vegetation Period

Studies of changes in weed occurrence in flax during vegetation period showed that the occurrence of monocotyledonous weeds (weeds m^{-2}) in spring and during harvest was constant while average occurrence of dicotyledonous weeds was by 19.9% lower at harvest day than two weeks after flax germination. Most of dicotyledonous weeds species occurred with lesser frequency at harvest day than two weeks after flax germination. The highest average percentage of weeds fade-outs was observed for *Thlaspi arvense* L. .(50,7 %), *Viola arvensis* Murr. (45,2 %), *Anthemis arvensis* L. (44,8 %) and *Sinapis arvensis* L. (28,2 %). Grasses (*Elymus repens* (L.) Gould and *Echinochloa crus-galli* (L.) P.B.) and some dicotyledonous species (*Capsella bursa-pastoris* (L.) Med.) were more frequent at harvest day than in spring.

Flax and Weeds Onthogenesis

Onthogenesis o flax and weeds is determined by environmental conditions, mainly by the weather and also by the type of soil. To higher extent this is true for weeds which react stronger to persisting soil and climatic conditions than flax. Even in unfavorable moisture conditions (drought) many of weed species mature before flax harvest, these are especially weeds with short vegetation period: *Capsella bursa-pastoris* (L.) Med., *Stellaria media* Vill., *Taraxacum officinale* Web. and *Poa annua* L. However, this is also true for *Chenopodium album* L., *Euphorbia helioscopia* L. and *Elymus repens* (L.) Gould. Among species that failed to bloom one should mention *Fallopia convolvulus* (L.) Á. Löve., *Polygonum aviculare* L. *and Echinochloa crus-galli* (L.) P.B.

Application Research

Herbicides Application in Fibrous Flax

The chemical control of flax cultivation is conducted according to the following principles:

- physical selectivity of herbicides (a physical separation of seeds from the herbicide, e.g. *linuron, lenacil*) [1],
- morphological resistance of flax to herbicides (e.g. *MCPA*, *bentazone*),
- physiological (biochemical) resistance of flax (*chlorsulfuron, thifensulfuron-methyl*) [3].

The herbicides recommended for flax belong to the class IV of toxicity to humans and bees i.e. they are little harmful for humans and practically harmless for bees. Unfortunately, they are usually harmful for algae (toxicity class I or II) and little less harmful for fish (toxicity class II-IV).

The herbicides used after flax emergence (a.i. *MCPA*, *bentazone*, *chlopyralid*, *thifensulfuron*, *sethoxydim*, etc.) are applied due to morphological resistance of protected plant to herbicides. The optimal time for applications of these substances is a "herring bone" stage of flax, i.e. when flax is 6-12 cm tall. In this stage of growth leaves of flax are set at sharp angle to the stem and covered with a thick layer of wax. This prevents the spray to stick on the crop.

Quite important in flax cultivation are the sulfonylurea herbicides mainly *chlorsulfuron* and *thifensulfuron-methyl*. These are applied mainly due to physiological and metabolic resistance of the crop because flax can detoxify these herbicides.

This type of selectivity offers application of herbicides to a wide range of time limits, when plants are 5-20 cm tall. This is important since the time of application of a herbicide may be earlier, when weeds are more susceptible to herbicides and this enables the reduction of herbicide doses (Table 1).

Table 1. The effect of time application on herbicide efficacy (ES Bukówka, ES Sielec Stary 1997-1999 average from 6 field trials)

No.	Treatment [g a.i. ha^{-1}]	Time of herbicide application – height of flax plants [cm]	Broad –leaved*/ biomass control [%]	Flax susceptibility to herbicides**/
1	*chlorsulfuron* [7,50]	5	86	1
2	*chlorsulfuron* [11,25]	12	88	1
3	*thifensulfuron methyl* [20,46] + *chlorsulfuron* [2,40]	5	83	1
4	*thifensulfuron methyl* [35,44] + *chlorsulfuron* [2,72]	12	84	1

*/ broad-leaved weed species dominating in the field trials: *Chenopodium album* L., *Falllopia convolvulus* (L.) Love, *Viola arvensis* Murr., *Lamium amplexicaule* L. *Thlaspi arvense* L.,

**/ according to 1-9 scale, where: 1 – no injury, 9 – total damage.

The research at the Institute of Natural Fibres is based on environmentally-friendly technologies of herbicides application in fiber flax on the following principles:

- the wide knowledge on weed communities in fiber flax in Poland (data from period 1967-2007) – relational database – an application of MS Access 97 [6].
- adjuvant application,
- decrease of herbicides doses,
- herbicides mixtures,
- optimal time of application – low herbicide doses,
- rotation of herbicides,
- application of herbicides according to the map of weed distribution in the field (precise treatments).

One of the factors enabling the reduction of doses of herbicides is application of adjuvants, which increase the biological activity of herbicides. Reduction of herbicide dose is beneficial for the environment and for the quality of flax fiber.

Biological Evaluation of Adjuvants in Flax

The research conducted at the Institute of Natural Fibres since 1981, tested the possibility of adjuvants application to enhance the effect of herbicides in fiber flax. The following adjuvants have been tested: linseed oil, ethoxylated fat amin, paraffin oil, post-refinement fatty acids, rapeseed oil, ethoxylated isodecyl alcohol.

Among advantages resulting from combined use of adjuvants and herbicides one should mention.:

- reduction of active ingredient (a.i.) of herbicides,
- improvement of herbicide solubility,
- reduction of herbicide a.i. volatilization,
- reduction of spraying mixture drift,
- enhancement of spray homogeneity,
- better penetration of herbicide a.i.,
- improvement of herbicide selectivity,
- broadening the spectrum of herbicides,
- control of spraying mixture pH,
- mitigation of negative effect of weather,
- reduction of residual effect,
- lower amounts of herbicide residues.

Among mentioned factors showing the advantages of using adjuvants the most important is the reduction of herbicide dose without compromising its efficacy. The advantage should be evaluated in both economical and ecological aspect. The adjuvant used "instead" of herbicide is cheaper and less toxic to the environment.

Lenmix 800 EC is a new adjuvant formulated at the Institute of Natural Fibers. The Lenmix 800 EC active ingredient is: 80 % raw linseed oil, 10 % non-ionic surfactant, 6 %

cation surfactant – A, 4 % cationic surfactant – B. The aim of the study was to evaluate the possibility of application of a new, linseed oil-based, adjuvant to increase the herbicide activity in fiber flax.

The adjuvant Lenmix 800 EC (1,0 l ha^{-2}) improved the performance of herbicides. Chlorsulfuron applied at lower doses (9 g ha^{-1}) in mixture with Lenmix 800 EC controlled efficiently the majority of dicotyledonous weeds infesting flax [5]. The graminicides haloxiphop-R – (104 g ha^{-1}) and chisalophop-P-ethyl (75g ha^{-1}) applied in mixture with Lenmix 800 EC showed good effect on grass weeds *Elymus repens* and *Echinochloa cruss-galli.*

Increasing Flax Resistance to Drought

Basic Research

Evaluation of Fiber Flax Water Requirements

The results of pot trials carried out in EF Petkowo showed that the most important period in fiber flax development is the period from phase of "fast growth" till beginning of bloom (phase 32 – 61 in BBCH scale). The studies also shown a high water requirements of fiber flax and strong effect of water conditions in the environment on the yield of flax straw, seed and fiber as well as on the quality of obtained raw material (fiber Ns). Despite the average transpiration area of flax leaf and stem is small (3800 mm^2), the number of guard cell on the leaf, however, is high comparing to other plants. It ranges from 3500 to 6350 on cotyledons, 2400-4000 on the leaves, and 12800-21800 on the stem. On average, the consumption of water in the middle of vegetation period is 5 g per plant daily. The flax requirement for water in the whole vegetation period (90-100 days) is 8 million l ha^{-1}. For comparison this value for oats is only 2 million l ha^{-1} and for rye – 800 000 l ha^{-1} (1/10 of that of flax).

Resistance of Fiber Flax Cultivars to the Stress of Drought

The important direction of the research in the world in terms of increasing the resistance of crops to drought stress is breeding work aiming at obtaining cultivars resistant to water deficiency in the environment. It is necessary to develop the methods for evaluation of cultivated crops resistance to drought stress and ability of giving yields in unfavorable environmental conditions.

The research on evaluation of resistance to drought in flax genotypes collected at INF has been carried out since 2002. A complex evaluation of moisture requirements for fibrous flax cultivars allows for selection of xerophil and hydrofil varieties. Among 25 tested genotypes of Linum, the highest resistance to water deficiency in soil was found for fiber cultivars Nike, Luna, Artemida, Diane, Caesar Augustus and oil flax cultivar Oliwin. A correlation was found between the growing season length and resistance to water deficiency in the soil: late maturing cultivars were usually characterized by higher resistance to drought.

Application Research

Research conducted at Institute of Natural Fibres in Poznan since 2002 have focused on increasing resistance of flax to drought. Studies are carried out in two directions:

- application of antiperspirants in flax to improve water management in plants,
- use of flax genotypes (from INF collection) characterized by better resistance to drought in breeding.

In studies investigating methods of increasing flax resistance to water deficiency in the environment, the preparations classified as antiperspirants have been evaluated.

In field and pot experiments (2002-2007) antiperspirants with the following action were tested for their usefulness in conditions of controlled drought:

- coating substances (biologically active substances (b.a.s): di-1-P menten 96%, paraffin, glycerin, colloids obtained from flax, triadimefon)
- desiccating substances (b.a.i. dikwat, ammonium glufosynat, ethyl karfentrazon – in very small doses – *ppm*)
- herbicides (b.a.i. glyphosate in *ppm* doses)
- specific substances (b.a.i. ASA – treatment developed at INF)
- fertilizers - K (K_2SO_4)
- growth regulators – Atonik SL [b.a.i. natrium para-nitrofenolan (0,3%) natrium orto-nitrofenolan (0,2%)5 + natrium nitroguajakolan (0,1%)]
- biopreparations (b.a.i. chitosan acetate, algae extract).

In discussed studies, the best results in increasing resistance of flax to drought were obtained when, ASA and Atonik SL were used (Fot. 3). Spraying fiber flax with ASA (1,0 kg ha^{-1} and 0,1 kg l ha^{-1}), Atonik SL (0,5 l ha^{-1}) and Reglone Turbo 200 SL (50 ppm solution) showed a positive effect on growth and development of fiber flax cultivated in controlled drought conditions. All mentioned treatments increased statistically technical and total length of straw which consequently increased yields of straw and what is most important – flax fiber yield [2].

Research conducted at INF encompass also compounds of poly(aspartic acid) which increases the root mass which has positive effect on flax resistance to drought. Compounds of PKA also cause better use of mineral fertilizers by plants.

In world research concerning increasing plant resistance to drought, a special attention should be put to results of prof. Parka and his team of Connecticut University (USA). The team has developed genetically modified tomato characterized by higher resistance to drought in 2005. In these studies he used gene AVP_1, obtained from Arabidopsis thaliana. This common weed can be also found in numerous countries of Europe. The AVP_1, gene used in genetic modification of tomato facilitates synthesis of auxin responsible for root growth. The auxin also increases concentration of enzymes causing gathering of ions in vacuole. A better water management in the cell (cells take and retain more water) makes the new tomato cultivar resistance to drought higher by 70% as compared to classical cultivars.

Incrasing Flax Resistance to *Fusarium* Wilt

Results of 3-year experiments cycle showed that the highest percent of healthy plants was found in combinations where seeds were treated with 0.1% chtitosan acetate and also with Polyversum, Bion 50 WG and Biosept 33 SL (Figure 1). These substances increased the condition of flax by approximately 29% compared to the control, where no treatment was used (Figure 1). When comparing to standard seed treatment – Oxafun T, this difference is much smaller – less than 5%, but still in a favor of biblically active substances.

The main yield in fiber flax is sowing seeds and straw which is than processed into fiber. The evaluation of the effect of biologically active ingredients and bio-preparations on straw yield showed that the highest yield was obtained when Biosept 33 SL, chitosan acetate, Bion 50 WG and Polyversum were used (Figure 2). In those combinations the yield of straw was higher than that of control by 20 dt/ha (Figure 2).

These treatments also cased the highest increase of seed yield (Figure 3). The observed yield increase surpassed the control by about 2 dt/ha (Figure 3).

Taking into account relatively short vegetation period and poorly developed root system, flax requires fast acting and easy to assimilate fertilizers. In addition to straw and seed yield, also the highest possible quality of fiber is very important. It is also known that magnesium, zink, boron and sulfite compounds show positive effect on quality of fiber. Microelements as coenzymes act in enzymatic processes and in mycotoxin detoxication.

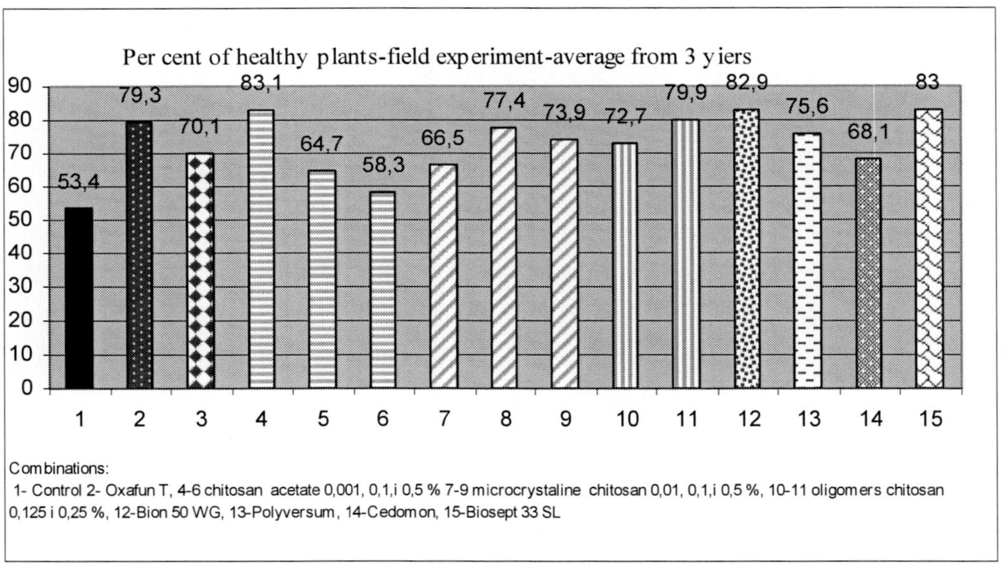

Figure 1. The healthiness of flax in pot experiment-average from 3 years.

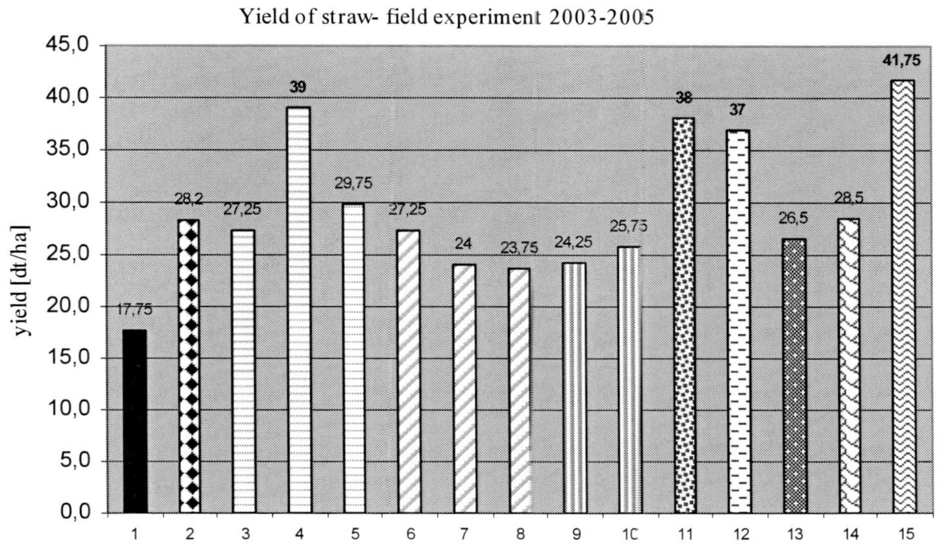

Combinations:
1- Control 2- Oxafun T, 3-5 chitosan acetate 0,001, 0,1,i 0,5 % 6-8 microkrystaline chitosan 0,01 0,1,i 0,5 %, 9-10 oligomers chitosan 0,125 i 0,25 %, 11-Polyversum 12-Bion 50 WG, 11-Polyversum, 13 Cedomon, 14-PSR 21, 15-Biosept 33 SL

Figure 2. The effect of straw yield- awerage from 3 years.

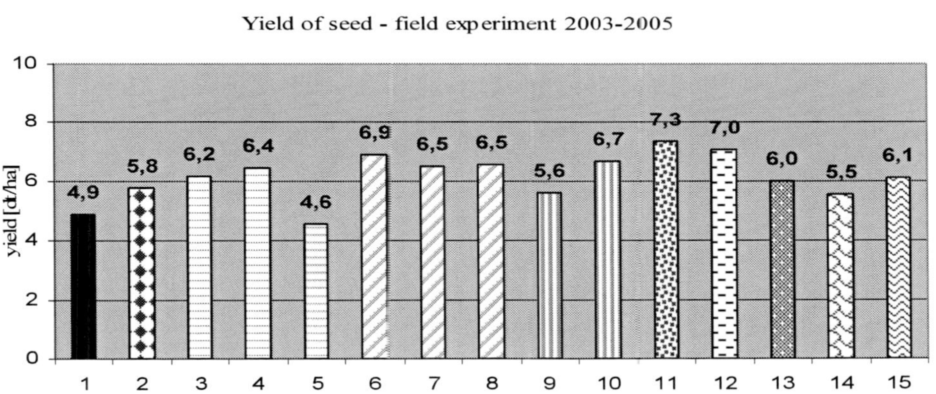

Combinations:
1- Control 2- Oxafun T, 3-5 chitosan acetate 0,001, 0,1,i 0,5 % 6-8 microkrystaline chitosan 0,01, 0,1,i 0,5 %, 9-10 oligomers chitozans 0,125 i 0,25 %, 11 Polyversum, 12-Bion 50 WG, 13-Cedomon, 14-PSR 21, 15-Biosept 33 SL

Figure 3. The effect of seed yield- average from 3 years.

The above issues were investigated at the Institute of Natural Fibres. Three chelate fertilizers were tested: Lubofos Len I, II and III; all produced by Zakłady Chemiczne Lubon Sp. z o. o.

It was found that using fertilizers of Lubofos Len type, one can obtain higher yields of flax straw, seed and fiber. The most efficient was application of Lubofos Len III/2 at 450 kg/ha and Lubofosem Len III/1 at 350 kg/ha (Figure 4).

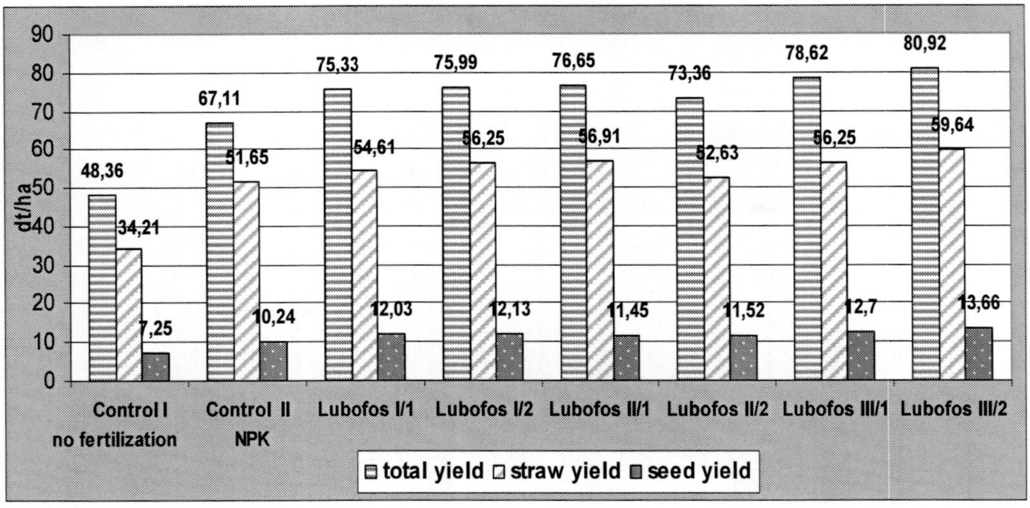

Figure 4. Influence of Lubofos type fertilizers on yield of flax in comparison to standard fertilization, 2004-2006 average(pot experiment).

Those fertilizers also had a beneficial effect on fiber quality, improving its strength, divisibility and fineness.

Positive effect of fertilizers on yields and quality of fiber allowed to conclude that microelements contained in those fertilizers show also beneficial effect on flax health and increase its resistance to *Fusarium* wilt.

As it was already mentioned before, in provocation conditions the cultivars and breeding material of flax from INF collection are tested for resistance to diseases. The INF holds a rich collection of flax serving as a crossing component source for breeding. Among 565 cultivars of fiber and oil flax cultivars tested so far, 99 were selected as very resistant to *Fusarium* wilt. These cultivars are most often used by breeders for crossings which gives the inheritance of resistance to *Fusarium* wilt. Most of them were considered by breeders as perspective ones, as they yield high fiber and seed. The final effect of close cooperation of breeders and phytopathologists was developing 6 cultivars of fiber flax that are very resistant to Fusarium wilt: Nike, Modra, Selena, Luna, Artemida i Temida.

Conclusions

1. Application of biological and conventional (chemical and mechanical) methods of plant protection integrated with breeding and cultivation techniques is a precondition for increasing flax resistance to biotic and abiotic stress.
2. Environmentally friendly methods of fiber flax protection should be based on three directions of activities: management of the environment, preventive measures and intervention treatments
3. New direction in breeding for resistance of flax is resistance to drought stress.
4. Among biologically active substances stimulating the natural resistance to *Fusarium* wilt the most effective one turned to be chitosan acetate in 0.1% concentration, Bion 50 WG, Polyversum and Biosept 33 SL.

5. Lubofos-Len fertilizers, containing optimal composition of microelements for flax, improve health condition of this plant.

REFERENCES

[1] Aldrich,R.J. 1997. Ekologia chwastów w roślinach uprawnych, *Solpress. Opole* pp. 461.

[2] Byczyńska M.and Heller K. 2003. Increasing resistance of flax plants to drought stress. Proc. International Conference of FAO European Research Network on Flax and Allied Fiber Plants for Human Welfare. 8-11.09.2003, Cairo:15-16.

[3] Heller K.1995. Weed and flax resistance to sulfonurea herbicides. *Natural Fibres XXXIX INF* Poznań:71-78.

[4] Heller K. 1998. Dynamika zbiorowisk chwastów segetalnych upraw lnu włóknistego w Polsce na przestrzeni lat 1967-1996. *Wyd. Inst. Włók. Nat. Poznań*: 105.

[5] Heller K., Rólski St. 2005. Lenmix 80 EC – new, linseed oil based, adjuvant for increasing herbicide activity in corn In: Development in production and use of new agrochemicals. *Chemistry for agriculture vol. 6*. Ed. Górecki H., Dobrzański Z., Kafarski P. Ed. Czech-Pol Trade. Czech Republic: 59-62.

[6] Heller K.,1996. Application of Computer Relational Database MS Access to the Evaluation of Results of Experiments with Fiber Flax. *Natural Fibres XL INF*, pp.31-44.

In: Renewable Resources: Obtaining, Processing... ISBN 978-1-6060692-534-8
Editors: R. Kozlowski, G. Zaikov and F. Pudel

Chapter 4

THE EFFECT OF SOWING DENSITY ON SEED AND PANICLE YIELDS OF INDUSTRIAL HEMP

H. Burczyk[1], M. Strybe[2], W. Koncewicz[1], N. Lukaszewska[1], G. Oleszak[2] and P. Baraniecki[1]
[1]Institute of Natural Fibres, Poznan, Poland
[2]Experimental Station INF in Petkowo, Poland

ABSTRACT

The predicted increase of cultivation area of industrial hemp in Poland requires optimalization of the production techniques for seed used both for sowing and for food applications (e.g. seeds and oil). The basic factor determining high yields is, apart from mechanization of the process, finding the optimum sowing density (plant number per m²).

Moreover, industrial hemp can be used for the production of essential oil that is present is the flowers. Therefore there is the need to determine also the optimum density and harvesting time for panicles to obtain the highest yield of the essential oil.

The field study was conducted at Experimental Farm of the Institute of Natural Fibres in Petkowo in years 2004-2006, where four sowing densities and three harvesting periods were examined. In the first two harvesting periods the yields of pannicles were determined, in the third also seed yield was checked.

The results indicate that the optimum sowing density for hemp cultivated for seeds was 10-15kg/ha. Similar amount of the sowing seeds results in the highest panicle yields when harvested during full flowering period. The efficiency of essential oil per area unit is correlated with the panicle yields. The yields of air-dried hemp were correlated with the sowing density.

Keywords: fibre hemp, sowing density, harvesting period, panicle, seed and straw yields.

I. Introduction

Hemp seed, so far, is obtained as a by-product from plants grown for textile fiber . However, as the demand for hemp products grows in textile, construction, automotive, food, etc. industries, the area of cultivation will grow and also will grow the demand for sowing seeds. *(Burczyk 2003), (Burczyk I in. 2005)* Hence, there is a need for further development of methods for sowing seeds production and seeds for food purposes (oil, bakery). *(Burczyk H. Kaniewski R.., 2005).*

In cultivation of hemp for the above mentioned purposes, seed-bearing panicles are a ballast worsening the conditions of stem retting. Thus the panicles can be used as a source for obtaining essential oils by steam distillation.

Recently, the interest in use of essential oils in cosmetics, repellents, food additives, etc, increases.

Additionally, essential oils obtained from hemp panicles can be an important factor improving the economical account of hemp cultivated for industrial purposes.

Therefore, to know the effect of important agronomical factors, such as: sowing density and time of harvest of hemp grown for seeds.

II. Materials and Methods

To precise agronomic recommendations the experiments with monoecious cultivar Silesia have been conducted at Institute of Natural Fibres Experimental Station in Pętkowo (wielkopolskie district). The experiments were set by randomized blocks, split-plot system with four replications. The plot area was 9 m2, raw spacing 37,5 cm. The following factors were tested:

- sowing density: 5, 10, 15, 20 kg/ha
- harvest dates: I – beginning of panicle formation, II – full bloom stage, III – end of seed formation.

The following observations were carried out during the growing period: hemp development stages, number of plants per unit area, plant height and technical length of straw and hemp panicles.

The experiments were set on black earth soil classified to IIIa class and wheat very good agricultural soil suitability complex. The average content of nutrients varied for P_2O_5 11-25 mg, for K_2O 10.5-24.5 mg and MgO 2.5-4.8 mg/100 g of soil. The pH measured in KCl varied from 5 - 6.2.

Evaluation of essential oils content in hemp panicles (fresh and dry) was carried out in the Laboratory of Department of Bast Fiber Plants Processing Technology according to Polish Farmakopea protocols *(2002).*

Hemp panicles were obtained form main shoot and lateral shoot. Samples were manually homogenized just before examination.

Each of 50 g samples of fresh homogenized material was water-distilled in Deryng apparatus with closed water cycle, for 3 h.

Distilled essential oil was gathered form water surface in distillation receiver. Volume of obtained essential oil was read after distillation process finished. Simultaneously hemp panicles were examined for water concentration during drying to constant weight in 100-105°C. Quantity of essential oils was evaluated by comparison with dry matter samples and presented in volume-weight percentage.

Biomass (panicles and stalks together) yield was evaluated with 18% moisture and seed yield with 8% moisture respectively.

Plant ginning was done manually using ginning combs.

Obtained results were analyzed statistically by variance analysis for split-plot system. Synthesis of multi-year results was carried out according to variance analysis for orthogonal experiments. *(Łubkowski 1968).*

The analysis of variance was supplemented with analysis of linear and square regression.

III. RESULTS

1. Influence of Sowing Density and Harvest Date on Seed Yield

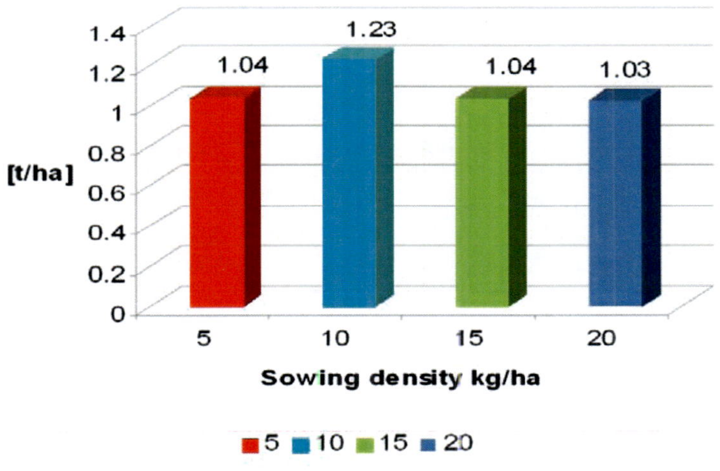

Figure 1. Influence of Sowing density on seed yield.

Mature seeds were harvested only at IIIrd harvest date. The highest seed yield occured at Sowing density of 10 kg per ha. Increase of Sowing density caused seed yield to decrease.

2. Influence of Sowing Density and Harvest Date on Panicles Yield

Analysis of variance show a significant influence of Sowing density and harvest date on panicles yield. The highest panicles yield occured at IIIrd harvest date combined with sowing density of 5 kg per ha. The increase of sowing density caused significant yield reduction.

Table 1. Influence of Sowing density and harvest date on seed yield dt/ha

Harvest date	Sowing density kg/ha				Average
	5	10	15	20	
I	14.6	15.1	13.8	15.1	14.7
II	39.5	37.6	31.6	31.3	35.0
III	44.9	39.5	39.8	39.4	40.9
Average	33.0	30.8	28.4	28.6	

LSD (lod 0,05) for harvest date2,87.
LSD (lod 0,05) for sowing density 1,84.
LSD (lod 0,05) for interaction 3,19.

3. Influence of Sowing Density and Harvest Date on Straw Yield

Statistical analysis showed that time of harvest and sowing density (to a lesser extent) has an effect on yields of straw.

The highest yields were obtained when plants were harvested on II date. Delaying harvest up to full maturity of seeds leads to straw yield reduction. Increasing amount of sown seeds over 10 kg/ha has no effect on yield increase.

Table 2. Influence of sowing density and harvest date on straw yield dt/ha

Harvest date	Sowing density kg/ha				Average
	5	10	15	20	
I	81.1	94.3	88.9	91.0	88.8
II	103.3	112.3	120.4	118.9	113.7
III	87.7	100.8	96.2	100.9	96.4
Average	90.7	102.5	101.8	103.6	

LSD (lod 0,05) for harvest date 5,57.
LSD (lod 0,05) for sowing density4,3.
LSD (lod 0,05) for interaction 7,46.

4. Influence of Sowing Density and Harvest Date on Essential Oils Content

Essential oils content in fresh hemp panicles (variety Silesia) for examined combinations of sowing density and harvest date was in range from 0,23 to 0,34%. The same examination for dry panicles show range of 0,17 to 0,25%. Therefore, essential oils in fresh panicles show twofold content than in dry panicles.

Examinations show that sowing did not influence the essential oils content.

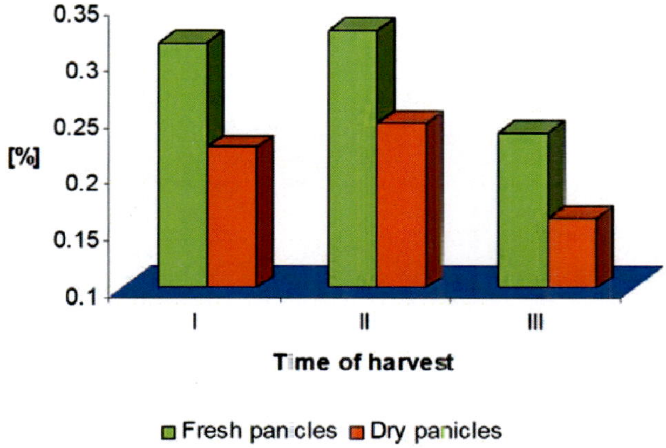

Figure 2. Influence of harvest date on essential oils content in hemp panicles.

5. Morphological Survey

5.A. Influence of Sowing Density and Harvest Date on Plant Losses

Real plant density determined before harvest was smaller than estimated. Increase of sowing density caused a linear increase of plant losses (Figure 4) in all harvest dates. Analysis of regressions indicates a 8 % increase of plant losses in IInd date, 10 % in IIIrd date of harvest. The highest level of plant losses (33,75 %) was noticed at sowing density 20 kg per ha at IIIrd harvest date.

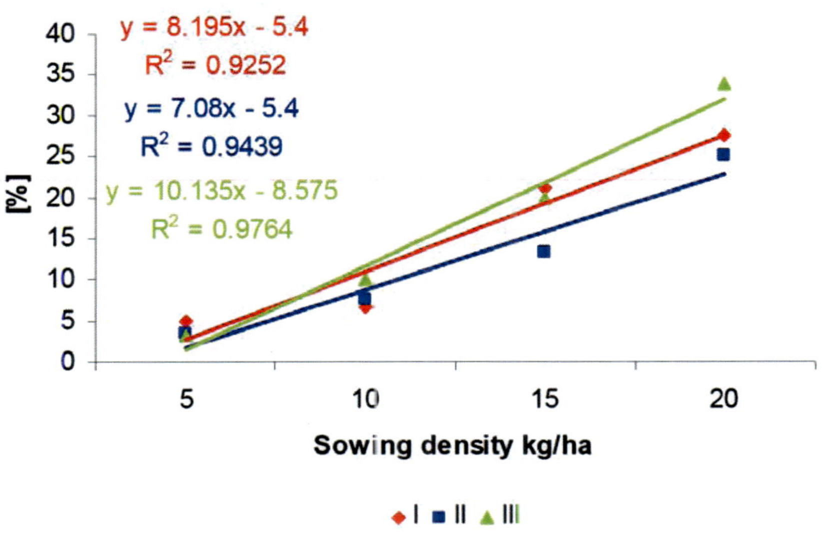

Figure 4. Influence of sowing density and harvest date on plant losses.

5.B. Influence of Sowing Density and Harvest Date on Technical Length and Stalk Diameter

Morphological survey show that increase of sowing density independently from harvest date caused an increase of technical length (Figure 5) and decrease of stalk diameter (Figure 6).

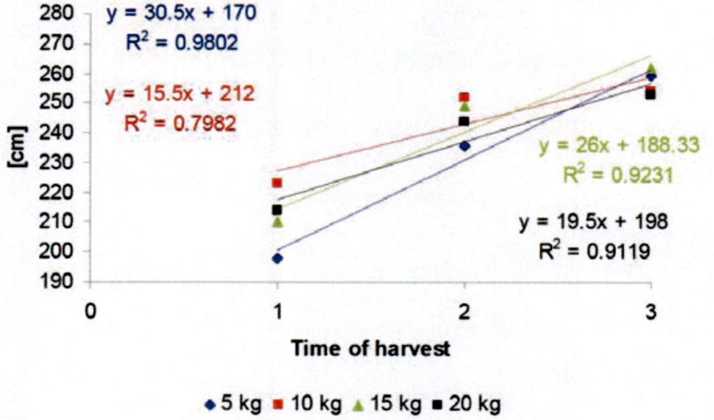

Figure 5. Influence of Sowing density and harvest.

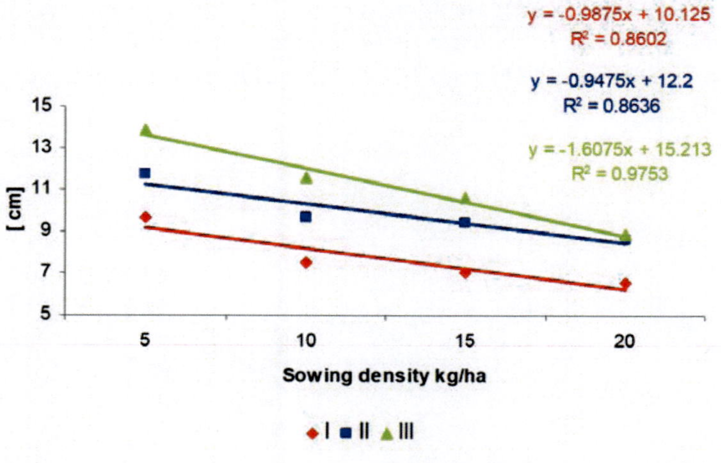

Figure 6. Influence of Sowing density and date on technical length harvest date on stalk diameter.

IV. Discussion

Different information can be found in literature as refers to sowing. It depends mainly on the use and kind of the material produced in the field, latitude, environmental conditions, cultivar, etc. *(Bocsa I., Carus M., 1997)* .Plant density is very important for development and growth of hemp plants. Contemporary grown Polish hemp cultivars have different biological and economical qualities compared to foreign ones. Therefore, is is necessary to know the optimum sowing density for conditions of Poland.

In France hemp for fiber is harvested after 20th August and for seeds – between 5-20 September *(Van der Werf, 1994)*. In Ukraine, the optimum harvest time is when 75%-80% of seeds reach maturity – mid September *(Golobarodko, 1994)*.

In studies of Grabowska and Koziara *(2005)* it was also found that excessive number of plants per m2 causes high fade outs, weaker plants condition and lower yields of straw.

In authors' experiments it was basically confirmed that optimum sowing density is 10 kg/ha. Similar results were reported by Jaranowska *(1968)*. On the other hand, Loch *(1996)* recommends higher sowing densities for seeds (20 kg/ha).

In hemp cultivated for industrial properties, the straw is the utilized raw material. Panicles are a by-product, which causes quality decrease in processing of straw or in retting. Initial studies showed that panicles are a good source of essential oils used for different purposes and traded. The results showed here indicate that optimum time to harvest hemp for essential oils is the flowering stage. The sowing density had no significant effect on the content of essential oils in panicles.

Results obtained for Poland show that it is possible to use hemp panicles to obtain 3-5 dm3 of essential oils from seed plantations. This can be an option to improve economical balance of hemp cultivated for sowing or technical seeds.

Conclusions

1. The results of field experiment show that the highest yield was obtained at sowing rate 10 kg/ha
2. Time of harvest and sowing density have effect on yields of hemp panicles. The optimum yield of panicles to be used for obtaining essential oils was during maturing of seeds at sowing density 5 kg/ha.
3. The highest content of essential oils in hemp panicles was observed during flowering.
4. Time of harvest is the main factor affecting yields of straw. To a lesser extent similar effect is shown by sowing density. Delaying time of harvest up to full seed maturity reduces yields.
5. Essential oils obtained from panicles can be a significant factor improving economical effect of industrial hemp plantations grown for seed and fiber.

References

Bocsa I., Karus M.: „Der Hanfanbau", Heidelberg: C.F. Muller, 1997.

Burczyk H.: Production of hemp sowing seed in Poland. *Journal of Industrial Hemp*. 2003, vol. 8(1), 81-88.

Burczyk H., Kaniewski R.: New Technology of Harvesting Hemp Grown for Seed. *Journal of Industrial Hemp*. 2005, vol. 10(1), 49-60.

Burczyk H., Kowalski M., Pławuszewski M.: The trends and methods of hemp breeding In Poland. *Journal of Natural Fibres*, 2005, vol. 2(1), 25-33.

Farmakopea Polska, 2002, tom VI, 151.

Goloborodko P.A. i in.: „*Sprawocznik konoplewoda*", „*Urożaj*", Kiew, 1994.

Grabowska L., Koziara W., The effect of Nitrogen Dose, Swing Density and Time of Harvest on Development and Yields of Hemp Cultivar Bialobrzeskie, *Journal of Natural Fibres*, vol 2,nr 4, 1-18, Haworth Press, 2005.

Jaranowska B.: „Wpływ czasu i gęstości siewu na wysokość i jakość plonu konopi jednopiennych na i dwupiennych uprawianych na glebach mineralnych", *Prace IPWŁ, Rocznik XV*, 25-36, 1968.

Loch L.: „Hanf – ein alter und neuer nachwachsender Rohstoff in Deutchland", Holttsee, 1996.

Łubkowski Z.: "*Metodyka doświadczalnictwa rolniczego*", PWRiL, W-wa, 1968.

Van der Werf H.: Crop physiology of fiber hemp (Cannabis sativa L.), *Doctoral thesis*, Wageningen, 1994.

In: Renewable Resources: Obtaining, Processing... ISBN 978-1-6060692-534-8
Editors: R. Kozlowski, G. Zaikov and F. Pudel

Chapter 5

BIOMODIFICATION OF HEMP AND FLAX FIBRES

J. Marek[1]***, V. Antonov***[1] ***and H. Fischer***[2]
[1]INOTEX Ltd. Czech Republic
[2]Faserinstitut Bremen e.V., Germany

Use of natural biocatalysts - enzymes for modification of natural bast fibres for wide range of applications. Enzymes as natural retting agents do all work needed during field dew-retting process of bast plants like hemp and flax. Recent studies and pilot trials clearly show that enzymatically activated processing of hemp and flax is able to exchange traditional process of bast plant treatment improving reproducibility, fibres yield and quality. It allows to prepare bast fibres for utilisation in high added value applications, like composites and plastics for automotive, building or other technical use. Enzymatically treated bast fibres can be used for the production of OE-spun yarns too. Enzymes seems to be a key to new possibility of linseed fibres utilization. Several special enzymatic agents for bast fibres elementarization have been developed by Inotex and tested for flax and hemp fibres separation.

Widely used conventional technologies of bast fibres treatment causes certain limitations in the processability and weaknesses of final quality of bast fibrous textiles. These problems have initiated the recent developments of new, cleaner production orriented processing methods to improve the position of bast fibres on the textile market. Also bio-technologies as a new, expanding discipline, based on use of microbiological procedures and enzymatic catalysis seems to be future alternative of bast noncellulosic compounds decomposition and fibre quality enhancement. In this case we focuse on biotechnology area with enzymes as biocatalysts.

Enzymes have great potential in catalysis of chemical compounds decomposition and synthesis. Enzymes tested and used in different areas of textile industry bring many surprising advantages and enable to achive results unreachable up to now. This could be a chance for growing and processing of bast fibres too.

Conventional bast fibres processing technologies in Central Europe are dew-retting as a first fibre releasing operation versus water retting widely used 50 years ago. All work during this retting period is done by different enzymes which come from natural microorganisms acting on the plant surface. Nevertheless dew-retting is negatively influenced by weather seasonal variation. Important part of under- and over-retted fibres is its unpleasant consequence.

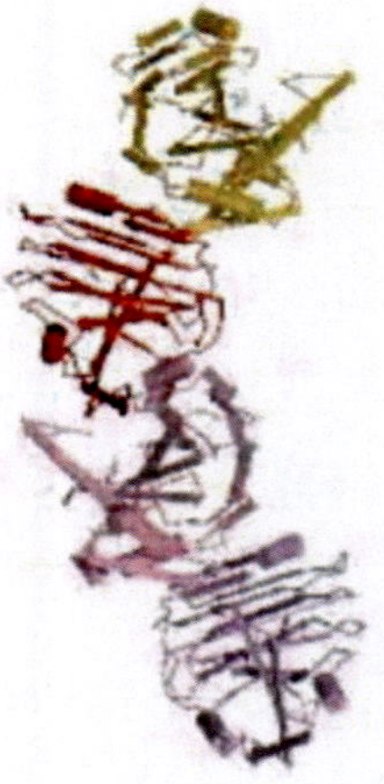

Enzyme - xylanase

For the production of wet-spun yarns alkaline scouring and combination of peroxide and chlorite bleaching are conventionally used.

As to experience outgoing from our research, development and many practical tests and trials there are several new possibilities of enzymatic technologies using of easily available commercial enzymes produced and sold by INOTEX®. as natural retting agents.

Enzymatic activities which seem to be useful for bast fibres processing are all pectinase activities, cellobiohydrolase, endoglucanase, xylanase, arabinase, rhamnogalacturonase, ligninperoxidase, Mn-dependent peroxidase, laccase, etc. These enzymes decompose polysaccharide polymer chains namely pectin and hemicelluloses (xylans, arabans, galactans,..) very effectively up to their total decomposition to oligomers and glucose or other relevant monomers. This leads to weakening or disruption of their bond with other biopolymers like lignin or cellulose. Lignin molekules thus released from lignocellulosic complex structure are therefore removed. We can take advantage of such mechanism in bast fibres processing, where the reduction of noncellulosic compounds content and thereby reaching desirable fineness, touch and colour is highly required. Selected commercial enzymatic products of INOTEX® corresponding to above mentioned activities are shown in the table.

Table. Commercially available enzymes of INOTEX®

Product name	Main substrates degraded by enzymes
Texazym BFE	pectin
Texazym SCW	pectin, hemicelluloses
Texazym DLG	hemicelluloses, cellulose, lignin
Texazym SER-3	pectin, hemicelluloses
Texazym SER-4	pectin, hemicelluloses, cellulose
Texazym SER-5	pectin, hemicelluloses, lipides
Texazym LOOK	lignin

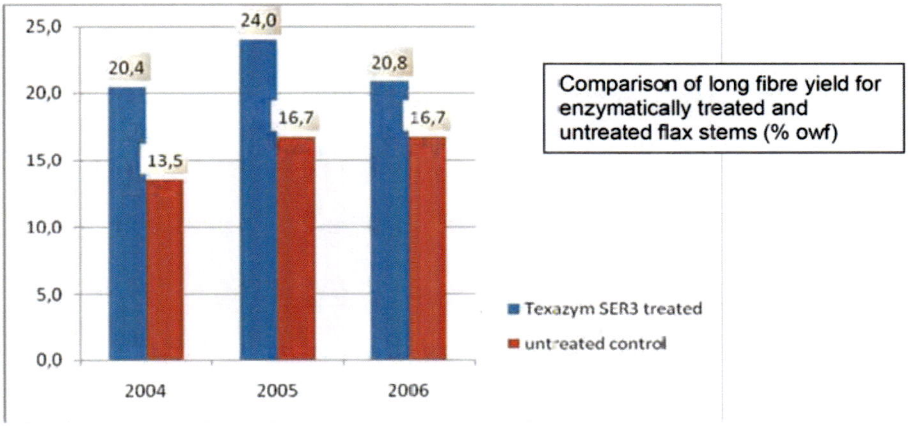

Enzymes can help in different stages of bast fibres processing like retting, cottonization, roving pretreatment and others. They are able to prepare tailor-made material for many different applications. Ring- and OE-spun yarns, automotive composites or building insulations are among them.

The crucial moment for final flax or hemp fibres quality and yield is in the dew-retting stage. The longer is retting period on the field the higher risk of material quality and/or yield losses. We have already run tests of enzymatic spraying of flax stems in first days of dew-retting for 3 years. Hemp stalks were enzymatically sprayed last year as well. These tests were broadly large trials because each test of after-pulling out application was done on cca 2,5 tons of stems.

There could be the opportunity for shortening of retting time because all enzymatically sprayed flax stems had scutching quality after 11-14 days of dew-retting.

Enzymes can facilitate the obtaining of good quality fibres separated from each other with minimum impurities and shives content as well. This is highly desirable for example in textile spinning and higher added-value composites. Enzymes are able to modify fibre surface with the aim of better compatibility with polymer matrix and tailor fibre lengths and diameters to maximum homogeneity and strength enhancement of final composite.

Enzymatic spraying of flax stems

Door panel made from natural fibres reinforced composite

Fineness improvement of bast fibres with enzymes leads to reduction of water absorption capacity due to reduction of microspaces among fibres and reduction of hemicelluloses content which are mainly responsible for undesirable absorptivity of natural fibre reinforced composites. New possibility is utilization of existing fibre dyeing devices for enzymatic cottonisation of bast fibres for spinning in blends with cotton and viscose using OE- or ring-spinning technology. Enzymes enable processing of waste-like bast materials like flax scutching tows, improve separation of hemp fibres and processability of linseed fibres. Enzymes allow better, more gentle and easier following mechanical processing of fibres and improve spinnability of such enzymatic pre-cottonised fibres.

Next opportunity for enzymes is by the linen eco-yearns production and by pre-treatment of flax roving for the production of wet spun long linen and tow linen yarns. There were compared different enzymes (Texazym DLG, BFE, SCW, etc.) as to their ability to process flax roving in laboratory and semi-industrial conditions. According to our rich experiences on this field following from commercial use of Texazym DLG enzymes seems to improve roving spinnability and yarn quality. Texazym SCW treatment showed the feasebility of „eco“ way of roving processing and preserved yarn tensile strength highest. Texazym SCW treatment enhances whiteness level significantly (if bleached) and maintain original natural flax colour (if not bleached).

Flax yarn

Enzymes have great potencial for the bast fibres area i.e. flax, linseed flax, hemp and other. Tests and trials carried out in past 7 years in INOTEX and cooperative companies have proved that enzymes are able to improve bast fibres quality: to increase long fibres yield significantly, to repair low-quality fibres, „eco-yarns“ can be produced. Enzymes can help to increase bast fibres growing areas and competitive ability of fiber producers

It seems that only complex solution of relations between growing and processing sphere and utilization of new technologies of biochemical and mechanical processing enable to create suitable conditions for utilization of traditional and perspective domestic renewable fibre sources and to open them ways to new qualities and to other industrial areas of use.

In: Renewable Resources: Obtaining, Processing... ISBN 978-1-6060692-534-8
Editors: R. Kozlowski, G. Zaikov and F. Pudel

Chapter 6

APPLICATION OF BIO-TECHNOLOGY IN PRE-TREATMENT OF COTTON TEXTILES TO ENHANCE QUALITY

A. I. Wasif and S. K. Laga*
Textile and Engg. Institute, Ichalkaranji,
Maharashtra, India

ABSTRACT

Textiles have undergone chemical processing since time immemorial. Human ingenuity and imagination, craftsmanship and resourcefulness are evident in textile products throughout the ages. We are to this day awed by beauty and sophistication of textiles sometimes found in archaeological excavations. After fabricating the mansions of fashion and comfort, textiles are now moving towards high-tech era of performance. This has brought up diversification and sea change in the chemical processing which was unchanged while years. Application of biotechnology in chemical processing of textiles is one of the revolutionary ways to enhance the textile wet processing pinnate right from pre-treatments to finishing.

In the present study, scouring of 100% cotton knit fabric has been carried out with pectinases enzymes alone at various concentrations (1, 3, 5, 7, 10 and 15 gpl) at various temperatures (40o C, 60o C and 80o C) for a period of 2, 4, 6, 8, 10 hours, as well as in combination of cellulases enzyme at various concentrations [0.1, 0.3, 0.5 % owf (cellulases) and 3, 5, 7 gpl (pectinases)] at 80o C for a period of 2, 3, 4 hours. The properties of bio-scoured fabric are compared with these of conventional scoured one. Encouraging results in terms of absorbency, whiteness, abrasion resistance and washing and rubbing fastness properties are obtained in case of bio-scouring. Further it reduces volume of effluent as well as BOD, COD, TDS and pH. It saves a substantial amount of energy and time. Bio-scouring causes savings of thermal energy is 50% while electrical energy while saving of electrical energy is 40%. Bio-scouring waste water has 40-45% less BOD and 60% less TDS as compared to conventional scouring waste water, and also 60% lower in TDS in waste water from bio-scouring.

* E-mail : swapan_laga@yahoo.co.in

1. INTRODUCTION

The enzymatic treatment of cotton can be subdivided into three major topics. The first one concerns the cellulosic part of the cotton and consequently the enzymes used are mainly cellulases. The effects achieved by these enzymes are pilling reduction, increase in softness, and amelioration of handle, surface structure and fabric appearance. These effects are summarized as "bio-polishing", a term created by the Danish company NOVO Nordisk. In the second application of enzymes in cotton finishing, the stone wash process of denim material is replaced by the use of cellulases. By the enzyme treatment, the amount of stone material required can be reduced or even completely replace and, unlike the stone wash process, the enzyme treatment can be prolonged without damaging the textile material. The third area of application concerns the removal of natural fibre-adhering material (pectin) as preparatory procedure for the subsequent processing steps. Pectinases, or a combination of pectinases and cellulases, are the enzymes used for degradation of the pectin material. The influence of interfering factors like surfactants, electrolytes and dyestuff materials on the quality of bio-finish processes has been monitored via weight loss determination. Cellulase treatment is not just for cotton but also for synthetic cellulosic fibres like Lyocell/tencel and rayon, where it improves softness, drapeability, pilling tendencies and post laundry appearance. On cellulose acetate, it is described as giving little benefit. Using enzymes for the modification of Lyocell in fibre blends. Using enzymes for the modification of Lyocell in fibre blends, the enzyme product always has to be adjusted to the needs of the accompanying fibre [1, 2, 3, 4].

Effective dyeing and printing of textiles requires uniform absorbency. Furthermore, no disturbing amounts of dirt, sizes or adhering natural material should remain on the fibers. Sixty percent of all the problems occurring in the dyeing and finishing of cotton originate from a wrong or uneven pre-treatment [5, 6].

Therefore, the pre-treatment is the most important step in cotton finishing. Consequently, the adhering natural material and the sizes have to be removed. Surface waxes produce a water-repellent fibre surface. Seed husks, pectins, hemicelluloses and sizes fix part of the dyestuff and therefore cause uneven dyeing. Furthermore, when using sodium hypochlorite as a bleaching agent, this material, if not removed, leads to the formation of AOX in the effluent. Therefore, prior to the bleaching process an alkaline boiling off is performed to reduce the amount of cotton adhering material [7, 8].

Regarding the enzymatic removal of natural material-adhering cotton, Bach and Schollmeyer examined the degradation of the cotton pectin by pectinolytic enzymes. The pectin was either extracted from the cotton and used as a substrate in homogeneous solution, or degraded directly on the cotton surface a heterogeneous solution, or degraded directly on the cotton surface in heterogeneous solution, or degraded directly on the cotton surface in heterogeneous reaction. The degree of degradation was determined by measuring the amount of reducing groups being released into the solution. The measuring the amount of reducing groups being released into the solution. The measuring the amount of reducing groups being released into the solution. The consequences of this degradation were published by Bach and Shcollmeyer in 1993. The degree of whiteness after enzymatic degradation of the pectin was lower compared to alkaline boiling-off. However, combining the enzymes treatment (a simultaneous treatment of pectinase and cellulase) or the alkaline boiling-off with in combination with the enzyme treatment. Resessner confirmed that the degree of whiteness of a cotton sample only treated with enzymes was lower by 8-10% than the degree of whiteness of an alkaline boiled off material. The effect of the enzymatic treatment was also determined by measuring the wetability of differently pre-treated fabrics. The differences were documented by the TEGEWA drop test. Desized fabrics were either enzyme or alkali treated. Differences in wetability occurring in this treatment step were removed by an additional peroxide bleaching process. The combined process, including an enzyme treatment, delivered results comparable to the alkali treatment [9].

Factors influencing scouring are the nature of the substrate, the kind of enzyme used, the enzyme activity, the use of surfactants and mechanical impact. It was observed that, during pectinase scouring, much less wax was removed compared with the alkaline scouring. If the treatment was combined with surfactant treatment, results equivalent to alkaline scouring could be achieved. A water treatment at 100°C is reported to increase the effectiveness of the subsequent scouring of cotton fabric with a combination of pectinase, protease and lipase, results equivalent to alkaline scouring could be achieved. A water treatment at 100°C is reported to increase the effectiveness of the subsequent scouring of cotton fabric with a combination of pectinase, protease and lipase, whereas the use of these enzymes alone showed very little effect. A new pectinase (Bio-Prep, NOVO Nordisk A/S) was screened that is stable under alkaline condition, i.e. the optimum conditions for removal of cotton adhering substances. Pectinase action does not create full removal of cotton adhering substance. Pectinase action does not create full removal of cotton adhering substance. Pectinase action does not create full wetability alone. Surfactants, Ca-binding salts, emulsifiers and high treatment temperatures complete the removal of the Ca-pectate/wax complex. The relative colour depths of caustic soda-scoured and environmentally friendly bio-prepared fabrics do not show significant differences [10, 11].

The effect of scouring with extraction. Whereas caustic soda treatment resulted in the highest deterioration on a molecular level. But led to a high level of whiteness, the solvent extracted sample showed superior tensile strength and bio-scoured samples the best softening effect [12, 13]. Textiles have undergone chemical processing since time immemorial. Human ingenuity and imagination, craftsmanship and resourcefulness are evident in textile products throughout the ages. We are too this day awed by beauty and sophistication of textiles sometimes found in archaeological excavations. After fabricating the mansions of fashion and comfort, textiles are now moving towards high-tech era of performance. This has brought up diversification and sea change in the chemical processing which was basically unchanged while years. Application of bio-technology in chemical processing of textiles is one of the revolutionary ways to enhance the textile wet processing pinnate right from pre-treatments to finishing.

In the present study, scouring of 100% cotton knit fabric has been carried out with Pectinases enzymes alone at various conc. (1, 3, 5, 7, 10, 15 gpl) at various temp. (40°,60°,80° C) for a period of 2,4,6,8,10 hrs, as well as in combination of cellulases enzyme at various conc. (0.1,0.3,0.5 % owf (Cellulases) and 3,5,7 (Pectinases) gpl) at 80° C for a period of 2,3,4 Hrs. The properties of Bio-scoured fabric are compared with these of conventional scoured one. Encouraging results in terms of absorbency, whiteness, abrasion resistance and washing and rubbing fastness properties are obtained in case of bio-scouring. Further it reduces volume of effluent as well as BOD, COD, TDS and pH. It saves a substantial amount of energy and time. Bio-scouring causes savings of thermal energy is 50% while electrical energy while saving of electrical energy is 40%. Bio-scouring waste water has 40-45 % less BOD and 60% less TDS as compared to conventional scouring waste water. and also 60% lower in TDS in waste water from Bio-scouring.

2. EXPERIMENTAL

Materials Used

(a) Fabric

The 88 x 72 double jersey cotton knitted fabric (180 gm/m2) was used for the study. The count of the yarn was 40s.

(b) Chemicals and auxiliaries used

The following chemicals and auxiliaries were used for the study:

Table 1. List of chemicals and auxiliaries used

Name of Chemicals and Auxiliaries	Manufacturer	Function
Didawin EWN 200%	Sarex Overseas, Mumbai	Nonionic wetting agent
Vacuum Salt	Laboratory Grade	Thermo stabilizer For α-Amylase Enzyme
Acetic Acid (2N)	Laboratory Grade	Maintaining The acidic pH
Pectinase, 8000 EGU	Rossari Biotech India Pvt. Ltd., Mumbai	Synergetic mixture of Protease and Pectinase Enzymes for Bio scouring of Cotton
Pectinase, 6000 EGU	Bayer ,Germany	Alkaline Pectinase for bio scouring of cotton
Cellulase	Novozymes	Acid Cellulases
Pectinase 4000 EGU	Sarex Overseas, Mumbai	Lipases
Sodium Hydroxide (Flakes)	Laboratory Grade	For Conventional scouring of cotton
Hydrogen Peroxide 50%	Laboratory Grade	For Bleaching of Cotton
Baystabil DBSX	Bayer ,Germany	Silicate free stabilizer
Oxalic acid	Laboratory Grade	After treatment of KMnO4 bleached Cotton Fabric
Sodium carbonate	Laboratory Grade	For fixation of reactive dye and maintaining alkaline pH

(c) Dye used :

Jakofix brilliant red HE3B (C.I. Reactive Red 24)

3. METHODS

3.1. Wetting

Grey fabric was treated with wetting agent. The procedure is as follows.

WETTING TREATMENT→COLD RINSE→ HOT WASH AT 80°C FOR 20 MIN. → COLD RINSE→ DRY

Recipe for wetting treatment:

DIDAWIN EWN 200% (WA)	= 3GPL
Emulsifier	= 0.5GPL
Temperature	= 70°C - 80°C
M:L Ratio	= 1:30

3.2. Conventional scouring

The pre wetted fabric was conventionally scoured using 2.5% sodium hydroxide, 1.5 %sodium carbonate and 0.5 % detergent at 100 °C for a period of 2 hrs,4 hrs,6 hrs. and 8 hrs.

Process sequence:

SCOURING TREATMENT→ HOT WASH AT 95°C FOR 10 MIN. → HOT WASH AT 95°C FOR 10 MIN →COLD RINSE→ DRY

3.3. Bio-scouring

i) Bio-scouring using pectinase 4000, pectinase 6000 and pectinase 8000:

Prewetted fabric was scoured using 1,3,5,7,10,15 gpl using pectinases and 1 gpl DIDAWIN EWN at a pH of 5.5,7.0 and 8.5 at various temperature (40°C, 60°C and 80°C) for a period of 2,4,6,8 and 10 hrs.

Process sequence:

BIO-SCOURING TREATMENT→ COLD RINSE→ HOT WASH AT 95°C FOR 10 MIN . → COLD RINSE→ DRY

ii) Two Step Bio-scouring with Cellulases Followed By Pectinases:

Prewetted fabric was first treatment with cellulase at various concentrations viz. 0.1,0.3 and 0.5% and then scoured using 3,5 and 7 gpl using pectinases and 1 gpl DIDAWIN EWN at various pH viz. 5.5,7.0 and 8.5 at various temperature (40°C, 60° C and 80°C) for period of 2,4,6,8 and 10 hrs.

Process sequence:

CELLULASES TREATMENT AT 50°C FOR 30MIN. →BIO-SCOURING→ HOT→WASH AT 80°C FOR 10 MIN. → COLD RINSE → DRY

3.4. Bleaching with Hydrogen Peroxide

Conventional as well as bio-scoured knitted fabric was bleached using 0.7% hydrogen peroxide (50% strength), 1% Baystabil DBSX, at a pH 10.5 at 80°C temperature for a period of 2 hrs.

Process sequence

PEROXIDE BLEACHING→ COLD RINSE → HOT WASH AT 80°C FOR 10 MIN. → COLD RINSE → DRY

3.5. Dyeing

Bio-Scoured, Peroxide Bleached Fabric As Well As Well Conventionally Scoured and Peroxide Bleached Fabric were dyed With Jakofix Brilliant Red HE3B to produce 1% Shade.

4. Testing and Analysis

Abrasion resistances have been measured using Martendel abrasion resistance tester, Model No. PRM 5LL/DB by ASTM 4966 method. Weight per square meter has been determined using cutting die equipment by IS 1964-1970 method. Fabric Thickness has been measured using Presser foot equipment, by IS 7702-1975 method. Residual wax content has been measured by IS 4390 -1967 method. Ash content has been determined by AATCC 78-1989 method.

Carboxylic group content has been carried out by IS 1560-1974 method. Absorbency has been measured by IS 2349-1963 method. Indices, K/S value and colour strength have been determined by using Macbeth colourEye CER 3000. Spectrophotometer with Colorlab+ Software. pH of effluent has been determined by using pH meter. TDS of effluent has been determined by evaporating the effluent. BOD of effluent has been determined by dilution followed by titration method. Washing fastness was measured as per ISO-3 recommendation using Laundero meter. Rubbing fastness was measured as per IS-766-1956 using Crock meter.

5. Results and Discussion

Bio-scouring of cotton knits with pectinases enzymes

Effect of concentration of enzyme, temperature, pH and duration on the properties of bio-scoured fabric have been studied.

(a) Effect of concentration of pectinase on absorbency

The effect of concentration of pectinase 4000, pectinase 6000 and pectinase 8000 on absorbency has been studied by varying the concentration from 1gpl to 15 gpl. at 80°C for 4 hrs. The results are tabulated in the Table 2,

Table 2. Effect of concentration of pectinase on absorbency

Concentration of Enzyme (gpl)	Absorbency [sec]		
	Pectinase 4000 EGU	Pectinase 6000 EGU	Pectinase 8000 EGU
1	40	35	35
3	22	20	20
5	15	14	14
7	7	5	5
10	1	1	1
15	1	1	1

(b) Effect Of Temperature of Scouring bath On Absorbency

The effect of temperature on absorbency has been studied by varying temperature from 40OC to 80OC keeping concentration of enzyme 10 gpl.and treatment time 4 hrs. The results are tabulated in the Table 3.

Table 3. Effect of temperature on absorbency

Temperature of Scouring Bath [°C]	Absorbency [sec]		
	Pectinase 4000 EGU	Pectinase 6000 EGU	Pectinase 8000 EGU
40	> 120	> 120	> 120
60	6	6	6
80	1	1	1

(c) Effect Of pH On Absorbency

The effect of pH on absorbency using pectinase 4000, pectinase 6000 and pectinase 8000 liquid have been studied by varying the pH from 5.5 to 8.5 keeping concentration of enzyme 10 gpl temperature and duration 4 hrs. The results are tabulated in Table 4.

Table 4. Effect of pH of scouring bath on absorbency

pH of Scouring Bath	Absorbency [sec]		
	Pectinase 4000 EGU	Pectinase 6000 EGU	Pectinase 8000 EGU
5.5	1	1	1
7.0	2	2	2
8.5	1	1	1

(d) Effect of duration of treatment on absorbency

The effect of duration of treatment on absorbency using using pectinase 4000, pectinase 6000 and pectinase 8000 have been studied by varying treatment time from 2 hr to 10 hrs. keeping concentration 10 gpl and temperature 80°C. The results are tabulated in Table 5.

Table 5. Effect of duration of treatment on absorbency

Duration of Treatment [Hours]	Absorbency [sec]		
	Pectinase 4000 EGU	Pectinase 6000 EGU	Pectinase 8000 EGU
2	13	13	12
4	1	1	1
6	1	1	1
8	1	1	1
10	1	1	1

(e) Effect of scouring on fabric shrinkage:

The effect of scouring on fabric shrinkage has been studied in both waleswise and coursewise.

i) Wales Way shrinkage:

Effect of scouring on fabric shrinkage (Wales way) have been studied keeping concentration of enzyme 10 gpl. pH 5.5 temperature 80°C and duration 4 hrs.The results are tabulated in Table 6.

Table 6. Effect of scouring on fabric shrinkage (Wales way)

Treatment	Shrinkage [%]
Conventionally Scoured	6.0
Pectinase 4000	2.2
Pectinase 6000	2.4
Pectinase 8000	2.5

ii) Course Way shrinkage

Effect of scouring on fabric shrinkage (course way) have been studied keeping concentration of enzyme 10 gpl. pH 5.5 temperature 80°C and duration 4 hrs.The results are tabulated in Table 7.

Table 7. Effect of scouring on fabric shrinkage (Course way)

Treatment	Shrinkage [%]
Conventionally Scoured	6.0
Pectinase 4000	2.8
Pectinase 6000	2.9
Pectinase 8000	3.0

(f) f) Effect of scouring on abrasion resistance of fabric

Effect of scouring on the abrasion resistance of fabrics has been studied. keeping concentration of enzyme 10 gpl. pH 5.5 temperature 80°C and duration 4 hrs.The results are tabulated in Table 8.

Table 8. Effect of scouring on abrasion resistance of fabric

Treatment	Difference in Thickness (mm)	Weight Loss [%]
Conventionally Scoured	0.36	1.45
Scouring with pectinase 4000	0.14	0.86
Scouring with pectinase 6000	0.16	1.00
Scouring with pectinase 8000	0.20	1.16

(g) Effect of scouring on fabric weight loss

Effect of scouring on fabric weight loss of fabrics has been studied keeping concentration of enzyme 10 gpl. pH 5.5 temperature 80°C and duration 4 hrs. The results are tabulated in Table 9.

Table 9. Effect of scouring on fabric weight loss

Treatment	Weight Loss [%]
Conventionally Scoured	6.25
Scouring with pectinase 4000	0.8
Scouring with pectinase 6000	1.0
Scouring with pectinase 8000	1.2

(h) Effect of scouring on absorbency

Effect of scouring on absorbency has been studied. keeping concentration of enzyme 10 gpl. pH 5.5 temperature 80°C and duration 4 hrs. The results are tabulated in Table 10.

Table 10. Effect of scouring on absorbency

Treatment	Absorbency (Sec.)
Conventionally scoured	1
Scouring with pectinase 4000	1
Scouring with pectinase 6000	1
Scouring with pectinase 8000	1

i) Effect of scouring on ash content of fabric:

Effect of scouring on ash content of fabric has been studied keeping concentration of enzyme 10 gpl. pH 5.5 temperature 80°C and duration 4 hrs. The results are tabulated in Table 11.

Table 11. Effect of scouring on ash content of fabric

Treatment	Ash Content [%]
Unscoured Sample	0.7
Conventional Scoured	0.2
Scouring with pectinase 4000	0.3
Scouring with pectinase 6000	0.4
Scouring with pectinase 8000	0.5

(j) Effect of scouring on the wax content of fabric:

Effect of scouring on wax content of fabric has been studied keeping concentration of enzyme 10 gpl. pH 5.5 temperature 80°C and duration 4 hrs. The results are tabulated in Table 12.

Table 12. Effect of scouring on wax content of fabric

Treatment	Wax Content [%]
Unscoured Sample	0.7
Conventional Scoured	0.1
Scouring with pectinase 4000	0.3
Scouring with pectinase 6000	0.3
Scouring with pectinase 8000	0.3

(k) Effect of scouring on -COOH group content of the substrate:

Effect of scouring on -COOH group content of fabric has been studied keeping concentration of enzyme 10 gpl. pH 5.5 temperature 80°C and duration 4 hrs. The results are tabulated in Table 13.

Table 13. Effect of scouring on -COOH group content of fabric

Treatment	-COOH Group Content
Conventionally Scoured	0.90
Scouring with pectinase 4000	0.80
Scouring with pectinase 6000	0.80
Scouring with pectinase 8000	0.80
Conventionally Scoured and H_2O_2 Bleached	0.95
Scouring with pectinase 4000 and H_2O_2 Bleached	0.90
Scouring with pectinase 6000 and H_2O_2 Bleached	0.85
Scouring with pectinase 8000 and H_2O_2 Bleached	0.85

(l) Properties of bio-scoured and bleached fabric:

Whiteness index, yellowness index and brightness index of Bioscoured and H_2O_2 Bleach fabric is shown in table 14.

Table 14. Properties of bio-scoured and H_2O_2 bleached fabric

Treatment	Indices		
	Whiteness Index (%)	Yellowness Index (%)	Brightness Index (%)
Conventionally Scoured and H_2O_2 Bleached	83.716	7.371	76.649
Scouring with pectinase 4000 and H_2O_2 Bleached	84.888	17.005	67.316
Scouring with pectinase 6000 and H_2O_2 Bleached	88.050	7.092	75.602
Scouring with pectinase 8000 and H_2O_2 Bleached	89.66	6.890	77.520

(m) Effect of scouring on washing and rubbing fastness of dyed fabrics:

Effect of scouring on wash and rubbing fastness of dyed fabrics has been studied. The results are tabulated in Table 15,16 and 17.

Table 15. Effect of Bio- scouring on Wash Fastness Rating

Treatment	Wash Fastness Rating [Staining Scale] 1 % Shade
Scouring with pectinase 4000 and H_2O_2 Bleached	3-4
Scouring with pectinase 6000 and H_2O_2 Bleached	4
Scouring with pectinase 8000 and H_2O_2 Bleached	4
Scouring with Cellulase and pectinase 4000	3-4
Scouring with Cellulase and pectinase 6000	3-4
Scouring with Cellulase and pectinase 8000	3-4

Table 16. Effect of scouring on Dry Rubbing Fastness

Treatment	Dry Rubbing Fastness Rating [Staining Scale] 1 % Shade
Scouring with pectinase 4000 and H_2O_2 Bleached	4
Scouring with pectinase 6000 and H_2O_2 Bleached	4
Scouring with pectinase 8000 and H_2O_2 Bleached	4
Scouring with Cellulase and pectinase 4000	4
Scouring with Cellulase and pectinase 6000	4
Scouring with Cellulase and pectinase 8000	4

Table 17. Effect of scouring on Wet Rubbing Fastness

Treatment	Wet Rubbing Fastness Rating [Staining Scale] 1 % Shade
Scouring with pectinase 4000 and H_2O_2 Bleached	3
Scouring with pectinase 6000 and H_2O_2 Bleached	3-4
Scouring with pectinase 8000 and H_2O_2 Bleached	3-4
Scouring with Cellulase and pectinase 4000	3
Scouring with Cellulase and pectinase 6000	3-4
Scouring with Cellulase and pectinase 8000	3-4

5.2. Two Step Bio-Scouring with Cellulase Followed by Pectinase 4000, Pectinase 6000 and Pectinase 8000

(a) Effect of Concentration of Cellulase On Absorbency

The effect of concentration of Cellulase on absorbency has been studied by varying the concentration of cellulase from 0.1% to 0.5%. keeping concentration of enzyme 10 gpl. pH 5.5 temperature 80°C and duration 2 hrs. The results are tabulated in the Table 18.

Table 18. Effect of Concentration of Cellulase on absorbency

Concentration of Cellulase (owf)	Absorbency (sec.)		
	Pectinase 4000 EGU	Pectinase 6000 EGU	Pectinase 8000 EGU
0.1	2	2	2
0.3	1	1	1
0.5	1	1	1

(b) Effect of Concentration of pectinases On Absorbency:

The effect of concentration of pectinase 4000, pectinase 6000 and pectinase 8000 on absorbency has been studied by varying the concentration from 5gpl to 10 gpl. keeping concentration of enzyme 10 gpl. pH 5.5 temperature 80°C, duration 2 hrs.and 0.3% cellulase (Cellosoft L) The results are tabulated in the Table 19.

Table 19. Effect of Concentration of pectinases on Absorbency

Concentration of Enzymes (owf)	Absorbency (sec.)		
	Pectinase 4000 EGU		Pectinase 4000 EGU
5	3	2	1
7	1	1	1
10	1	1	1

(c) Effect Of duration of Treatment On Absorbency:

The effect of Treatment time on absorbency using Cellulases followed by Pectinase 4000, pectinase 6000, pectinase 8000 have been studied by varying duration from 2 to 4 hrs. keeping concentration of enzyme 7 gpl. pH 5.5 temperature 80°C and the results are tabulated in the Table 20.

Table 20. Effect of Treatment time on Absorbency

Treatment Time [Hours]	Absorbency (sec.)		
	Pectinase 4000 EGU	Pectinase 6000 EGU	Pectinase 8000 EGU
2	1	1	1
3	1	1	1
4	1	1	1

(d) Effect of scouring on fabric shrinkage

i) Wales Way shrinkage:

Effect of scouring on fabric shrinkage (Wales way) have been studied. keeping concentration of Pectinase enzyme 10 gpl. pH 5.5 temperature 80°C duration 2 hrs. and 0.3% cellulase. The results are tabulated in Table 21.

Table 21. Effect of scouring on fabric shrinkage (Wales way)

Treatment	Shrinkage [%]
Scouring With pectinase 4000 EGU	2.4
Scouring With pectinase 6000 EGU	2.5
Scouring With pectinase 8000 EGU	2.6
Scouring With Cellulase and pectinase 4000 EGU	2.6
Scouring With Cellulase and pectinase 6000 EGU	2.8
Scouring With Cellulase and pectinase 8000 EGU	2.9

ii) Course Way shrinkage:

Effect of scouring on fabric shrinkage has been studied. keeping concentration of Pectinase enzyme 10 gpl. pH 5.5 temperature 80°C duration 2 hrs. and 0.3% cellulase. The results are tabulated in Table 22.

Table 22. Effect of scouring on fabric shrinkage (Course way)

Treatment	Shrinkage [%]
Scouring With pectinase 4000 EGU	2.8
Scouring With pectinase 6000 EGU	2.9
Scouring With pectinase 8000 EGU	3.0
Scouring With Cellulase and pectinase 4000 EGU	2.9
Scouring With Cellulase and pectinase 6000 EGU	3.0
Scouring With Cellulase and pectinase 8000 EGU	3.1

(e) Effect of scouring on abrasion resistance of fabric:

Effect of scouring on the abrasion resistance of fabrics has been studied. keeping concentration of Pectinase enzyme 10 gpl. pH 5.5 temperature 80°C duration 2 hrs. and 0.3% cellulase. The results are tabulated in Table 23.

Table 23. Effect of scouring on Abrasion resistance of fabric

Treatment	Difference in Thickness (mm)	Weight Loss (%)
Scouring With pectinase 4000 EGU	0.16	0.86
Scouring With pectinase 6000 EGU	0.18	1.00
Scouring With pectinase 8000 EGU	0.2	1.16
Scouring With Cellulase and pectinase 4000 EGU	0.19	1.15
Scouring With Cellulase and pectinase 6000 EGU	0.21	1.17
Scouring With Cellulase and pectinase 8000 EGU	0.23	1.19

(f) Effect of scouring on fabric weight loss:

Effect of scouring on weight loss of fabrics has been studied keeping concentration of Pectinase enzyme 10 gpl. pH 5.5 temperature 80°C duration 2 hrs. and 0.3% cellulase. The results are tabulated in Table 24.

Table 24. Effect of scouring on fabric weight loss

Treatment	Weight Loss [%]
Scouring With pectinase 4000 EGU	0.6
Scouring With pectinase 6000 EGU	0.8
Scouring With pectinase 8000 EGU	1.2
Scouring With Cellulase and pectinase 4000 EGU	1.1
Scouring With Cellulase and pectinase 6000 EGU	1.2
Scouring With Cellulase and pectinase 8000 EGU	1.4

(g) Effect of scouring on absorbency:

Effect of scouring on absorbency has been studied. keeping concentration of Pectinase enzyme 10 gpl. pH 5.5 temperature 80°C duration 2 hrs. and 0.3% cellulase. The results are tabulated in Table 25.

Table 25. Effect of scouring on absorbency

Treatment	Absorbency (sec.)
Scouring With pectinase 4000 EGU	1
Scouring With pectinase 6000 EGU	1
Scouring With pectinase 8000 EGU	1
Scouring With Cellulase and pectinase 4000 EGU	1
Scouring With Cellulase and pectinase 6000 EGU	1
Scouring With Cellulase and pectinase 8000 EGU	1

(h) Effect of scouring on ash content of fabric

Effect of scouring on ash content of fabric has been studied. Keeping concentration of Pectinase enzyme 10 gpl. pH 5.5 temperature 80°C duration 2 hrs. and 0.3% cellulase. The results are tabulated in Table 26.

Table 26. Effect of scouring on ash content of fabric

Treatment	Ash Content [%]
Scouring With pectinase 4000 EGU	0.3
Scouring With pectinase 6000 EGU	0.4
Scouring With pectinase 8000 EGU	0.4
Scouring With Cellulase and pectinase 4000 EGU	0.4
Scouring With Cellulase and pectinase 6000 EGU	0.4
Scouring With Cellulase and pectinase 8000 EGU	0.4

(i) Effect of scouring on the wax content of fabric:

Effect of scouring on wax content of fabric has been studied. keeping concentration of Pectinase enzyme 10 gpl. pH 5.5 temperature 80°C duration 2 hrs. and 0.3% cellulase. The results are tabulated in Table 27.

Table 27. Effect of scouring on wax content of fabric

Treatment	Wax Content [%]
Scouring With pectinase 4000 EGU	0.3
Scouring With pectinase 6000 EGU	0.3
Scouring With pectinase 8000 EGU	0.3
Scouring With Cellulase and pectinase 4000 EGU	0.3
Scouring With Cellulase and pectinase 6000 EGU	0.3
Scouring With Cellulase and pectinase 8000 EGU	0.3

(j) Effect of scouring on -COOH group content of the substrate:

Effect of scouring on -COOH group content of fabric has been studied. Keeping concentration of Pectinase enzyme 10 gpl. pH 5.5 temperature 80°C duration 2 hrs. and 0.3% cellulase.

The results are tabulated in Table 28.

Table 28. Effect of scouring on -COOH group content of fabric

Treatment	-COOH Group Content [%]
Scouring With pectinase 4000 EGU	0.8
Scouring With pectinase 6000 EGU	0.8
Scouring With pectinase 8000 EGU	0.8
Scouring with pectinase 4000 EGU and H_2O_2 Bleaching	0.90
Scouring with pectinase 6000 EGU and H_2O_2 Bleaching	0.85
Scouring with pectinase 8000 EGU and H_2O_2 Bleaching	0.85
Scouring With Cellulase and pectinase 4000 EGU and H_2O_2 Bleaching	0.95
Scouring With Cellulase and pectinase 6000 EGU and H_2O_2 Bleaching	0.95
Scouring With Cellulase and pectinase 8000 EGU and H_2O_2 Bleaching	0.95

(k) Properties of bio-scoured-bleached fabric:

Whiteness Index, Yellowness Index and Brightness Index of Bioscoured keeping concentration of Pectinase enzyme 10 gpl. pH 5.5 temperature 80°C duration 2 hrs. and 0.3% cellulase and Bleached with Hydrogen Peroxide are shown in Table 29.

Table 29. Properties Of Bioscoured and Bleached Fabric

Treatment	Indices		
	Whiteness Index (%)	Yellowness Index (%)	Brightness Index (%)
Scouring with pectinase 4000 EGU and H_2O_2 Bleaching	84.888	17.005	67.316
Scouring with pectinase 6000 EGU and H_2O_2 Bleaching	88.050	7.092	75.602
Scouring with pectinase 8000 EGU and H_2O_2 Bleaching	87.05	7.09	75.88
Scouring With Cellulase and pectinase 4000 EGU and H_2O_2 Bleaching	85.326	12.365	71.325
Scouring With Cellulase and pectinase 6000 and H2O2 Bleaching	86.669	5.023	77.025
Scouring With Cellulase and pectinase 8000 EGU and H_2O_2 Bleaching	86.70	5.10	77.30

(l) Effect of scouring on washing and rubbing fastness of dyed fabrics:

Effect of scouring on wash and rubbing fastness of dyed fabrics has been studied Bio-scouring has been carried out keeping concentration of Pectinase enzyme 10 gpl. pH 5.5 temperature 80°C duration 2 hrs. and 0.3% cellulase. The results are tabulated in Table 30, 31 and 32.

Table 30. Effect Of Scouring on Washing Fastness of Dyed Fabrics

Treatment	Wash Fastness Rating [Staining Scale] 1 % Shade
Scouring with pectinase 4000 EGU and H_2O_2 Bleaching	3-4
Scouring with pectinase 6000 EGU and H_2O_2 Bleaching	4
Scouring with pectinase 8000 EGU and H_2O_2 Bleaching	3-4
Scouring With Cellulase and pectinase 4000 EGU and H_2O_2 Bleaching	3-4
Scouring With Cellulase and pectinase 6000 EGU and H_2O_2 Bleaching	3-4
Scouring With Cellulase and pectinase 8000 EGU and H_2O_2 Bleaching	3-4

Table 31. Effect of scouring on Dry Rubbing Fastness of Dyed Fabrics

Treatment	Dry Rubbing Fastness Rating [Staining Scale] 1 % Shade
Scouring with pectinase 4000 EGU and H_2O_2 Bleaching	4
Scouring with pectinase 6000 EGU and H_2O_2 Bleaching	4
Scouring with pectinase 8000 EGU and H_2O_2 Bleaching	4
Scouring With Cellulase and pectinase 4000 EGU and H_2O_2 Bleaching	4
Scouring With Cellulase and pectinase 6000 EGU and H_2O_2 Bleaching	4
Scouring With Cellulase and pectinase 8000 EGU and H_2O_2 Bleaching	4

Table 32. Effect of scouring on Wet Rubbing Fastness of Dyed Fabrics

Treatment	Wet Rubbing Fastness Rating [Staining Scale] 1 % Shade
Scouring with pectinase 4000 EGU and H_2O_2 Bleaching	3
Scouring with pectinase 6000 EGU and H_2O_2 Bleaching	3-4
Scouring with pectinase 8000 EGU and H_2O_2 Bleaching	3-4
Scouring With Cellulase and pectinase 4000 EGU and H_2O_2 Bleaching	3
Scouring With Cellulase and pectinase 6000 EGU and H_2O_2 Bleaching	3-4
Scouring With Cellulase and pectinase 8000 EGU and H_2O_2 Bleaching	3-4

5.3. Comparative study of characteristics of effluent

Effluent obtained from pre-treatment of cotton bath has been analyzed for pH, TDS and BOD. Total volume of the effluent has also been measured. The results are tabulated in Table 33, 34, 35 and 36.

Table 33. pH of the effluent

Treatment	pH
Conventional scouring	12.4
Scouring with pectinase 4000 EGU / pectinase 6000 EGU / pectinase 8000 EGU	5.5-8.4
Scouring with mixture of cellulase and pectinases	5.5

Table 34. TDS of the effluent

Treatment	TDS [ppm
Conventional scouring	7420
Scouring with pectinase 4000 EGU / pectinase 6000 EGU / pectinase 8000 EGU	3054
Scouring with mixture of cellulase and pectinases	3456

Table 35. BOD of the effluent

Treatment	BOD5 Days [ppm]
Conventional scouring	1940
Scouring with pectinase 4000 EGU / pectinase 6000 EGU / pectinase 8000 EGU	1120
Scouring with mixture of cellulase and pectinases	1120

Table 36. Volume of the effluent

Treatment	Volume of Effluent [Litres]
Conventional scouring	2800
Scouring with pectinase 4000 EGU / pectinase 6000 EGU / pectinase 8000 EGU	1800
Scouring with mixture of cellulase and pectinases	1800

5.4. Comparison of electricity consumption

Consumption of electrical energy and thermal energy for conventional and bio-scouring (including washing) have been calculated. The results are tabulated in Table 37 and 38.

Table 37. Electricity Consumption in %

Treatment	Electricity Consumption (%)
Conventional Scouring	100
Scouring with pectinase 4000 EGU / pectinase 6000 EGU / pectinase 8000 EGU	50
Scouring with mixture of cellulase and pectinases	25

Table 38. Thermal Energy Consumption (%)

Treatment	Thermal Consumption (%)
Conventional Scouring	100
Scouring with pectinase 4000 EGU / pectinase 6000 EGU / pectinase 8000 EGU	50
Scouring with mixture of cellulase and pectinases	25

5.5. Combined bioscouring and bleaching

(a) Effect of combined bioscouring and bleaching On Absorbency:

The effect of combined scouring (one step, two steps) and bleaching on absorbency has been studied. The results are shown in Table 39.

Table 39. Effect of combined scouring and bleaching on Absorbency

Treatment	Absorbency (Sec.)
Scouring with pectinase 4000 EGU and H_2O_2 Bleaching	1
Scouring with pectinase 6000 EGU and H_2O_2 Bleaching	1
Scouring with pectinase 8000 EGU and H_2O_2 Bleaching	1
Scouring with Cellulase and pectinase 4000 EGU and H_2O_2 Bleaching	1
Scouring with Cellulase and pectinase 6000 EGU and H_2O_2 Bleaching	1
Scouring with Cellulase and pectinase 8000 EGU and H_2O_2 Bleaching	1

(b) Effect of combined bio-scouring and bleaching on indices properties of bio-scoured and bleached fabric:

The effect of combined scouring (one step, two steps) and bleaching on indices has been studied. The results are shown in Table 40.

Table 40. Indices properties of bio-scoured and bleached fabric

Treatment	Indices		
	Whiteness Index (%)	Yellowness Index (%)	Brightness Index (%)
Scouring with pectinase 4000 EGU and H_2O_2 Bleaching	64.568	17.005	64.325
Scouring with pectinase 6000 EGU and H_2O_2 Bleaching	68.050	11.092	66.245
Scouring with pectinase 8000 EGU and H_2O_2 Bleaching	68.99	11.34	67.52
Scouring with Cellulase and pectinase 4000 EGU and H_2O_2 Bleaching	55.356	12.365	59.248
Scouring with Cellulase and pectinase 6000 EGU and H_2O_2 Bleaching	56.688	15.023	62.356
Scouring with Cellulase and pectinase 8000 EGU and H_2O_2 Bleaching	67.89	15.33	63.36

(c) Effect Of Combined bio-scouring and Bleaching On Wash And Rubbing Fastness Of Dyed Fabrics:

Effect of combined bio-scouring and bleaching on wash and rubbing fastness of dyed fabrics has been studied. The results are tabulated in Table 41, 42 and 43.

Table 41. Effect of combined bio-scouring and bleaching on Wash Fastness of Dyed Fabrics

Treatment	Wash Fastness Rating (Staining Scale) 1 % Shade
Scouring with pectinase 4000 EGU and H_2O_2 Bleaching	3-4
Scouring with pectinase 6000 EGU and H_2O_2 Bleaching	4
Scouring with pectinase 8000 EGU and H_2O_2 Bleaching	3-4
Scouring with Cellulase and pectinase 4000 EGU and H_2O_2 Bleaching	3-4
Scouring with Cellulase and pectinase 6000 EGU and H_2O_2 Bleaching	3-4
Scouring with Cellulase and pectinase 8000 EGU and H_2O_2 Bleaching	3-4

Table 42. Effect of combined bio-scouring and bleaching on Dry Rubbing Fastness of Dyed Fabrics

Treatment	Dry Fastness Rating (Staining Scale) 1 % Shade
Scouring with pectinase 4000 EGU and H_2O_2 Bleaching	4
Scouring with pectinase 6000 EGU and H_2O_2 Bleaching	4
Scouring with pectinase 8000 EGU and H_2O_2 Bleaching	4
Scouring with Cellulase and pectinase 4000 EGU and H_2O_2 Bleaching	4
Scouring with Cellulase and pectinase 6000 EGU and H_2O_2 Bleaching	4
Scouring with Cellulase and pectinase 8000 EGU and H_2O_2 Bleaching	4

Table 43. Effect of combined bio-scouring and bleaching on Wet Rubbing Fastness of Dyed Fabrics

Treatment	Wet Fastness Rating (Staining Scale) 1 % Shade
Scouring with pectinase 4000 EGU and H_2O_2 Bleaching	3
Scouring with pectinase 6000 EGU and H_2O_2 Bleaching	3-4
Scouring with pectinase 8000 EGU and H_2O_2 Bleaching	3-4
Scouring with Cellulase and pectinase 4000 EGU and H_2O_2 Bleaching	3
Scouring with Cellulase and pectinase 6000 EGU and H_2O_2 Bleaching	3-4
Scouring with Cellulase and pectinase 8000 EGU and H_2O_2 Bleaching	3-4

CONCLUSION

Optimum conditions of bio-scouring with pectinases require treatment with 10 gpl pectinases at pH 5.5 or 8.5 at 80°C for 4 hrs. Optimum conditions of two step bio-scouring with 0.3% Cellulase followed by Pectinases need treatment with 7 gpl pectinases at pH 5.5 at 80°C for 2 hrs. Conventional scouring produces excellent whiteness causing more shrinkage of fabrics. Conventional scouring shows more weight loss than of bio-scouring. Ash content as well as wax content of bio-scoured fabric are slightly higher than those of conventionally scoured fabric. The volume of effluent generated by bio-scouring is 40% less as compared to conventional scouring. Abrasion resistance of bio-scoured fabric is better than that of conventionally scoured fabric. The bio-scoured fabric produces slightly duller shades while wash fastness and rubbing fastness properties of bio-scoured bleached and dyed fabric are better than those of conventionally scoured one. Two step bio-scouring do not have significant influence on properties of fabric compared to bio-scouring by only Pectinases. Combined bio-scouring and bleaching shows less Whiteness index, Yellowness index and Brightness index.

ACKNOWLEDGEMENT

The authors are very much thankful to Dr. C.D. Kane, Executive Director, Textile and Engg., Institute, Ichalkaranji, India for his encouragement and support to carry our research work. Thanks are also due to M/S Zytex India Pvt. Ltd., M/S. Rossari Biotech Pvt. Ltd., M/S Sarex Overseas and Jay Chemicals for their support by providing precious chemicals, enzymes and dyes.

References

[1] Dr. Petry GMBH , *Application of Enzymes in the Textile Industry*, NCM 15 (1), 23-28 (2001)

[2] Fornelli Saverio , The personal enzyme horoscope! , ITB Dyeing/Printing/Finishing 3RD Q 41, 62-65 (1995).

[3] Nalankilli G., Sunder Edwin A,Bio-bleaching ,ATJ 11 (1),33-34(2002).

[4] Diller Gisela Buschie , Broughton Royal , Ren Xuehong ,Boimimicking of Enzymes for textile Processing , *National Textile center Annual Report* CO2-AE07 , (2002).

[5] Diller Gisela Buschie, Xiang Dong Yang ,Enzymatic Bleaching of Cotton Fabric with Glucose Oxidase, *Textile Res. J.* 71(5), 388-394 (2001).

[6] Abdalah Gus, Ramkumar S.S., *Surface characterization of enzyme treated fabrics*, Colourage XLVIII(4),15-18 (2001).

[7] Tzanov Tzanko , Costa Silgia, Calafell ,Gubitz George M., Cavaco-Paulo Artur, Enzymes for cotton fabric preparation and recycling of waste waters for dyeing,Annual2000, 65-72 (2000).

[8] Hsieh You-lo , Cram Lisa A. , Enzymatic Hydrolysis to Improve Wetting and Absorbency of Polyester Fabric, *Textile Res.J.* 68(5), 311-319 (1998).

[9] Diller Gisela Buschie, Traore M. K. , Influence of Direct and Reactive Dyes on the Enzymatic Hydrolysis of Cotton, *Textile Res.J.* 68(3), 185-192 (1998).

[10] Cortez Joao M., Ellis John, Bishop David P., Using Cellulase to Improve the Dimensional Stability of Cellulosic Fabrics, *Textile Res.J.* 72 (8),673-680 (2002).

[11] Buschle-Diller.G., And Zeronian S.H, Enhancing The Reactivity And Strength Of Cotton Fibres , *J. Appl. Polym. Sci.* 45, 967-979 (1992).

[12] Yadav, R P, Saxena R K Gupta R And Sheba W D: *Biotechnology Appl. Bio. Chem.*, 28, 243- 249 (1998).

[13] Dong H, Gao S, Han S And Cao S: *Biotechnol. Appl. Biochem.*, 30, 251-256 (1999).

In: Renewable Resources: Obtaining, Processing... ISBN 978-1-6060692-534-8
Editors: R. Kozlowski, G. Zaikov and F. Pudel

Chapter 7

REMOVAL OF SOLVENTS FROM PLANT MATERIALS

Hartmut Haida[1], Vera Miteva[1], Matthias Börner[1], Wolfgang Behns[1], Mirjana Kostova[2], Stefan Heinrich[1] and Lothar Mörl[1]

[1]Otto-von-Guericke-University of Magdeburg
Institute of Process Equipment and Environmental Technology
P. O. Box 4120, 39016 Magdeburg (Germany)
[2]University of Chemical Technology and Metallurgy Sofia
8, St. Kliment Ohridski blvd. 1756 Sofia (Bulgaria)

ABSTRACT

The removal and recovery of residual solvents from the extracted porous plant material can be performed e. g. by a fluidized bed process. Experimental investigations of this fluidized bed process also using a model substance are described in this chapter. The carbohydrates concentration in the fluidizing matter was measured by a flame ionization detector. The test results and the results of model calculations are compared. The required data for the design of industrial plants can easily be determined by such model based experimental investigations.

INTRODUCTION

The liquid-solid-extraction is a very often applied process using suitable solvents (hydrocarbons) to produce the active ingredients from plant materials. An important task for such processes in reference to the economic efficiency is the removal and recovery of residual solvents from the extracted porous plant material. There the solvent and the plant material should not be damaged thermally.

This task can be performed e. g. by a fluidized bed process. In the fluidized bed the residual solvents can be transfered into the vapor phase. An inert gas or the solvent vapour itself is used as fluidizing matter for the extracted plant materials. In both cases the knowledge of the fluidization and drying behaviour is decisive for the feasibility and profitability of the process.

In the presentation experimental investigations of this fluidized bed process using a model substance are described. The test results and the results of model calculations are compared.

It will be shown that the required data for the design of industrial plants can easily be determined by such model based experimental investigations.

EXPERIMENTAL INVESTIGATIONS

The experimental investigations are carried out using fluidized bed pilot plants with a pivoting gas distributor plate according to Figure 1 and Figure 2. The diameter of the fluidizing chamber is 100 mm resp. 200 mm. The carbohydrates concentration in the fluidizing matter was measured by a flame ionization detector besides to the characteristic temperatures and pressure drops.

At first Gamma-alumina (γ-Al_2O_3) was used as test material because of the defined adsorption properties (particle size d_p = 1,8 mm, density ρ = 1040 kg/m³, porosity ε = 0,68). These particles were loaded by tetrachloromethane CCl_4. The load was analysed before and after the fluidization usind a drying oven.

Figure 1. Fluidized bed pilot plant ND 100 with flame ionization detector.

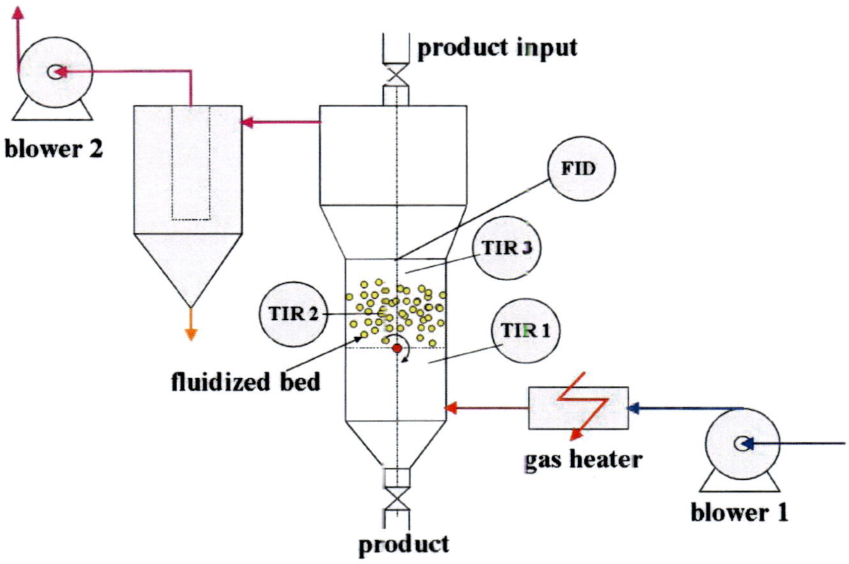

Figure 2. Scheme of the fluidized bed pilot plant ND 100.

MODELING

The modeling of the solvent removal (evaporation and desorption) is based according to the schematic representation in Figure 3.

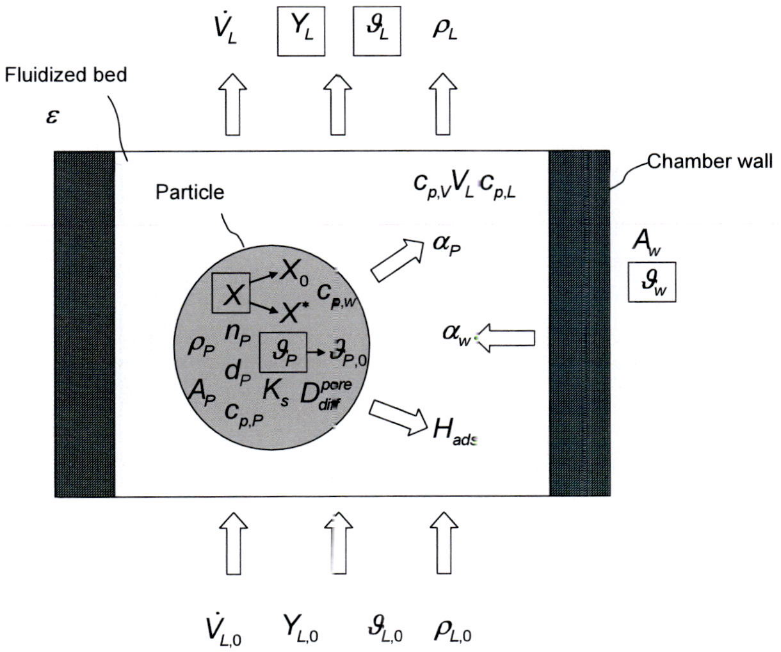

Figure 3. Schematic representation of the solvent removal model.

The mass and (thermal) energy balances for the fluidizing gas and the particles are transfered in a system of differential equations which can be solved by MATLAB.

RESULTS

Figure 4 shows e. g. the recorded measured data of the flame ionization detector.

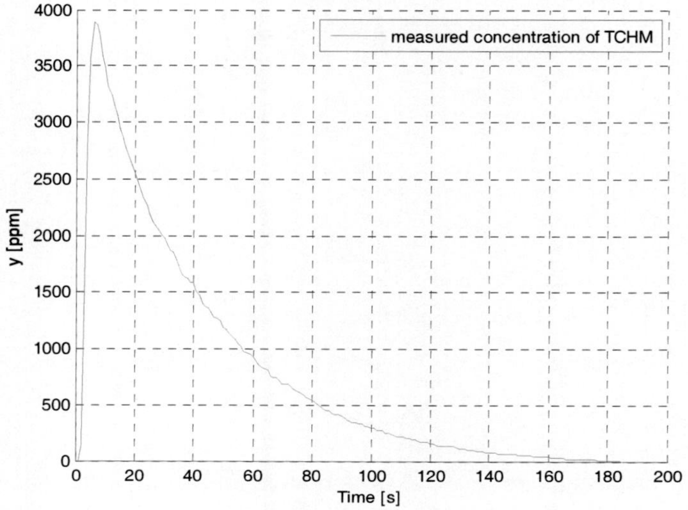

Figure 4. Measured carbohydrates concentration in the fluidizing gas after the fluidizing chamber.

As a first approximation the area under the graph in Figure 4 can be seen as an equivalent of the desorbed mass of carbohydrates. This mass depending on the time can be calculated by integration – Figure 5.

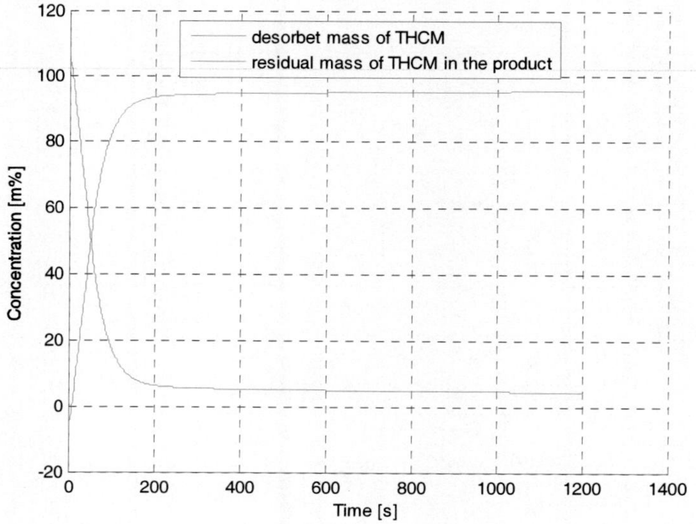

Figure 5. Desorbed and residual mass of carbohydrates in the solids.

The particle temperatures, the temperatures of the fluidizing gas and the corresponding loadings of the particles can be calculated by the model equations. Figure 6 represents e. g. the measured inlet and outlet temperatures of the fluidizing gas and of the fluidized bed in comparison to the calculated temperatures of the outgoing fluidizing gas and of the solid particles. In Figure 7 the calculated particle load and the analyzed loads of samples are collected as an example.

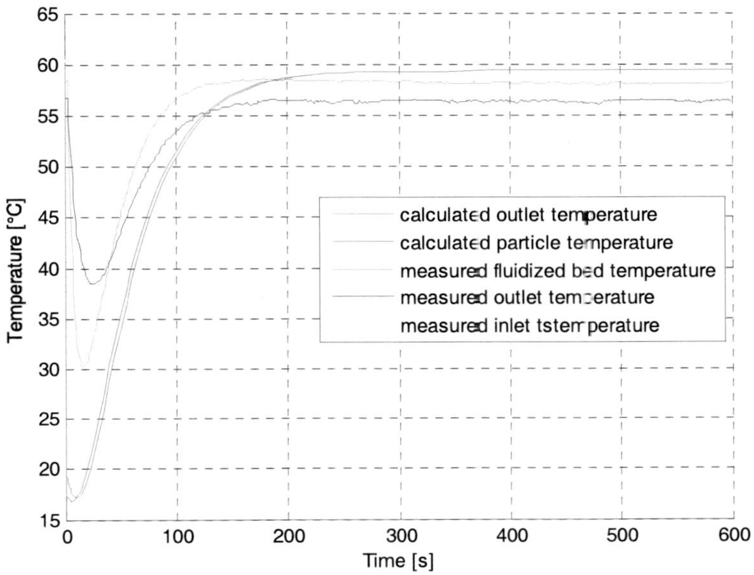

Figure 6. Comparison of measured and calculated temperatures.

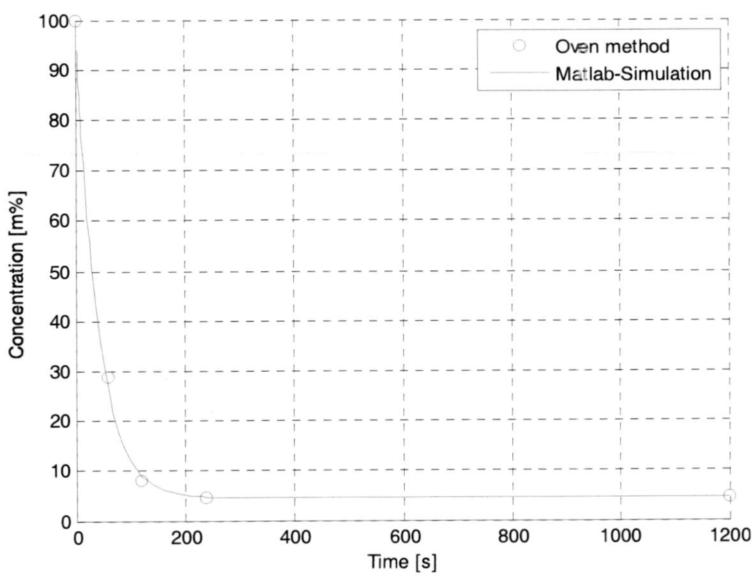

Figure 7. Calculated and measured values of the particle load.

CONCLUSION

The so far performed experimental investigations and the comparison with the model based calculations confirm that fluidized bed processes are basically suitable for the removal of residual extraction agents from plant materials. This process is completed in a very short time. The current results must be supported with other plant materials. So fluidized bed plants for industrial applications could be designed.

In: Renewable Resources: Obtaining, Processing... ISBN 978-1-6060692-534-8
Editors: R. Kozlowski, G. Zaikov and F. Pudel

Chapter 8

DETECTION OF ESTROGENIC ACTIVITIES IN ENVIRONMENTAL SAMPLES, URINE AND MILK BY A NOVEL YEAST BIOSENSOR

Martina Körner[1], Steffen Uhlig[2], Katja Heide[1], Kirsten Simon*[2], Christian Kaiser[1], Kirstin Kunath[2], Kristina Florschütz[1], Kristin Suckau[1], Elmarie Fischer[1] and Gotthard Kunze[1]

[1]Leibniz Institute of Plant Genetics and Crop Plant Research (IPK), Corrensstr. 3, D-06466 Gatersleben, Germany

[2]quo data GmbH, Kaitzer Str. 135, D-01187 Dresden, Germany

ABSTRACT

The *Arxula adeninivorans* yeast estrogen screen (A-YES) assay has been developed and established for the detection of estrogenic activities. The *in vitro* assay is based on recombinant yeast cells engineered to co-express the human estrogen receptor α (hERα) and the phytase (*phyK*) reporter gene under the control of an *Arxula*-derived promoter that had been modified by the insertion of estrogen response elements (EREs). With its superior characteristics in handling, analysis time, detection and determination limits it is applicable for a wide range of matrices like wastewater, urine and milk. 120 samples from 60 wastewater treatment plants (effluent, indirect discharge, municipal wastewater) in Sachsen-Anhalt, Germany, were repeatedly analyzed to investigate the reproducibility of the A-YES assay and the influence of matrix effects on the measured values. The analyses were performed according to a nested design. According to the outcome of this analysis, the average estrogenic effect in the wastewater treatment plant effluent was equivalent to 5 ng l^{-1} 17ß-estradiol (E2), whereas in untreated municipal wastewater the average was equivalent to 20 ng l^{-1} E2. Matrix effects appeared to be relatively low for effluent water, as the relative inhibition was typically around 20%. Untreated wastewater was however, around 70%. The repeatability standard deviation (SD) is approx. 2.7 ng l^{-1} E2 and 3.8 ng l^{-1} E2 for effluent water and untreated wastewater. This corresponds to a detection limit between 6 and 8 ng l^{-1} E2. The in-house reproducibility SD ranges from

* E-mail: k.simon@quodata.de

3.7 to 5.4 ng l^{-1} E2. Because the latter was only slightly higher than the repeatability SD, the A-YES can be considered very stable over time.

1. Introduction

Estrogens are a group of steroid compounds, named for their importance in the estrous cycle. It's function is as primary female sex hormone. Beside the uniquely natural estrogen like 17ß-estradiol, there is a continuous increasing number of estrogen acting products on the market. Examples are the birth control pill and in a few medical and industrial products. All these estrogen acting products caused an accumulation of estrogens and estrogenic acting compounds in the environment, especially in sewage water where they are not completely degraded. A prevalence of all these estrogenic compounds, however, can disrupt reproduction of organisms, especially in aquatic environments. It is known that the low concentrations of 17ß-estradiol induce vitellogenin synthesis in primary hepatocytes and it promotes feminization in fish species. Furthermore, estrogens are considered to be weak carcinogens for mammalian species (Tag et al. 2006).

The permanent improving concentration of estrogenic substances and their drastic effects on the environment indicate a need to monitor estrogenic activity in wastewater treatment installations, in food, feed and animal products. A number of different methods are available. So far, analytical methods such as LC-MS or GC-MS systems have been preferred. However, these systems detect estrogenic compounds separately and cannot give total estrogenicity content for the samples. In addition extensive sample preparation is necessary, for example extraction and clean-up procedures (Tag et al. 2006).

In order to detect the estrogenicity, the measuring method must be able to detect all compounds with estrogenic activity as sum parameter. This assumption fulfils only biological measuring systems. One of the most sensitive biological screening techniques is the E-Screen which is based on the proliferation of MCF-7 breast cancer cells. This method however, is not useful for measuring the estrogens and estrogenic substances because the cancer cells are very sensitive to the hazardous substances in sewage, urine and milk. Its use thus requires a time-consuming purification of the samples before measurement. The E-screen additionally requires more than 4 days to complete. Other biological methods such as the enzyme-liked receptor assays (ELRA) are only useful in a laboratory setting and for detection of 17ß-estradiol at high concentrations (>20 ng l^{-1}).

A further possibility is to measure the estrogenicity by using microorganisms. Microorganisms which contain all these mentioned proteins are necessary for the estrogen induced pathway in human and animal cells. These assays based on recombinant *Saccharomyces cerevisiae* strains provide direct measurement of the estrogenic activity in samples. The organisms have the DNA sequence of the human estrogen receptor (hER) and an expression plasmid carrying a reporter gene. This reporter gene is under the control of a promoter that contains estrogen-responsive elements (ERE). The production of the encoded reporter protein is induced in the presence of estrogens. The use of the intracellular localized reporter enzymes as measuring signal is however relatively problematic, because of the long measuring time and the inadequate reproducibility. In addition the microorganisms host *S. cerevisiae* is known to be sensitive to environmental variation, urine and milk (Tag et al. 2006, Hahn et al. 2006).

2. Detection of Estrogenic Activities in Environmental Samples, Urine and Milk

To establish an estrogen assay with high resilience to toxic molecules and environmental fluctuations in wastewater, urine and milk, we have developed the *Arxula adeninivorans* yeast estrogen screen (A-YES) assay. The *in vitro* assay based on recombinant *A. adeninivorans* cells engineered to co-express hERα and the phytase (*phyK*) reporter gene under the control of an *Arxula*-derived promoter that had been modified by the insertion of estrogen response elements (Figure 1).

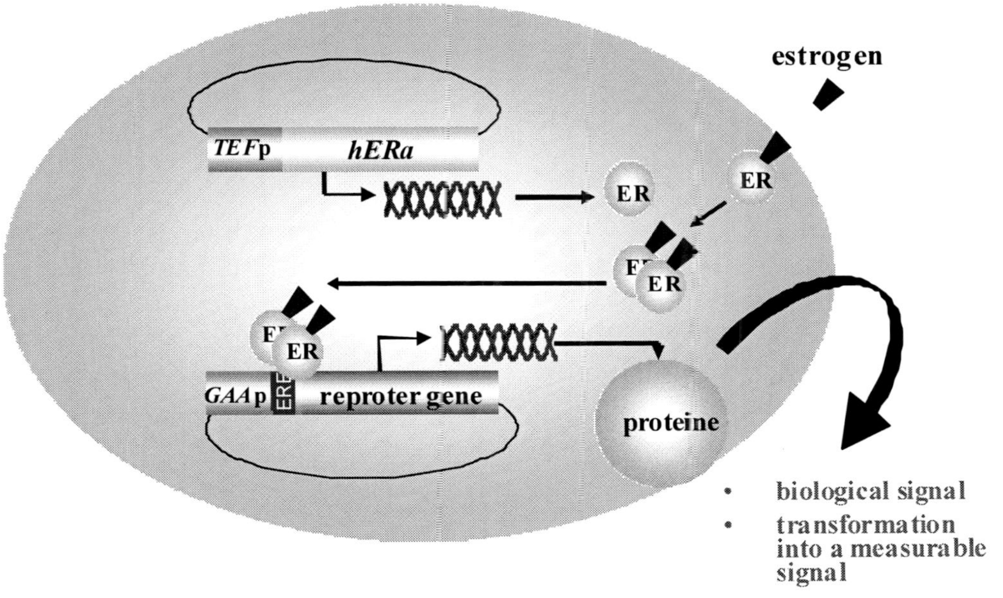

Figure 1. Measuring principle of the A-YES assay for the detection of estrogenic activity in environmental samples, urine and milk. The microbial component of the assay comprises transgenic *A. adeninivorans* cells containing a receptor gene cassette with *TEF1* promoter - *hERα* gene – *PHO5* terminator and a reporter gene cassette with *GAA-ERE* promoter – *reporter* gene (*phyK*) – *PHO5* terminator. The constitutive *TEF1* promoter is arranged for the permanent expression of the *hERα* gene and therefore the synthesis of the recombinant estrogen receptor (hERα) is independent of the estrogen concentration. In the presence of estrogen or estrogen analogues, however, hERα forms a hERα-estrogen-dimer complex, which bind to the ERE-region of the *GAA* promoter located in the reporter gene expression cassette. The cassette (*GAA-ERE* promoter – *reporter* gene (*phyK*) – *PHO5* terminator) is activated, the *phyK* gene is expressed and phytase is synthesized. Since this enzyme contains a native secretion signal sequence it is secreted and accumulates extra-cellularly. The recombinant phytase level is then quantified by a simple biochemical method (Hahn et al. 2006).

For the biochemical phytase detection *p*-nitrophenolphosphate is used as substrate which is digested in presence of phytase to *p*-nitrophenolate and phosphate (Figure 2).

In order to eliminate or reduce an activation of the reporter gene cassette by non-estrogen substances, a control strain with reporter gene cassette (without receptor gene cassette) was constructed and involved in the A-YES assay. In contrast to the estrogen measuring strain (contains both cassettes) it is not able to synthesized phytase in presence of estrogens.

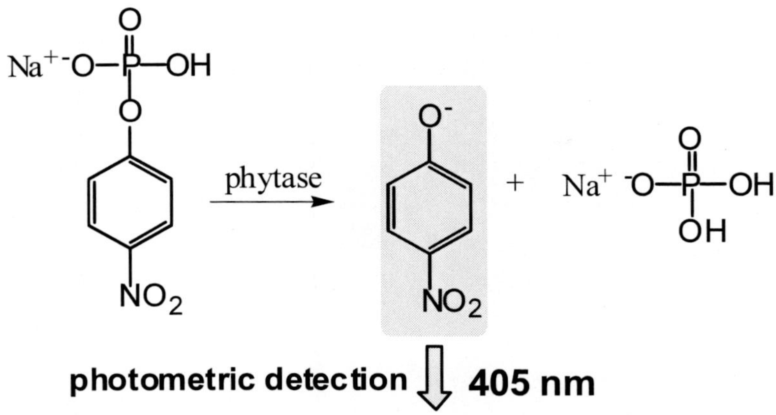

Figure 2. Biochemical detection of phytase activity with the A-YES assay. For this reaction *p*-nitrophenolphosphate is used as substrate. In presence of phytase the substrate is digested to *p*-nitrophenolate and phosphate. Since *p*-nitrophenolate exhibits a yellow colour at basic pH value, the concentration can be detected by a photometric measurement at 405 nm.

The measuring procedure for estrogenic activities by the A-YES assay includes three working steps. In the first step lyophilized yeast cells of the measuring strain plus the control strain are incubated with the respective measuring samples in microtiter plates. Subsequently the biochemical reaction follows. The yeast cells of both strains were centrifuged and the supernatants transfer into new microtiter plates for this purpose. After adding of *p*-nitrophenolphosphate the phytase activity is measured (Figure 3).

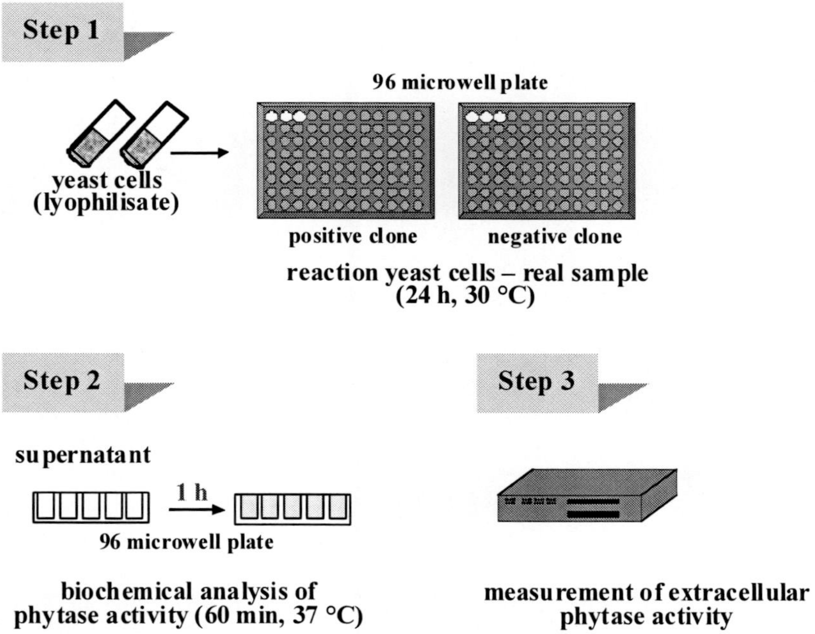

Figure 3. Working steps of the A-YES assay for detection of estrogenic activities.

At the biochemical reaction of *p*-nitrophenolphosphate and phytase it gets a colour reaction which can be measured by a microtiter plate reader. The concentration of *p*-nitrophenolate is therefore a measurement of the estrogenic activity in the measuring sample (Figure 3 – Simon et al. 2008).

In order to reduce the influence of non-estrogenic compounds in samples on the yeast physiology, a correct measured value (y_{ij}) was determined. Therefore the value (M-E_{405nm}) of the phytase activity based on the measuring strain (with reporter gene cassette and receptor gene cassette) and the control value (C-E_{405nm}) based on the control strain (with reporter gene cassette but without receptor gene cassette) were detected. Subsequently the corrected measuring value (y_{ij}) was determined by dividing the control value (C-E_{405nm}) by the measuring value (M-E_{405nm}), i.e.

$$y_{ij} = \text{M-E}_{405nm} \text{ x M-E}_{405nm} \text{(water control) / C-E}_{405nm}.$$

with a detection limit <5 ng l^{-1} 17ß-estradiol in water the A-YES assay is classified as the most sensitive method for detection of estrogenic activities. It measure solid results within 26 hours.

Since the A-YES assay detect all estrogens and estrogen analogues which are able to bind to hERα as sum parameter the specificity is not only limited on the natural estrogen 17ß-estradiol. With identical sensitivity the artificial estrogen 17α-ethynylestradiol can be measured by the A-YES assay. All other previously tested estrogens and estrogen analogues did not, or only binded weakly to hERα. At this reason the sum parameter obtained by the A-YES assay includes particularly 17ß-estradiol and 17α-ethynylestradiol. Since measuring samples like wastewater, urine and milk contain predominantly 17ß-estradiol, the A-YES assay was calibrated with 17ß-estradiol and all measuring values were relating to 17ß-estradiol as equivalent estrogenicity to 17ß-estradiol (Table 1).

Table 1. Specificity of the A-YES assay. The estrogenicity of various estrogens and estrogen analogues were measured by the A-YES assay for this purpose. To compare the obtain data the EC_{50} value (effective concentration of the respective substance causing 50% of the maximal phytase signal) and the relative estrogenicity relation to 17ß-estradiol were calculated (REP)

	EC_{50} [ng l^{-1}]	REP		EC_{50} [ng l^{-1}]	REP
17β-estradiol	~60	1.000000	17β-estradiol-3-sulfate	~550000	0.000109
17α-ethynylestradiol	~55	1.090909	17β-estradiol-3-β-D-glucuronide	~850000	0.000071
estrone	~500	0.120000	17β-estradiol-17-β-D-glucuronide	~12000000	0.000005
17α-estradiol	~1200	0.050000	estron-3-β-D-glucuronide	~27000000	0.000002
estriol	~25000	0.002400	estron-3-sulfate	~3400000	0.000018
2-hydroxyestradiol	~8200	0.007317	cumestrol	~16000	0.010000
4-hydroxyestradiol	~15000	0.004000	campesterol	~2800	0.021429
2-hydroxyestron	~25000	0.002400	bisphenol A	~650000	0.000092
4-hydroxyestron	~3500	0.017143			

With its superior characteristics in handling, analysis time and detection and determination limits, it is applicable for a wide range of matrices like wastewater, urine and milk. 120 samples from 60 wastewater treatment plants (effluent, indirect discharge, municipal wastewater) in Sachsen-Anhalt, Germany, were repeatedly analysed to investigate the reproducibility of the A-YES assay and the influence of matrix effects on the measured values. The analyses were performed according to a nested design. According to the outcome of this analysis, the effective average estrogenic effect in the wastewater treatment plant effluent is equivalent to 5 ng l-1 17ß-estradiol (E2), whereas in untreated municipal wastewater the average is equivalent to 20 ng l-1 E2. Matrix effects appeared to be relatively low for effluent water, as the relative inhibition is typically around 20%. Untreated wastewater is however, around 70%. The repeatability standard deviation (SD) is approx. 2.7 ng l-1 E2 and 3.8 ng l-1 E2 for effluent water and untreated wastewater, respectively. This corresponds to a detection limit between 6 and 8 ng l-1 E2. The in-house reproducibility SD ranges from 3.7 to 5.4 ng l-1 E2. The latter is only slightly higher than the repeatability SD, therefore the A-YES can be considered very stable over time (Figure 4).

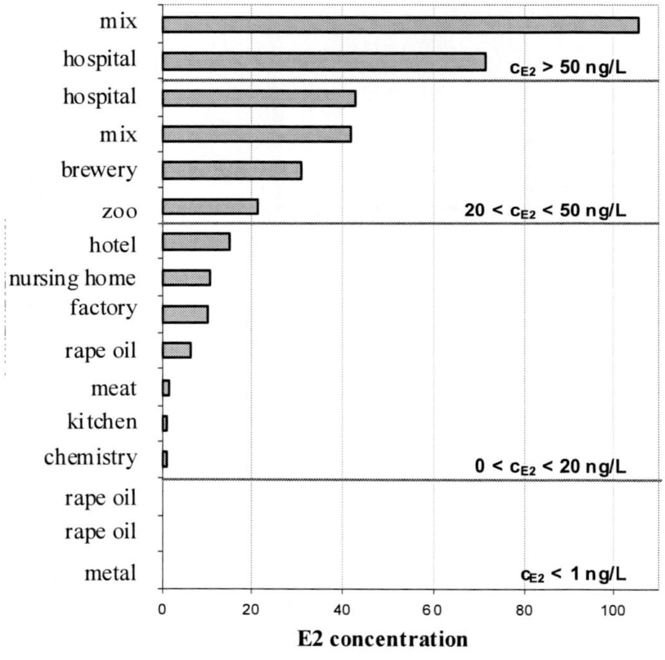

Figure 4. Detection of estrogenic activities of wastewater from various indirect discharges in Sachsen-Anhalt (Germany). All samples were measured by the A-YES assay as described in the text. The presented results are the corrected measuring value (y_{ij}).

In parallel the A-YES assay was applied to detect the estrogenicity in urine and milk. However, a pre-treatment of the samples must be occurring because the estrogens (containing in both types of samples) are available in a bounding form, which reduced the sensitivity of the A-YES assay. This is the reason why an enzymatic treatment has to transfer the estrogens in a soluble form. Subsequently a heat shock step for inactivation of all enzymes and bacteria in the sample must follow and two filtration steps to separate the un-soluble compounds (Figure 5).

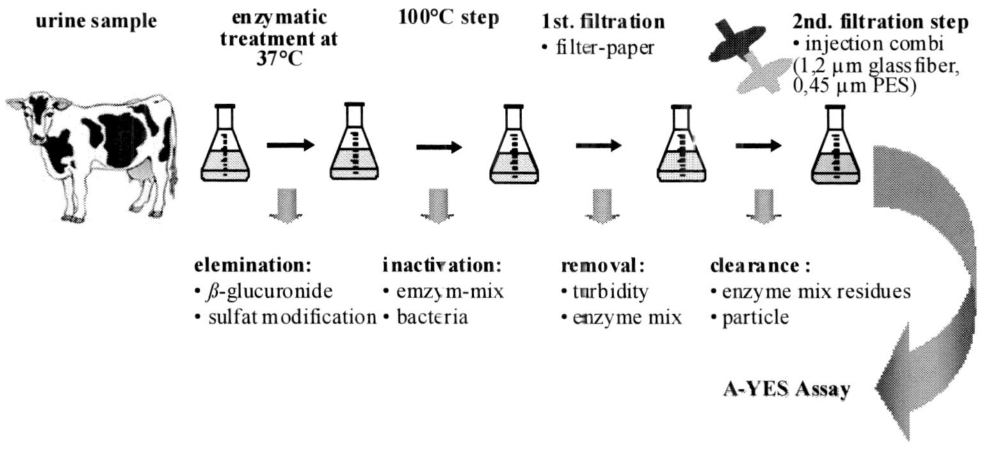

Figure 5. Pre-treatment of urine samples for the detection of estrogenic activities by the A-YES assay. The pretreatment includes (1) the enzymatic treatment of the sample to transfer the estrogens, complete in the solved form, (2) the heat shock step for inactivation all enzymes in the sample including bacteria, (3) two filtration steps for the separation of the unsolved compounds. Subsequently the estrogenic activities can be detected by the A-YES assay.

First measurements of estrogenic activities in cattle urine confirm the functional efficiency of the A-YES assay. In collaboration with the "Referenzlabor für Rückstände beim Bundesamt für Verbraucherschutz und Lebensmittelsicherheit" cattle urine samples were analysed concerning estrogens by using the A-YES assay and GC-MS analysis. Both methods provide nearly identical results. However it must be consider, that the A-YES assay measure the sum parameter estrogenicity which includes all estrogens and estrogen analogous which are able to bind to hERα whereas the GC-MS analysis detect the concentration of each single estrogen. With a detection limit of ~8 ng l^{-1} and a measuring area between 10 and 80 ng l^{-1} the A-YES assay fulfill all requirements for an application in this field (Figure 6).

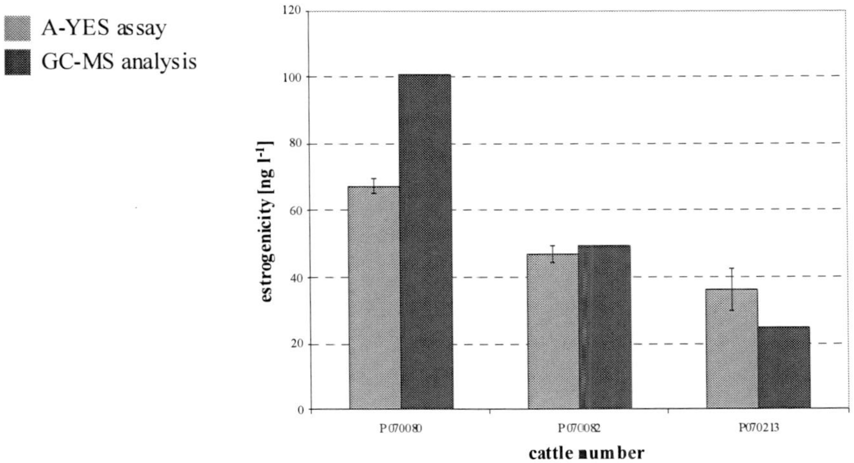

Figure 6. Detection of estrogenic activities in cattle urine. All samples were pre-treated and measured by the A-YES assay as described in the text. The presented results are the corrected measuring value (y_{ij}).

Conclusion

As summary, the A-YES assay is a suitable assay for the analysis of environmental samples, urine and milk concerning the estrogenic activity. In order to obtain reproducible results, the measurements of the samples have to be complemented by measurement of controls and of spiked samples.

In the future the assay will be expanding on further hormone classes like androgens, progesterons and glucocorticoides. Therefore, yeast strains are constructed to contain the androgen, progesteron and glucocorticoide receptors plus the reporter gene cassette including the respective responsive elements. With this assay an overview of the hormone composition in environmental samples, urine and milk will be realised.

References

Hahn T, Tag K, Riedel Km, Uhlig S, Baronian K, Gellissen G, Kunze G (2006) A novel estrogen sensor based on recombinant *Arxula adeninivorans* cells. *Biosens Bioelectron* 21: 2078-2085.

Simon K, Uhlig S, Körner M, Kunze G (2008) Mit mikrobiologischem Testsystem gegen die Wasserverschmutzung. *GIT Labor-Fachzeitschrift* 4: 374-375.

Tag K, Hahn T, Riedel K, Kunze G (2006) Assays and microbial biosensors for detection of endocrine disruptors. In: Biosensors and bioassays based on microorganisms (eds Riedel K, Kunze G, Baronian KHR) *Reseach Signpost*, Kerala, India.

In: Renewable Resources: Obtaining, Processing... ISBN 978-1-6060692-534-8
Editors: R. Kozlowski, G. Zaikov and F. Pudel

Chapter 9

BIODEGRADATION OF FILM POLYMER COATING ON THE BASIS OF CHITOSAN

E. I. Kulish*[*1], *V. P. Volodina*[2], *V. V. Chernova*[*], *S. V. Kolesov*[1] and *G. E. Zaikov*[3]

[1]The Bashkir State University, Ufa
[2]The Institute of Organic Chemistry of the Ufa Scientific Centre, The Russian Academy of Science, Ufa
[3]Institute of Biochemical Physics, Russian Academy of Sciences, Moscow

ABSTRACT

The present paper deals with enzymatic biodegradation of film chitosan coatings which can be used for protecting open wounds (burning, surgical ones) and the means of their modification for extending the service life.

Keywords: chitosan, polysaccharide, enzymatic destruction, collagenase, lirase, intrinsic viscosity.

One of the advantages of using chitosan (ChT) for producing various medical materials – films, threads, coatings, gels – is its capability for biodegradation. However, rapid biodegradation under the influence of enzymes may become the factor limiting the usage of such materials by reducing their service life under the conditions of medical application. The present paper deals with enzymatic biodegradation of film chitosan coatings which can be used for protecting open wounds (burning, surgical ones) and the means of their modification for extending the service life. To provide the prolongation of ChT action there are at least two possible ways. The first one is to change its supermolecular structure, to increase the packing

[*] E-mail: alenakulish@rambler.ru

density and correspondingly to decrease ChT units accessibility for enzymes. The second way is the transformation of ChT into water- insoluble form.

Chitosan (TU 6-09-05-296-76) produced by the company "Chimmed" (Russia) was used as the object of investigation. It is produced by deacytelation of crab chitin, the degree of deacytelation being ~75% and M_η=120000. Polyvinyl alcohol (PVA) of the 16/1 brand with M_η=25000 was also used. The enzyme preparation "Collagenase KK" obtained from hepatopacreas of industrial species of crab and produced by SUE "Immunopreparat" (Ufa) was used as enzyme. ChT films as well as films of its mixtures with PVA were obtained by the method of casting of the polymer solution onto the glass surface. Acetic acid (AcOH) was used as the solvent, its concentration varying from 1 to 70% of mass. The polymer mass concentration in the initial solution was 2% of mass. The film thickness in all the experiments was maintained constant and equal to 0.1 mm. The films were dried until reaching constant weight. For modelling the process of biodegradation of a ChT film specimen on wound surface the film was placed on the base moistened with collagenase solution and was held at 35°C for 1.5 hour. The biodegradation degree was determined according to the data of viscosimetry in the buffer solution consisting of 0.3 M AcOH and 0.2 M sodium acetate. The transformation into the basic form was carried out by holding the film in 0.1 N NaOH solution for 1 hour with subsequent washing in distilled water. The thermomodification of films was carried out by heating at 120°C. The swelling of film specimens was determined by the gravimetric method.

The principal possibility of ChT biodegradation process under the influence of the enzymatic complex "Collagenase KK" has been established earlier [3]. In addition to ascertaining the destruction fact itself it has been found out that the conformational and supermolecular state of ChT in the film and correspondingly the rate and degree of ChT biodegradation are greatly determined by its aminogroups protoning degree which in its turn depends on the concentration of the acid used. The variations in ChT state in AcOH solutions are demonstrated by the values of intrinsic viscosity of ChT which hasn't been in contact with enzyme yet (Figure 1, curve 1).

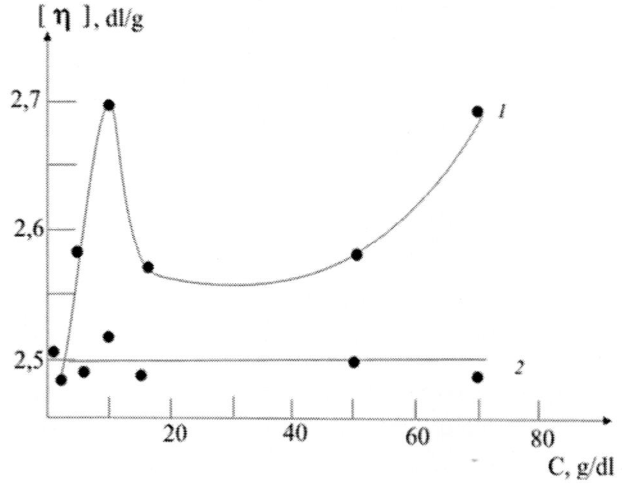

Figure 1. The dependence of the limit value of viscosity of ChT films in the salt (1) and basic (2) forms which were not subjected to biodegradation, on the concentration of acetic acid in initial solution.

For all ChT films the type of dependence of viscosity variation on the time of holding in contact with enzymatic complex is the same (Figure 2).

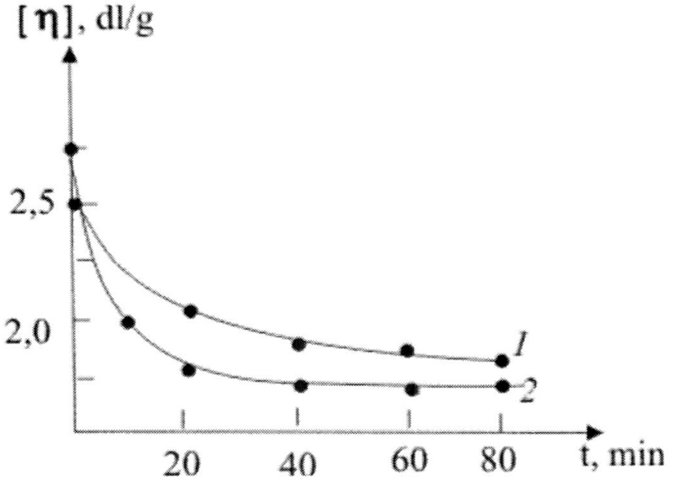

Figure 2. The dependence of the limit value of viscosity of ChT solutions obtained from film specimens in the salt form made of 1% (1) and 10% (2) acetic acid on the time of contact with enzyme.

The most significant drop of intrinsic viscosity takes place in the initial (15-20 min) period. Further enzymatic exposure of specimens doesn't greatly affect the degree of intrinsic viscosity drop. One can also note that both at short and at long periods of holding the films in contact with enzyme solution one and the same trend persists – the least variation of polymer molecular mass is observed in the film obtained from 1% acid while the greatest one – in the films from 10% and especially 70% AcOH (Table 1).

Table 1. The characteristics of ChT film specimens obtained from initial solution with different acetic acid content. (The time of holding in collagenase solution is 1 hour. Molecular mass of initial chitosan is 126.*10^5 (c [η]=3.0,dl/g)

Δ[η] in the process of biodegradation, dl/g	Concentration of AcOH in the initial solution, dl/g.						
	1	5	10	15	20	50	70
For salt films	1.03	1.28	1.33	1.15	1.08	1.12	1.59
For basic film	0.51	0.46	0.63	0.51	0.60	0.65	0.72

ChT films obtained in the salt form (ChT acetate) have good solubility in water and high moisture-absorbing ability (Figure 3). High moisture-absorbing ability of ChT is its undeniable advantage, however, good solubility of such films promotes the accelerated mechanical destruction of a polymer film on wound surface.

Chitosan films holding for 1 hour in 0.1N NaOH solution is accompanied by ChT transformation from the salt form into the basic one [4]. This transformation results in a considerable increase in the density of ChT macromolecules packing with a corresponding decrease in the degree of accessibility of chitosan units for interaction with enzyme.

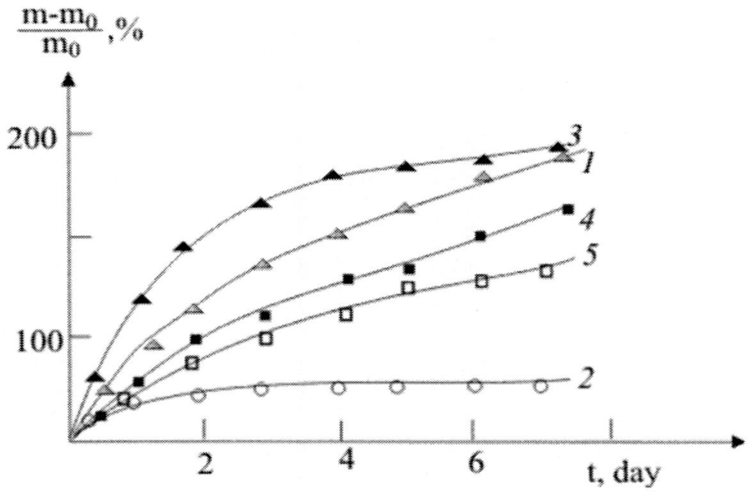

Figure 3. The kinetics of swelling in water vapours of ChT film specimens obtained from 1% acetic acid in the salt form (1), basic form (2), subjected to thermal treating (3) as well as the mixtures of ChT-PVA of the composition 80:20 (4) and 50:50 (5).

The variation of chemical and supermolecular ChT structure in the films transformed into the basic form results in a number of consequences. First, the transition of ChT films into the basic form is accompanied by a considerable decrease in, the degree of film specimens swelling (Figure 3, curve 2). Second, there occurs levelling in the values of ChT intrinsic viscosity (from the comparison of curves 1 and 2, Figure 1). This situation differs significantly from that which was observed in the case of films in the salt form where spread in viscosity values is great. Third, the general form of the curve of dependence of viscosity variation on the time of film contact with solution of the enzyme complex "Collagenase KK" (Figure 4) is the same as in the case of films obtained in the salt form, however, the degree of film specimens biodegradation is considerably lower (see Table 1).

As in the case of biodegradation of salt films the greatest degree of decreasing ChT molecular mass is observed in the case of basic films obtained from 70% AcOH but the general degree of biodegradation of ChT obtained in the basic form is considerably less than that of ChT in the salt form.

Significantly smaller extent of enzymatic degradation of basic films as compared with salt ones is their undeniable advantage because it provides the possibility of longer stay of the film on wound surface without destruction. However, mechanical properties of such films are not high: they are fragile, not strong enough and non-elastic. Besides, such films have low moisture-absorbing ability which suggests their low ability to absorb the liquid given off by wounds and doesn't make these films very suitable for practical use.

Considerable improvement of mechanical properties of the film polymer coating may be attained by adding another water-soluble polymer-PVA- to Chitosan at the stage of film formation. In this case the films formed are strong and elastic [5, 6]. It should be noted that under the chosen conditions PVA doesn't biodegrade. This is demonstrated by the fact that its relative viscosity and intrinsic one are constant for a long period of its contact with enzyme.

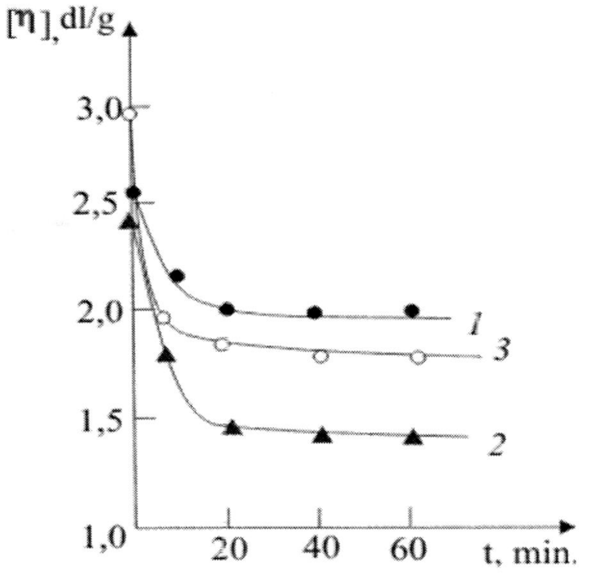

Figure 4. The dependence of the limit viscosity value of solutions of ChTobtained from film specimens formed from 1% acetic acid in the basic (1) and salt forms (2) and heated for 1 hour at 120°C, on the period of contact with enzyme.

However, in the case when the ChT-PVA specimens obtained in the form of films from 1% solution in AcOH begin to contact with enzyme, the ChT enzymatic destruction at all the investigated polymer relationships in the mixture takes place in a greater extent than for individual ChT (Table 2).

Table 2. The characteristics of ChT-PVA film specimens obtained from solutions in 1% acetic acid. (The time of holding in collagenase solution is 1 hour. The molecular mass of initial chitosan is 1.26*10^5 ([η] =3.0 dl/g).)

Variation in the process of destruction	PVA content in the mixture, % mas.				
	0	20	40	60	80
Δ[η] , dl/g	1.03	1.32	1.35	1.35	1.40
$\Delta M_\eta * 10^{-3}$	36	48	50	50	52

As mentioned in [7] irrespective of acid concentration in the initial solution, in the formed ChT-PVA film there takes place a decrease in the number of supermolecular formations and an increase in their size as compared with additive values. This fact is evidently due to the existing interpolymer interaction, i.e. the formation of ChT-PVA heteroaggregates which are known to be characterized by lower packing density and by great sizes.

Thus, the introduction of the second polymer is likely to change the supermolecular structure of the initial film, which is embodied in the kinetics of biodegradation of the film polymer composition. ChT accessibility for its interaction with collagenase in such loosely-packed aggregates is greater than in the case of ChT-ChT homoaggregates with high packing density.

It should be noted, however, that ChT-PVA film polymer coatings don't have high sorption ability despite their good solubility in water because individual PVA doesn't possess high moisture-absorbing ability (Figure 3, curves 4, 5).

To obtain films insoluble in water which preserve a high degree of swelling is possible by thermal modification of films. As known [8] the heating of chitosan films for 3 hour at the temperature about 120°C is accompanied by significant rearrangement of the chemical structure, which doesn't, however, affect their swelling ability (Figure 3, curve 3).

Since in the case of films insoluble in water it isn't possible to analyze the capacity for destruction, the specimens put through thermal treatment for 1 hour during which films still retain the capacity for dissolution in water, were taken for investigation. Nevertheless, in this case one can observe a pronounced tendency for increasing ChT intrinsic viscosity depending on the time of specimens heating (Figure 5, curve 1). The evident reason of increasing intrinsic viscosity is the process of ChT joining accompanying the modification. At increasing the time of thermomodification the film stability to the process of biodegradation also increases (Figure 5).

In fact, biodegradation of such thermomodified specimens proceeds up to reaching the viscosity values close to those which they have in basic films as well (Figure 4, curve 3). However, such thermomodified specimens, unlike basic films, are characterized by good mechanical properties.

Thus, the chemical modification of chitosan films carried out by treating the formed film with alkaline solution or by thermal treatment (joining) of the film makes it possible to obtain polymeric coating with a good complex of physical properties and increased stability to enzymatic degradation.

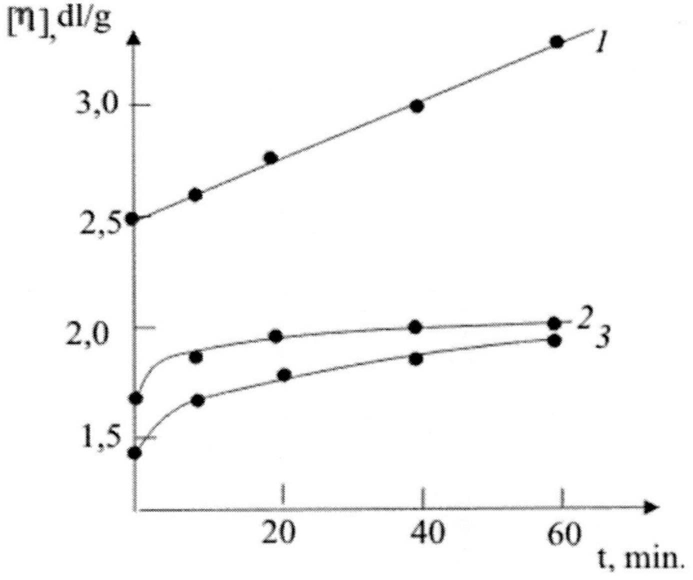

Figure 5. The dependence of the limit value of viscosity of ChT film specimens obtained from solution in 5% acetic acid on the time of heating at 120°C. 1- the films not subjected to contact with enzyme, 2- the films contacting with enzyme for 10 min, 3- the films contacting with enzyme for 30 min.

Thus, the following conclusions can be made from the given work:

(1) Thermomodified chitosan films are characterized by the absence of solubility in water, high moisture-absorbing ability, high strength and elasticity. At increasing the time of thermal treatment the stability of chitosan films to the process of enzymatic biodegradation also increases.
(2) Modified films in which ChT is in the basic form are characterized by low solubility in water as compared with non-modified films, and by low moisture-absorbing ability. The given films possess high stability to the process of enzymatic biodegradation and slightly depend on the prehistory of the film formation.
(3) Mixture polymeric coating "chitosan-polyvinyl alcohol" have high strength and elasticity, increased biodegradating ability and good solubility in water. Moisture-absorbing capacity of such films is somewhat lower than that of individual chitosan.

REFERENCES

[1] Chitin and Chitosan: Obtaining, properties and use. /Under the red. of Scraybyn K.G., Vikhoreva G.A., Varlamov V.P. M.: *Nauka*, 2002, (in Russian).
[2] Muzzarelli R.A.Chitin.Oxford : Pergamon Press, 1977, 309 p.
[3] Kulish E.I., Volodina V.P., Fatcullina R.R., Kolesov S.V. *Proceedings of the 8 th International Conference "Modern prospects in the Investigation of Chitin and Chitosan"*. M.,VNIRO. 2006. S.290.
[4] Ageev E.P., Vikhoreva G.A., Matushkina N.N., Pchelko O.M. // *Vysokomolec. soed.* A. 2000. V.42. №2. S.333.
[5] Mukhina V.R., Pastukhova N.V., Semchikov Yu.D., Smirnova Z.A., Kiryanov K.V., Zhernenkov M.N. //*Vysokomolec. soed.* 2000. V. 43 A. №10. S.1797.
[6] Nikolaev A.F., Prokopov A.A., Shulgina E.S., Golenyscheva S.A., Klubikova Z.E., Musikhin V.A. // *Plast.massy*.1987. №11, S.40.
[7] Kulish E.I., Kolesov S.V.. // J.P.Ch.2005.№9. S.1511.
[8] Zotkin M.A., Vikhoreva G.A., Kechekyan A.S. // *Vysokomolec. soed.* B. 2004. V.46. №2. S.359.

In: Renewable Resources: Obtaining, Processing... ISBN 978-1-6060692-534-8
Editors: R. Kozlowski, G. Zaikov and F. Pudel

Chapter 10

THE EFFECT OF ADDITION OF FLAXSEED OIL TO THE DIET OF LAYING HENS HOUSED IN THE DIFFERENT HOUSING SYSTEM ON THE NUTRITIONAL VALUE OF THE EGGS

Kozłowska Jadwiga[1], Biskupski Marcin[1], Gromek Marta[1], Wężyk Stanisław[2] and Połtowicz Katarzyna[2]

[1]Institute of Natural Fibres, Poznań, Poland

[2]National Research Institute of Animal Production at Balice/Kraków

ABSTRACT

Supplementing the diet for laying hens with components characterized with high content of n-3 polyunsaturated fatty acids (PUFA n-3) allows for controlling fatty acid composition of lipid in egg yolks.

The studies were aimed at determining the effect of addition of flaxseed oil in doses 0, 3 and 6% to the feed of laying hens on yolk fatty acid incorporation, egg quality as well as hen performance and egg production parameters.

The experiment was carried out for 25 weeks on Hy Line Brown laying hens housed in the two systems: individual cages and on the litter.

It was observed that the 3 and 6% addition of flaxseed oil had a significant effect of enriching the egg yolks in alpha-linolenic acid, total omega-3 polyunsaturated fatty acid content and better, pro-healthy fatty acid composition of egg yolk lipids. Average ALA content in yolk lipids from layers housed in cages, fed with 0, 3 and 6 % flaxseed oil supplementation, went up respectively 1.6, 3.3, 5.8% and the content of ALA in yolk lipids from layers housed on the litter, reached respectively 1.7, 5.5, 8.3%.

Feeding hens with the enriched diet did not result in any significant changes in egg quality, productivity and the feed consumption per egg.

Better results in terms of nutritional value and omega-3 PUFA concentration were obtained for hens housed on the litter system. This results from lower productivity and higher feed consumption per egg on the litter in comparison to the high efficiency cage system.

1. Introduction

The popularization of WHO/FAO recommendations on lowering cholesterol and change in consumption structure of fatty acids has strongly affected egg and meat markets. In the 90s in USA and Canada the egg consumption dropped by more than 15%. [1,2] While the demand for eggs and meat was decreasing the information on negative effects of excessive limiting the consumption of these products started to appear. Numerous studies on modifying the fatty acid profile of egg yolks through especially designed selection of feed for laying hens were carried out. [3-7] For enriching the laying hen diet in PUFA n-3 flaxseed is most commonly used. The studies concerning enriching eggs in unsaturated n-3 fatty acids by addition of higher doses of flaxseed results in lower production parameters, lower feed consumption and also worse quality features of eggs. An alternative to flaxseed as a source of unsaturated n-3 fatty acids is flaxseed oil, which is free from anti-nutritional components such as mucilaginous substances, cyanogenic glucosides, linatin or estrogen precursors. However few studies were conducted with the use of flaxseed oil as the enriching in PUFA n-3 component, mostly because of the oil susceptibility to oxidation [8]. The advances in the research on antioxidant additions to feeds allows for gradual introduction of unstable components.

2. Experimental [9]

2.1. Materials and Methods

The experiment was conducted from 26 to 34 week aged laying hens (layers) from Hy Line Brown strain housed in two housing systems: in cage and on the litter. The layers were divided into three groups: control (diet without flaxseed oil) and two experimental (diet with 3% and 6% addition of flaxseed oil). The experiment was carried out in the experimental farm of the National Research Institute of Animal Production at Balice/Kraków.

During the experiment the layers were fed with commercial diet based on commonly used industrial recipes. Balanced diets were modified regarding the fatty acid profile by introducing flaxseed oil of 53.62% α-linolenic acid (ALA) as a feed source of PUFA n-3. Three kinds of diet were administered: two with addition of flaxseed oil in the dose of 3 or 6% and the third – control, where the source of fat was lard. To the three diets the anti-oxidant containing 8.85% BHT (E-321 butylohydroxytoluen) and 19.6 % EQ (E-324-ethoksyquin-polietoksychinolina) was added at 250g/MT of fat.

In the diet, cold-pressed flaxseed oil obtained from Omega linseed cultivar was used. The analyses of fatty acid profiles of feed and egg yolks were performed by gas chromatography (Perkin Elmer Autosystem XL, with flame detector).

In each tested periods i.e. 26, 27, 28, 30, and 34 weeks of layers' age in each experimental group 10 eggs were picked randomly for the evaluation of egg and shell quality. The quality parameters of the egg content were determined: average egg mass (g), yolk mass (g), yolk colour (Points of La Roche's scale), and shell quality parameters: shell mass (g), thickness (μm), density (mg/cm^2) and strength (N).

Every day during the experiment individual (cage housing) and group (on the litter housing) laying efficiency was measured (%) as well as the amount of administered feed (g).

Based on the results average laying efficiency in percent, average feed consumption per layer and feed consumption per egg in each week of the experiment were calculated.

2.2. Results

After enriching the diet with 3 or 6% addition of flaxseed oil, a statistically significant increase ($P \leq 0.05$) in ALA content in yolk, correlated with the oil dose, was observed. Average ALA content in yolk lipids from layers housed in cages, fed with 6 % flaxseed oil supplementation, went up gradually from 4.33% after the first week of the experiment to 9.57% after 8 weeks (Figure 1., 4). Average content in yolk lipids from layers housed on the litter, fed with 6 % flaxseed oil supplementation, went up from 2.00% to 10.51%, as compared with the range of 1.13% - 2.12% observed in yolk lipids in the control group (Figure 2., 4.).

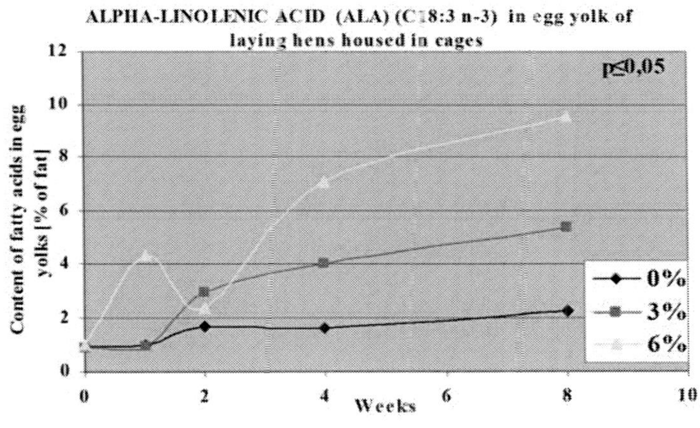

Figure 1. Alpha-linolenic acid (ALA) content in yolk lipids. Layers housed in cages during 8-week feeding with diet enriched in 0, 3, and 6% flaxseed oil.

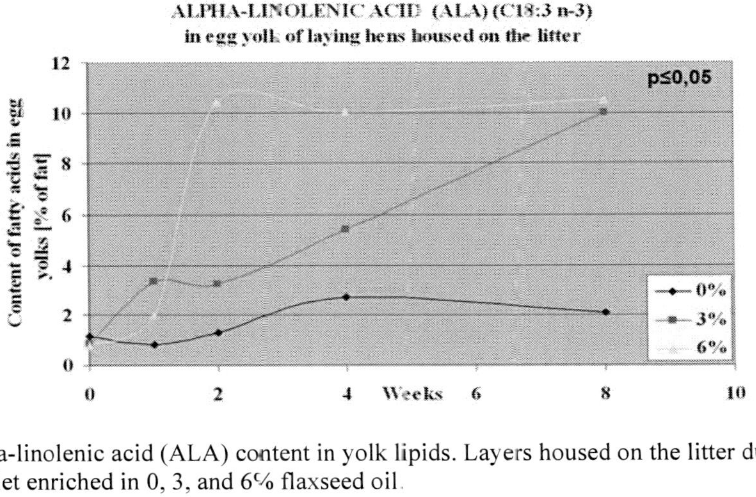

Figure 2. Alpha-linolenic acid (ALA) content in yolk lipids. Layers housed on the litter during 8-week feeding with diet enriched in 0, 3, and 6% flaxseed oil.

Total PUFA n-3 content in yolk lipids during the experiment for the 6 groups of layers is presented on the figure 3.

The main PUFA n-3 component in the enriched eggs was alpha-linolenic acid, which, according to the analyses of hens diet was mainly PUFA n-3 provided in the diet. After 8 weeks of supplementing layers' diet with the addition of 0, 3 and 6% flaxseed oil, the percentage share of ALA reached respectively 78, 85 and 86 % of total PUFA n-3 (figure 5). The result analysis shows that the total PUFA n-3 content in yolk lipids increased in the consecutive terms of tests when layers were fed with diets containing 3 and 6% flaxseed oil relatively the dose.

Period of feeding, weeks	Hy Line Brown laying hens								
	0% dose of oil			3% dose of oil			6% dose of oil		
	Cage	Litter	Average	Cage	Litter	Average	Cage	Litter	Average
	Content of fatty acids in egg yolks [% of fat] (PUFA n-3)								
0	2.43	2.68	2.56 a, x	2.06	2.54	2.30 a, y	3.12	2.87	3.00 a, z
1	2.61	2.51	2.56 b, x	2.66	4.67	3.67 ab, y	6.57	5.51	6.04 a, y
2	2.73	2.75	2.74 b, x	4.36	4.63	4.50 ab, y	4.52	11.72	8.12 a, y
4	2.70	3.64	3.17 c, x	7.58	6.67	7.13 b, x	8.37	11.75	10.06 a, x
8	2.87	2.68	2.78 c, x	7.04	10.96	9.00 b, x	10.66	12.61	11.64 a, x
Average	2.67	2.85	2.76 c	4.74	5.89	5.32 b	6.65	8.89	7.77 a
Average (1-8)	2.73	2.90	2.81 c	5.41	6.73	6.07 b	7.53	10.40	8.96 a

a - b Mean within a row with a common letter do not differ significantly P(≤0.05).
x – y Średnie wewnątrz kolumn oznaczone tą samą literą nie różnią się istotnie na poziomie P≤0.05.

Figure 3. Total PUFA n-3 content in yolk lipids in the yolks within 8 weeks of feeding laying hens with diet containing 0, 3, 6 % of flaxseed oil.

Period of feeding, weeks	Hy Line Brown laying hens								
	0% dose of oil			3% dose of oil			6% dose of oil		
	Cage	Litter	Average	Cage	Litter	Average	Cage	Litter	Average
	Content of fatty acid in egg yolks [% of fat] (ALA)								
0	0.94	1.13	0.69 a, y	0.90	0.85	0.75 a, w	0.97	0.78	0.78 a, u
1	1.02	0.84	0.95 c, y	0.91	3.37	1.35 b, u	4.33	2.00	1.75 a, z
2	1.66	1.28	1.65 c, xy	2.92	3.26	2.13 b, z	2.38	10.44	3.08 a, y
4	1.61	2.72	2.78 c, x	3.97	5.42	3.42 b, y	7.06	10.03	4.56 a, x
8	2.22	2.12	4.11 c, xy	5.37	10.00	5.30 b, x	9.57	10.51	6.36 a, x
Average	1.49	1.62	1.55 c	2.81	4.58	2.41 b	4.86	6.75	3.26 a
Average (1-8)	1.63	1.74	1.68 c	3.29	5.51	2.77 b	5.84	8.25	3.84 a

a - b Mean within a row with a common letter do not differ significantly P≤0.05.
x – y Mean within a column with a common letter do not differ significantly P≤0.05.

Figure 4. The average concentration of ALA in the yolks within 8 weeks of feeding laying hens with diet containing 0, 3, 6 % of flaxseed oil.

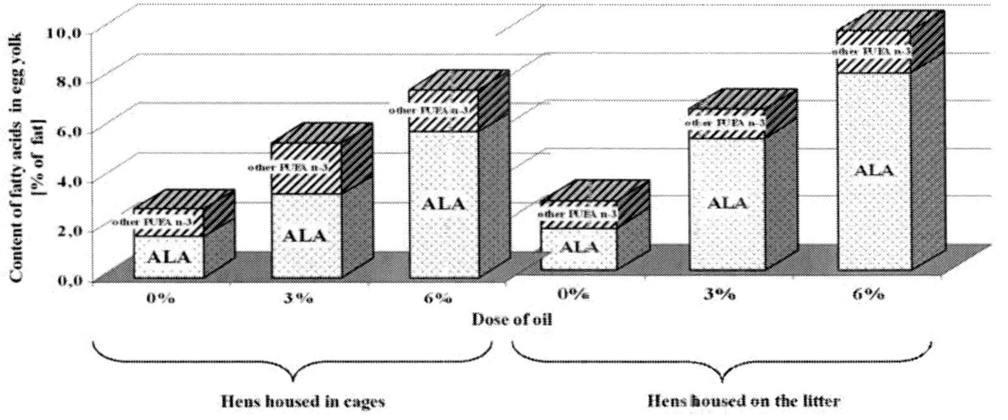

Figure 5. The average concentration of PUFA n-3 in yolk lipids during feeding of laying hens with diet enriched with flaxseed oil.

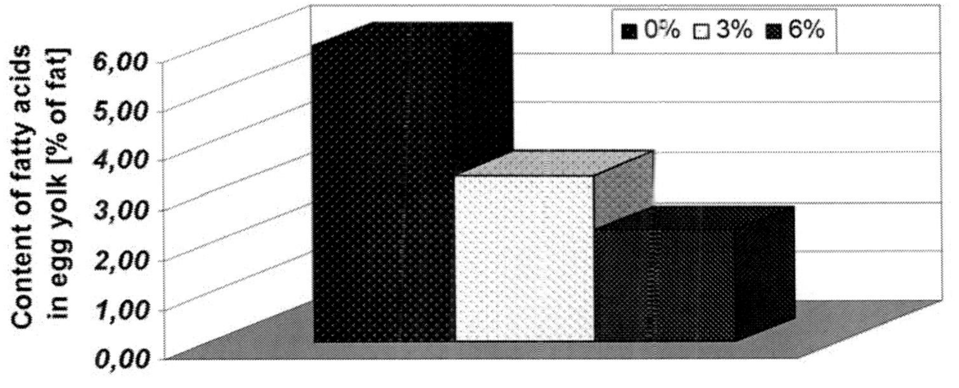

Figure 6. PUFA n-6:n-3 ratio in yolk lipid during feeding layers with diet enriched in 0, 3 and 6 % of flaxseed oil.

As Figure 6 illustrates, the ratio PUFA n-3:PUFA n-6 was much lower than in the control dose (0%) within 8 weeks of feeding layers with diet enriched in 3 and 6 % of flaxseed oil.

In yolks of eggs laid by layers fed with enriched diets (3 and 6 % of flaxseed oil) lower content of MUFA is observed as compared to the control group. The content of each of the remaining MUFA was insignificant in terms of diet (each acid below 0.5%).

The results of accumulation of SFA show that in the second half of the experiment for the yolks from layers fed with flaxseed oil a lower concentration of SFA with shorter carbon chains i.e. miristinic (C14:0) and palmitinic acid (C16:0), significantly different between the oil doses in the feed. The content of other SFA lowered only slightly or stayed the same in that period.

PUFA:SFA ratio in yolks of layers fed with flaxseed oil addition went up gradually, although not necessarily reaching the point of statistical significance, along with the increase of the oil dose. The average values of PUFA:SFA ratio in yolks for the whole experimental period were 0.60, 0.75, 0.84 for 0, 3 and 6% addition of flaxseed oil respectively and were considerably higher than for the control.

Important effect of flaxseed oil supplementation on average egg mass, yolk mass, yolk colour intensity, density and strength of shells was not observed in any of the testing periods (figure 7).

Egg and shell quality parameters	Average value								
	0% dose of oil			3% dose of oil			6% dose of oil		
	Cage	Litter	Average	Cage	Litter	Average	Cage	Litter	Average
Egg mass [g]	60,85	58,97	59,91 a	61,15	61,83	61,49 a	62,78	60,79	61,79 a
Yolk mass [g]	15,57	15,14	15,36 a	15,88	15,14	15,51 a	15,43	15,13	15,28 a
Yolk colour (Points of the La Roche's scale)	13,53	12,53	13,03 a	13,38	12,23	12,80 a	13,53	11,95	12,74 a
Shell mass [g]	5,49	6,05	5,77 a	5,75	5,71	5,73 a	5,74	5,74	5,74 a
Shell thickness (μm)	357,28	381,25	369,26a	373,05	364,93	368,99a	364,10	371,78	367,94a
Shell density (mg/cm2)	76,07	82,24	79,15 a	79,42	78,22	78,82 a	77,88	79,56	78,72 a
Shell strength (N)	28,66	33,63	31,15 a	34,50	27,37	30,93 a	29,24	28,78	29,01 a

a –b Mean within a row with a common letter do not differ significantly P≤ 0.05.

Figure 7. The average value of egg and shell quality parameters within 8 weeks of feeding hens with feed containing 0, 3, 6 % of flaxseed oil.

For layers housed on the litter, after 3 and 4 weeks of introducing the experimental diet, a temporary decrease of laying efficiency was observed when compared with the control group. Finally the average laying efficiency in the whole experimental period, i.e. between 25th and 34th week of layers' life, did not show considerable differences between the three groups Generally, in the whole experimental period, average daily feed consumption per layer in groups fed with diets supplemented with 0, 3 or 6% flaxseed oil reached respectively 119,85, 127.44, 128.75 g/day and was significantly different between 0 and 3 or 6% oil dose, without much difference between the 3 and 6% doses. Average feed consumption per egg for the whole experimental period did not vary considerably in the layers fed with all three diets. (figure 8).

Production results	Average value								
	0% dose of oil			3% dose of oil			6% dose of oil		
	Cage	Litter	Average	Cage	Litter	Average	Cage	Litter	Average
Laying rate [%]	96,63	83,22	89,93 a	96,34	82,69	89,52 a	94,95	81,88	88,42 a
Daily feed intake per hen [g]	119,85	127,90	123,88 b	127,44	128,03	127,74 a	128,75	127,96	128,36 a
Feed consumption per egg [g]	128,48	154,05	141,27 a	133,61	151,28	142,45 a	134,47	155,60	145,04 a

a - b Mean within a row with a common letter do not differ significantly P ≤ 0.05.

Figure 8. Results of productivity within 8 weeks of experiment - addition of 0, 3, 6% of flaxseed oil.

2.3. Discussion

Drawing from the results of the study, anti-oxidatively stabilized flaxseed oil is an efficient feed source of PUFA n-3 for laying hens. A lot of studies on using flaxseed in production of eggs enriched with PUFA n-3 indicate side effects of 10-20% addition of flaxseed to the diet, including mainly lower feed consumption and body weight, worsening of several egg quality parameters or liver damage in layers. Whereas with using flaxseed oil as a source of PUFA n-3, in most studies unfavourable changes in quality and production parameters were not observed.

This chapter also did not report the influence of 3 and 6 % dose of flaxseed oil on the egg quality or productivity values as compared to the control.

The comparison of the results for cage and on the litter housing systems indicates that layers housed on the litter cumulated in yolks more ALA than those housed in cages (5.2 and 3.6 of yolk lipids respectively) ($p \leq 0.01$). Higher ALA accumulation in yolks results from the production efficiency in both housing systems, i.e. in cage system the average laying efficiency was higher than in on the litter by 13.93%, at lower feed consumption by 21.46g/szt. and feed consumption per layer at 1.96g ($p \leq 0.01$).

Conclusion

(1) The addition of 3% and 6% of flaxseed oil to the layer diet resulted in significant increase of ALA content in eggs and also in better pro-healthy composition of yolk lipids.
(2) ALA content in yolk lipids after 8 weeks of feeding layers housed in cages with diets containing 0, 3 or 6% flaxseed oil reached 2.2; 5.4 and 9.6%, respectively and from layers housed on the litter reached respectively 2.1; 10.0 and 10.5%.
(3) Feeding layers with the enriched diet with 3% or 6% of flaxseed oil did not cause considerable changes in egg quality, eggs productivity and feed consumption per egg.
(4) More desirable results regarding nutritional value and PUFA n-3 concentration in yolks were achieved on the litter system, what was linked with lower eggs productivity and higher feed consumption in comparison to the high efficiency cage system.

References

[1] Agricultural Economics Report No.380.(1993). Potential Use of Flaxseed in Egg Production NDSU, Fargo, North Dakota. Prep by Stearns L.D, Petry T.A., Holstun J., Zetocha D.F., for the ND Oilseed Council and ND Agricultural Products Utilization Commission.

[2] Born F. (1996). Flaxseed in the Egg Industry. *Proceedings Flax Council of Canada Conference " Flax in next decade"* December 3-4, Winnipeg, Canada.

[3] Cunnane S.C. (1995). Metabolism and function of alfa-linolenic acid in humans, in *Flaxseed in Human Nutrition*, Ed. S.C. Cunnane, L.U.Thompson. AOCS Press. *Champaign,* s.99-127.

[4] Van Elswyk M.E. (1997). Comparison of n-3 fatty acid sources in laying hen rations for improvement of whole egg nutritional quality: a review. *Br. J. Nutr., Jul*, 78, Suppl 1, 61-69.

[5] Garcia A.U., Albala C. (1998). Lipid composition of eggs from hens fed with marine protein and fat products. *Archivos Latinoamericanos de Nutrition*, Mar., 48(1),71-76.

[6] Scheideler S.E., Jaroni D., Froning G. (1998). Strain and effects on eggs composition from hens fen diets rich in n-3 fatty acids. *Poltry Science*, 77(2), 192-196.

[7] Cunnane S.C.(2003a). Dietary Sources and Metabolism of α-linolenic acid, *in Flaxseed in Human Nutrition*. Ed. L.U. Thompson, S.C. Cunnane. AOCS Press. Champaign, s.63-91.

[8] Scheideler S.E. (2003a). *Flaxseed in Poultry Diets: Meat and Eggs in Flaxseed in Human Nurtition*, wyd.2. Ed. L.U. Thompson, S.C. Cunnane. AOCS Press. Champaign, 423-428.

[9] Kozłowska J., Wężyk S.(2005). Wpływ dodatku oleju lnianego do paszy kur nieśnych na profil kwasów tłuszczowych w żółtku jaja, Kraków, Poznań.

In: Renewable Resources: Obtaining, Processing... ISBN 978-1-6060692-534-8
Editors: R. Kozlowski, G. Zaikov and F. Pudel

Chapter 11

Bio-Damages of Materials. Adhesion of Microorganisms on Materials Surface

***K. Z. Gumargalieva*[1], *I. G. Kalinina*[1], *S. A. Semenov*[1], *M. I. Artsis*[2] *and G. E. Zaikov*[2]**

[1]N.N. Semenov Institute of Chemical Physics, Russian Academy of Sciences, 4 Kosygin Str., Moscow 119991, Russia
[2]N.M. Emanuel Institute of Biochemical Physics, Russian Academy of Sciences, 4 Kosygin Str., Moscow 119991, Russia

Abstract

Adhesion interaction of the most widely distributed species of microscopic fungi: Aspergillus niger, Trichoderma viride, Penicillium funiculosum, Aspergillus terreus to surfaces of materials (polymers, metals) is studied. The force of adhesion interaction was measured by the method of centrifugal detachment. On the base of analysis of kinetic curves the macroscopic characteristics of adhesion micro-organism–metal surface were obtained. We show on the example of Aspergillus niger that stochastic nature of adhesion of microorganisms cells is caused by heterogeneity of support surface, heterogeneity of conidium sizes, and distribution of them by forces of adhesion obeys the Gauss law. We established that structure of cellular wall of microscopic fungi was changed in dependence on age, and change of force of adhesion interaction correlated with changes in cellular wall structure of microscopic fungi, and dominating role in strengthening of adhesion interaction was played by increase of albuminous components concentration in surface layer of cell.

Keywords: adhesion, adhesion interaction, adhesion force, polymer surfaces, microorganism, Aspergillus niger, Aspergillus terreus, Trichoderma viride, Penicillium funiculosum.

INTRODUCTION

Professor N.M. Emanuel's set problems in the field of applied works were characteristic by fundamental approach. In particular, at the beginning of eighties of the last century the questions of polymer materials bio-damages traditionally related to micro-biological discipline were included by N.M. Emanuel into section of chemical destruction of polymer materials science. Actually, further investigations in this field showed that active destructing agents of materials under interaction with microorganisms were the products of their vital functions – metabolites representing famous chemical agents: water, salts, acids, alkali, ferments, toxins, i.e. finally bio-damage should proceed by laws of chemical destruction [1-10].

Investigation of result of bio-destruction is generally accepted, whereas it follows two stages: adhesion or microorganism connection to material surface with further growth of microorganism bio-mass as a result of substrate-support utilization. These two stages should predetermine further processes of material degradation.

At conditions of present-day ecologically fraught environment and fast adaptation of various microorganisms species to changing environment conditions from the one hand, and accumulation of various waste of synthetic origin from the other one, the investigations in the field of development of both fast degradating materials for the cases of utilization necessity and stable materials for cases of protection from bio-destruction are necessary [11-12].

Studying of adhesion of microorganisms cells is of great interest now, but these investigations in many cases are "exotic" in relation to selection of microorganisms' species. In work [13] they studied selective adhesion to micaceous plates covered by polyethyleneimine, extremorphil cultures grown at high temperatures and low phosphorescent; by method of scanning electron microscopy they affirmed that adhesion of extremorphil to polyethylenimine covering was preferred in comparison with adhesion to polylysine one. The main physical-chemical factors of material surface such as hydrophobicity, roughness, morphology of microorganisms cells are discussed in works [13-16]. These works and up to date state of the problem confirm the necessity of investigation of adhesion interaction nature on the interface material–cell's wall of microorganism with attempt of quantitative estimation of this process.

OUR RESULTS AND THEIR DISCUSSION

Adhesive cells of microorganisms act as aggressive bio-reagents as a result of evolving of exo-ferments or other low-molecular substances organizing so-called bio-film. That is why quantitative parameters of adhesion are determinative for rates of biofouling (biomass accumulation) and bio-destruction.

Interaction polymer surface–microorganism was determined by the value of adhesion force in dependence on contact time at various external conditions (temperature, moisture) to polymer materials of various degrees of hydrophilicity: polymethyl methacrylat, cellophane, polyethylene, acetylcellulose, epoxide resin and polyethylene terephthalate.

Conidiums of microscopic fungi Aspergillus niger were drifted on materials surface in doses from water suspension with definite titre by micro-batcher, then they were dried and aged at various thermo-moist conditions in the interval from 0 up to 24 hours. The value of conidiums Aspergillus niger adhesion force to surfaces of polymer materials was determined by method of centrifuge detachment [17]. Change of morphology, (form) and sizes of conidiums were studied by method of scanning electron microscopy "Tesla" BS300. Amount of flown off or residual conidiums on the surface of material as a result of action of centrifuge force (angle rate of rotor w = 15000 revos for 15 minutes) was counted in Goryaev's chamber.

Change of amount of residual conidiums $\gamma = N/N_0$ Aspergillus niger after centrifugation in dependence on time of preliminary ageing at given thermo-moist conditions ($T = 22^oC$, $\varphi = 30\%$) is presented in Figure 1. adhesion interaction of system polymer material–microorganism has kinetic character. Each system polymer material–microorganism is characterized by time of end of adhesion forces formation between conidiums and surface (approaching the plateau by kinetic curve) which depends on external conditions.

Presented in Figure 1 dependences are well described by equation (1):

$$\ln \gamma/\gamma_\infty = -K \cdot t \tag{1}$$

where γ, γ_∞ – part of particles remained on the surface of polymer material after preliminary ageing during the time t, and irreversibly adhesive at given conditions accordingly, K – constant of rate of formation of adhesion forces between conidiums and material surface. Values of parameters determined from equation for investigated polymers and also the value of force influencing on one cell are presented in Table 1. At $\omega = 15000$ min^{-1} the force influencing on each conidium was equal to $F = 1{,}2 \cdot 10^{-4}$ dyne/cell. As it is known [18] the value of force under adhesion corresponds to the value $\gamma = 50\%$. If we know F_{50}, the time of kinetic curve approaching the plateau and exponential course of kinetic curves of adhesion we may determine the values of adhesion forces for each polymer at given thermo-moist conditions.

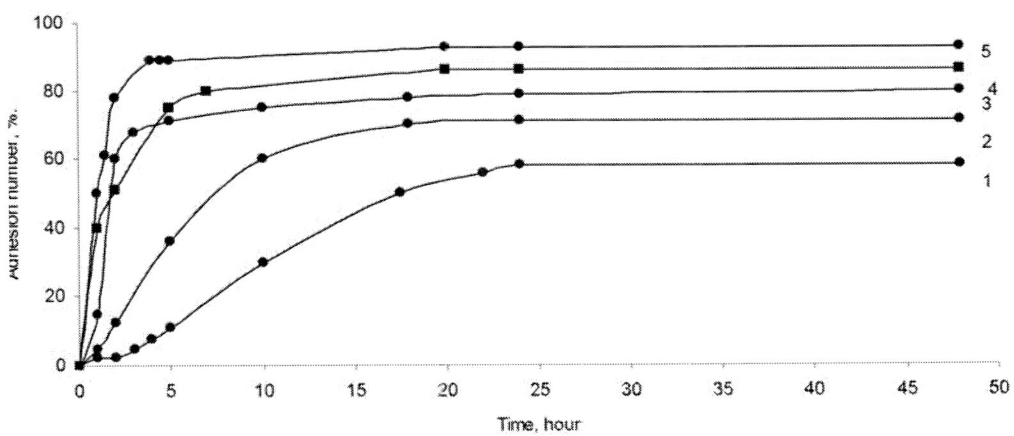

Figure 1. Kinetic curves of conidiums A niger adhesion to various polymer materials at temperature 22°C and relative moisture 30%. 1 – polyethylene, 2 – epoxide resin, 3 – polymethyl methacrylate, 4 – acetyl cellulose, 5 – cellophane.

Table 1. Parameters of adhesion of conidiums A niger to surface of polymer materials

Material	k, sec^{-1}	F, dyne/cell	γ_∞, %
Polyethylene	$1{,}66 \cdot 10^{-5}$	$3{,}3 \cdot 10^{-4}$	55
Epoxide resin	$2{,}20 \cdot 10^{-5}$	$6{,}6 \cdot 10^{-4}$	70
Polymethyl methacrylate	$1{,}0 \cdot 10^{-4}$	$1{,}1 \cdot 10^{-3}$	80
Cellophane	$1{,}0 \cdot 10^{-4}$	$1{,}6 \cdot 10^{-3}$	85
Acetyl cellulose	$1{,}3 \cdot 10^{-4}$	$2{,}5 \cdot 10^{-3}$	90

Thus, at fixed thermo-moist conditions each material may be characterized by proper set of values of K and F. By increase of adhesion of conidiums A. Niger the investigated polymer materials are formed the following raw: polyethylene, epoxide resin, polymethyl methacrylate, cellophane, lavsan, acetyl cellulose.

As it was mentioned above, in experiments the polymer materials with various degree of hydrophilicity were used. It is obvious from Figure 1 and Table 1 that hydrophilic polymers have high values of adhesion force and constants of rates of their formation, whereas for hydrophobic polymer materials these values are significantly lower. Obviously, the values of K and F are determined by the ability of material to strong adhesion of microorganisms connected with material moister capacity.

For more detailed studying of process the influence of external exploitation factors on adhesion process was investigated. Kinetic curves of adhesion on polyethylene and cellophane at various temperatures of conidiums ageing on polymers surface are presented in Figure 2. It is obvious that with temperature rise adhesion of conidiums Aspergillus niger to polymer materials and rate of formation of adhesion bond are decreased, that is characteristic for phenomena namely of physical nature.

The values of characteristics of adhesion interaction for polyethylene and cellophane at various temperatures and relative moisture are presented in Table 2.

Table 2. Adhesion parameters of interaction for polyethylene and cellophane at various thermo-moist conditions

T, °C	φ, %	Polyethylene			Cellophane		
		γ_∞, %	k, sec^{-1}	F, dyne/cell	γ_∞, %	k, sec^{-1}	F, dyne/cell
	0	70	$2{,}20 \cdot 10^{-5}$	$3{,}0 \cdot 10^{-4}$	-	-	-
10	30	85	$1{,}66 \cdot 10^{-5}$	$5{,}2 \cdot\cdot 10^{-4}$	90	$1{,}0 \cdot 10^{-4}$	$2{,}5 \cdot 10^{-3}$
	98	100	$9{,}70 \cdot 10^{-5}$	$1{,}9 \cdot\cdot 10^{-3}$	-	-	-
22	30	55	$1{,}66 \cdot 10^{-5}$	$3{,}3 \cdot\cdot 10^{-4}$	85	$1{,}0 \cdot 10^{-4}$	$1{,}6 \cdot 10^{-3}$
38	30	50	$1{,}66 \cdot 10^{-5}$	$1{,}3 \cdot 10^{-4}$	-	-	-

As it is obvious from Figure 2 and Table 2 the adhesion force (F) and constant of adhesion forces formation rate are observed at 98% relative moisture of air. Obviously, the conidiums of macroscopic fungi in air medium interact with surface of polymer material at the expense of both molecular forces and at the action of capillary forces of liquid condensed in gap between conidium and polymer surface under the action of forces of Coulomb interaction and other reasons.

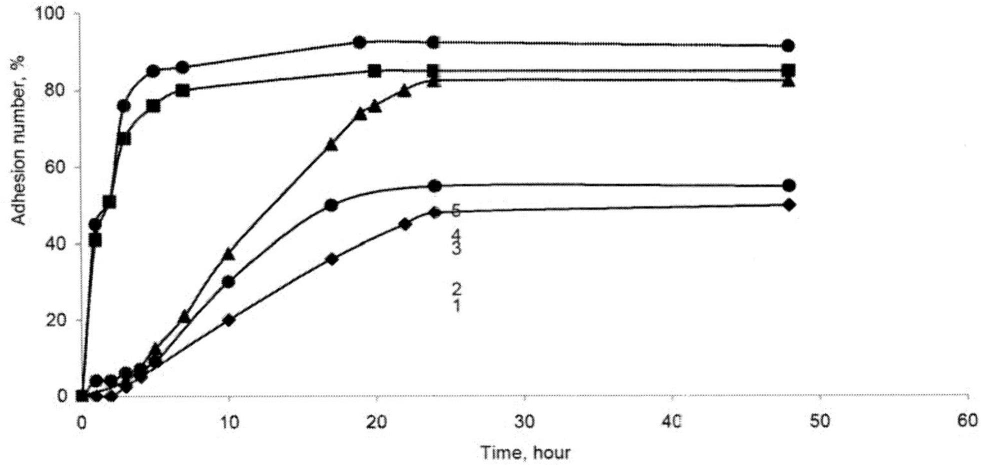

Figure 2. Kinetic curves of conidiums A niger adhesion to polyethylene and cellophane at relative moisture 30% and various temperatures: 1 – polyethylene, T = 38°C; 2 – polyethylene, T = 22°C; 3 – polyethylene, T = 10°C; 4 – cellophane, T = 22°C; 5 – cellophane, T = 10°C.

In the presence of moisture in air medium the condensation of vapor occurs between conidium and polymer surface. Capillary forces are the higher the higher the surface tension of liquid the vapor of which surround the region of interaction of conidiums with polymer surface and where moistening of polymer surface is better. Liquid interlayer between interacting objects (in given case between conidiums and polymer surface) excludes or to a great extent weaken the action of forces of electric nature. That is why results of experiments carried out with polyethylene at fixed temperature 10°C but at various air moistures which are illustrated in Figures 3 and Table 2 confirm physical character of adhesion interaction of system.

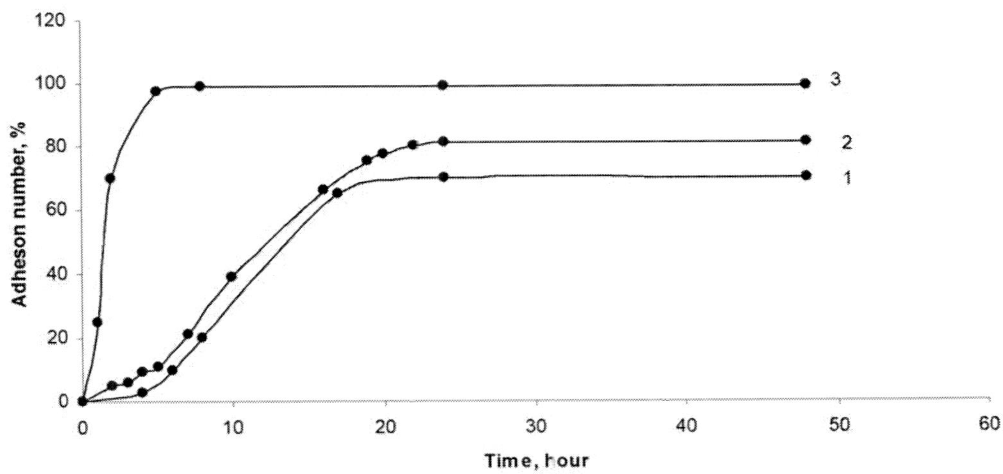

Figure 3. Kinetic curves of conidiums A niger adhesion to polyethylene at 10°C and relative moisture of environment 0 (1), 30 (2), 100% (3).

For confirmation of assumption the adhesion interaction of Aspergillus niger on the surface of polyethylene at 22°C and relative moisture of air $\varphi = 30\%$ was studied during days on raster electron microscope. Increase of interaction time leads to the change of geometric forms of conidiums Aspergillus niger and increase of contact square almost by 100%. Probably, by this fact we may explain the kinetic character of conidiums adhesion to polymer surfaces that allows quantitative description of microorganisms adhesion process to various surfaces and characterizing of it by kinetic parameters.

The process of microorganisms adhesion is also determinative under biofouling of metals in water and air mediums [19-22]. In these experiments the conidiums of Trichoderma viride fungi were used which were met in water medium and possessed optimal sizes for quantitative microscopic analysis. They determined the force causing the detachment of 50% of conidiums from general amount of all presenting on metal surface cells from integral curves of distribution of conidiums by adhesion forces characterizing the dependence of part of detached particles on force of detachment.

The force of detachment of 50% of conidiums was calculated specially:

$$F_{50} = \pi^3 / 675 \cdot R (p_c - p_{av}) \cdot \omega^2 \cdot r^3, \text{ dyne/cell}, \quad (2)$$

where ρ_{av} – density of medium (water); ρ_c – density of conidiums equal to 1,15 g/cm^3, R = 4cm – the radius of centrifuge rotor, ω – number of revos, r – conidiums radiuse equal to $(3 \pm 0,3)\cdot 10^{-4}$ cm. The adhesion of conidiums to metal surface was characterized by two parameters γ_∞ and F_{50}. The curves of conidiums Trichoderma viride adhesion to various metals are presented in Figure 4, the values of γ_∞ and F_{50} are presented in Table 3.

All metals by adhesion parameters may be divided into two groups: the first one is characterized by $F_{50} \geq 10^{-4}$ dynes/cell; and the second by F_{50} from 10^{-5} up to 10^{-7} dyne/cell.

Prehistory of surface plays significant role, for example for nickel treated by cold plastic deformation $\gamma_\infty = 79 \pm 8$ %, and $F_{50} = 2,0\cdot 10^{-4}$ dyne/cell, and for nickel aged for 1 hour at temperature 700°C and pressure 10^{-4} mm of mercury $\gamma_\infty = 50 \pm 2\%$, and $F_{50} = 8,2\cdot 10^{-4}$ dyne/cell.

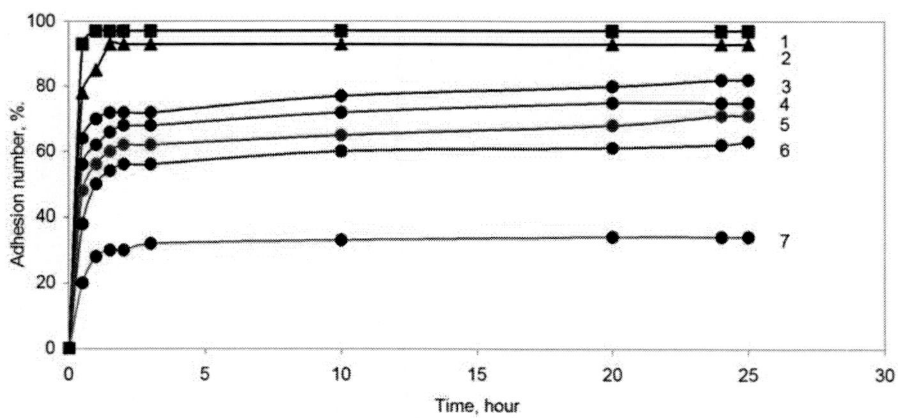

Figure 4. Curves of increase of conidiums Tr. Viride adhesion to metals: 1 – zinc, 2 – copper, 3 – aluminum, 4 – nickel, 5 – titanium, 6 – tantalum, 7 – molybdenum.

Table 3. Adhesion parameters o conidiums Trichoderma viride at T = 22°C and ω = 15000 revos in water medium to metals treated in solvent by a – boiling, b – at 22°C

No.	Metals	a		b	
		γ_∞, %	F_{50}, %	γ_∞, %	F_{50}, %
1	aluminum	95 ± 5	-	82 ± 5 (88B)	$3{,}0 \cdot 10^{-4}$
2	tungsten	0	$1{,}86 \cdot 10^{-7}$	0	$1{,}9 \cdot 10^{-7}$
3	gold	-	-	25 ± 5	$2{,}3 \cdot 10^{-5}$
4	copper	85 ± 3	$4{,}1 \cdot 10^{-4}$	93 ± 7	$5{,}8 \cdot 10^{-4}$
5	molybdenum	-	-	34 ± 5	$7{,}4 \cdot\cdot 10^{-5}$
6	nickel	62 ± 5	$1{,}67 \cdot 10^{-4}$	79 ± 8	$2{,}0 \cdot 10^{-4}$
7	platinum	-	-	53 ± 5	$2{,}2 \cdot 10^{-4}$
8	plumbum	96 ± 2	$7{,}00 \cdot 10^{-4}$	100 ± 1	$7{,}0 \cdot 10^{-4}$
9	silver	-	-	78 ± 5	$5{,}0 \cdot 10^{-4}$
10	tantalum	63 ± 8	$2{,}97 \cdot 10^{-4}$	65 ± 10	$2{,}5 \cdot 10^{-4}$
11	titanium	76 ± 5	$1{,}35 \cdot 10^{-4}$	71 ± 3	$9{,}0 \cdot 10^{-5}$
12	zinc	100	-	97 ± 3	$6{,}7 \cdot 10^{-4}$

Minimum adhesion is observed on gold, tungsten and molybdenum, the maximum is on zinc, plumbum and copper. Aluminum and titanium take intermediate position.

In the whole the raw of limited number of adhesion is the following: plumbum, zinc, copper, aluminum, nickel, silver, titanium, tantalum, platinum, molybdenum, gold, tungsten; and by decrease of values of force of 50% detachment of conidiums: plumbum, zinc, copper, silver, aluminum, tantalum, platinum, nickel, titanium, molybdenum, gold, tugsten.

Thus, conidiums Trichoderma viride reveal low level of adhesion to metals which at experiment conditions are not oxidized and not form oxides at all, for example gold.

Obtained results testify that the main characteristic of adhesion – the force of adhesion is increased in time reaching equilibrium value and is determined by nature of material surface and type of microscopic fungi. The experimental fact of distribution of adhesion forces in dependence on applied stress was also noticed. This effect was studied earlier and probably was caused by both heterogeneity of surface of material and heterogeneity of conidiums by their sizes.

The dependences of conidiums Aspergillus niger number remained on material surface after centrifugation (adhesion number γ) on angle rate of rotation for polyethylene and cellophane at temperature T = 22°C and moisture of environment φ = 98% are presented in Figure 5. From obtained data we calculated forces of 50% detachment of conidiums F_{50} by formula (2).

Results of experiments show that at time of conidiums ageing on polymer surface for 4 and 24 hours the adhesion force F_a (forces of 50% detachment of conidiums F_{50}) is increased for polyethylene in 2 times, and for cellophane in 1,5 times, and this fact obviously testifies to significant difference in constants of adhesion forces formation rates for the same system (Tables 1 and 2).

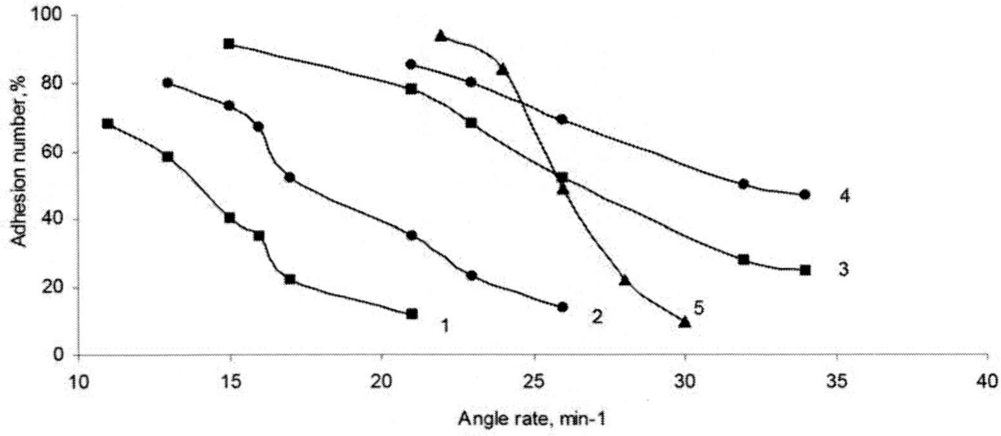

Figure 5. Distribution of adhesion number (γ) of conidiums A. niger in dependence on angle rate (ω) on polyethylene (1, 2) and cellophane (3, 4) at 22°C and φ = 98%. 1 and 3 – time of ageing is 4 hours; 2 and 4 – time of ageing is 24 hours; 5 – distribution of polystyrene particles in dependence on ω in integral form at the same conditions of adhesion.

For more detailed investigation of dependences of adhesion number on number of revos γ = γ (ω) (in implicit form it is dependence of adhesion number on applied force γ = γ (F) that represents the distribution by adhesion forces) the derivative of this function was determined for polyethylene γ' (F) = dγ/dF (F) (Figure 6). Dependence presented in Figure 6 is described by equation of the type corresponding to Gauss distribution:

$$\varphi (x) = A \exp (-\alpha x^2) \text{ in coordinates } \varphi = \varphi (x) \quad (3),$$

where A and α are constants.

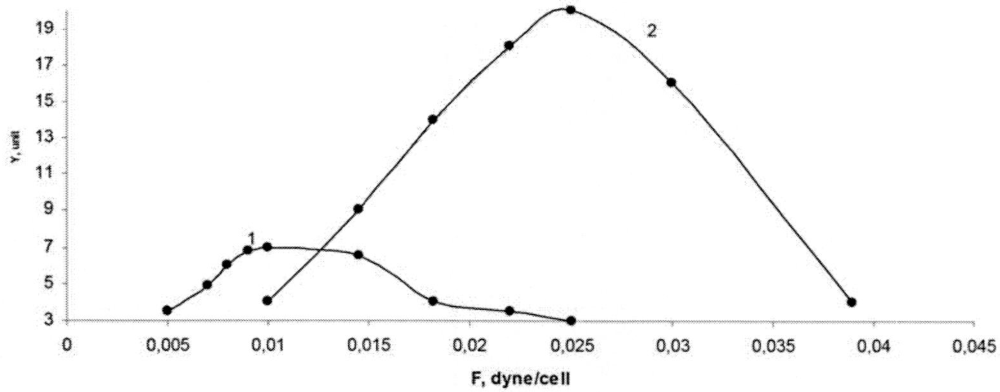

Figure 6. 1-Distribution of conidiums A. niger by forces of adhesion on polyethylene in differential form. Time of ageing is 24 hours. 2–distribution of polystyrene particles by adhesion forces in dofferentil form. T = 22°C, φ = 98%.

Such distribution may be caused by both heterogeneity of conidiums by sizes and heterogeneity of polymer surface. With this aim the sizes of conidiums A. niger were determined on raster electron microscope "TESLA BS 300". Measured values of conidiums diameters on electron micro-photographs demonstrate significant scattering by sizes (Figure 7) that should influence on scattering of adhesion force values.

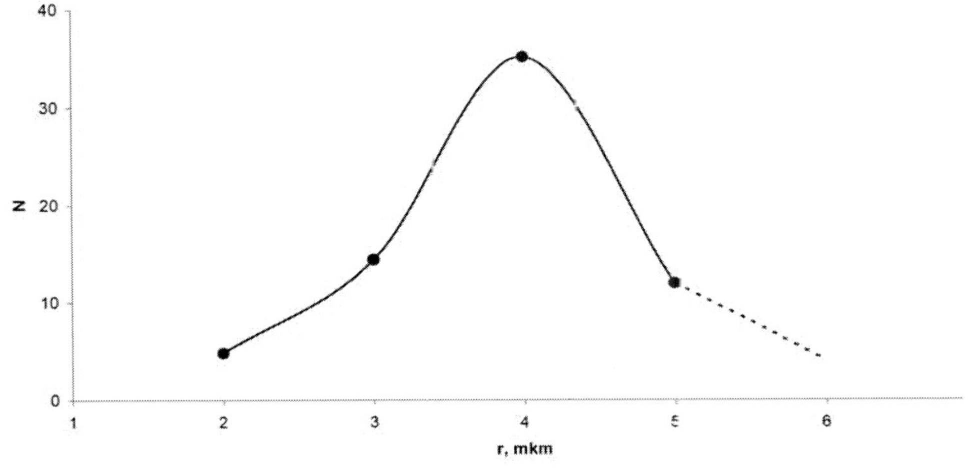

Figure 7. Distribution of conidiums A. niger by radiuses.

Let show that distribution of conidiums by sizes makes its contribution (i.e. is described by the same function) into distribution of adhesive conidiums by forces.

We present derivatives of adhesion number by radius and force:

$$\gamma\ (r) = d\gamma/d\ r\ (r);\ \gamma\ (F) = d\gamma\ /\ dF\ (F).$$

We should prove that from $\gamma = \gamma$ (r) it is consequent that $\gamma = \gamma$ (F). Let write the distribution by conidiums sizes in the following form:

$$d\gamma/dr = (d\gamma/dF) \qquad (dF/dr) = K\ d\gamma\ /\ dF \tag{4}$$

For determination of value of K we present as $d\gamma/dr$ from equation (2):

$$d\gamma/dr = (3\ F/r) \qquad d\gamma/dF\ ;\ K = 3\ F/r \tag{5}$$

It is consequent from experimental data (dependence F = F(r) and $\gamma = \gamma$ (r)) that for all values of radiuses (r) of conidiums at functioning force of detachment from $F_1 = 5,5 \cdot 10^{-3}$ dynes per cell up to $F_2 = 2,3 \cdot 10^{-2}$ dynes per cell (Figure 6) the values in right part of equation (5) will be

$(3\ F/r) \qquad d\gamma/dF = 4,6 \pm 0,3,$ and in left part of equation (5):

$d\gamma/dr = 4,7 \pm 0,3.$

In limits of experiment error these values are equal, i.e. distribution of conidiums by sizes is one of the reasons of real distribution by adhesion force. The heterogeneity of support should also contribute into distribution of particles by adhesion forces. For this we should have uniform by sizes particles.

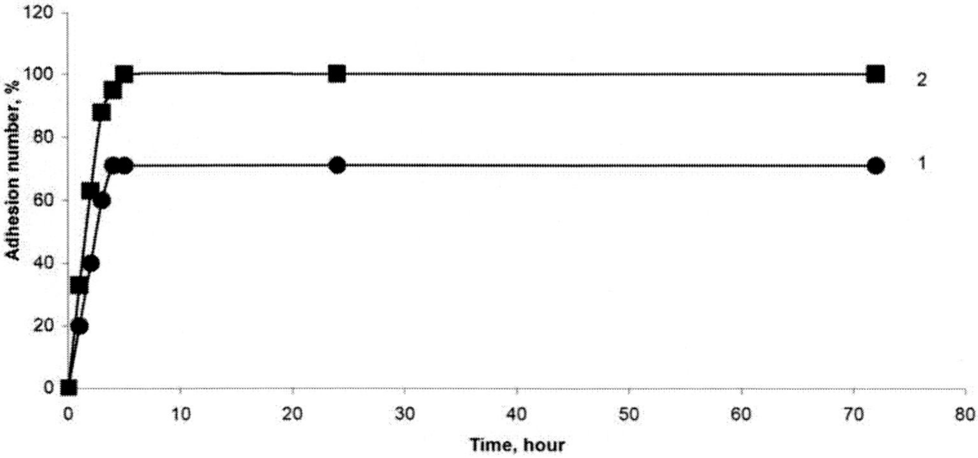

Figure 8. Kinetic curves of polystyrene particles adhesion to polyethylene (1) and cellophane (2) at T=22°C, $\varphi = 30\%$, $\omega = 15\,000\ min^{-1}$.

With this aim we carried out experiemtn with spherical particles of polystyrene with diameter (d) 2,8 mcm of "SERVA" firm. We plot kinetic curves of adhesion and distribution of particles by adhesion force by obtained experimental values (Figure 8).

By their character kinetic curves and curves of distribution of conidiums A. niger and polystyrene particles are similar, although there are some differences in parameters ofadhesion interaction (Figure 4 and 6). The constants of adhesion forces formation for polystyrene particles have high values in comparison with conidiums A. niger. Moreover, scattering of adhesion force values γ in the case of polystyrene particles is significantly lower.

Results of experiment give the basis to assume that adhesion interaction of system polystyrene particles–polymer is changed in time, distribution of particles by adhesion forces is described by Gauss equation. Consequently principle picture of polystyrene particles and conidiums A. niger adhesion to polymer surface is analogous that may be explained for polystyrene particles by the following:

1. increase of contact area of polystyrene particles with polymer surface causes the rise of adhesion force in time at the expense of deformation of surface itself;
2. capillary concentration of moisture in contact zone also promotes adhesion force;
3. energetic heterogeneity of the surface of polymer material causes distribution of polystyrene particles by adhesion forces.

Thus, adhesion of conidiums A. niger has stochastic nature caused by their heterogeneity by sizes and heterogeneity if polymer support, distribution of conidiums by adhesion forces obeys Gauss distribution. Quantitative characteristics of process of microscopic fungi interaction with solid polymer surfaces may serve as estimative criterion of processes of microdestruction of materials.

In fact, the adhesion process also should be determined by parameters of external wall of conidiums (structure of cell wall) and by their ability to evolve various organic substances (exo-ferments, organic acids, etc.) [9, 10]. For establishment of correlation between adhesion properties of conidiums micro-destructors of polymer materials and structure of cell wall of cultures of various ages the samples obtained by the method of wash-out from surface of conidiums Aspergillus niger, A. Terreus, Penicillium funiculosum, Trichoderma viride agar, deposited on membrane filters "Synpor" (thickness of layer ~ 0,01mm) and dried in exiccator over $CaCl_2$ were studied by method of IR-spectroscopy with Fourier analysis.

Concentration of analyzed substances was estinated according to Lambet-Buger-Ber law with the help of the following equation:

$$C = D/l \cdot E,$$

where l – the depth of beam penetration, E – extinction coefficient.

For comparative quantitative analysis of IR-spectra the absorption bands significantly changed in intensity in the course of growth process should be chosen. As such band the following ones were selected: 1545 – 1550 cm^{-1} (amide 2) – deformational vibrations of NH-group, 1275 – 1280 cm^{-1} – stretching vibrations P=O of phospholipids, 831 – 835 cm^{-1} – deformational non-planar vibrations of CH-group of α-glycanes.

The values of optical densities and concentrations of analysed substances for four types of conidiums of microscopic fungi are presented in Table 4. In the case of some types of conidiums in the course of growth noticeable quantitative changes in structure of cell walls occur, and for other types significant changes were not noticed.

Table 4. Characteristic of IR-spectra of conidiums of fungi of various ages

Microscopic fungi	ν, cm^{-1}	Conidiums age, days					
		3		15		30	
		D	C,%	D	C,%	D	C,%
Aspergillus niger	1550 – 1560	0,01	5	0,08	58	0,07	56
	1275 – 1278	0,84	71	0,20	34	0,20	34
	831 – 835	0,73	24	0,14	8	0,15	10
A. terreus	1550 - 1560	0,10	37	0,07	30	0,08	32
	1275 – 1278	0,54	47	0,56	54	0,55	50
	831 – 835	0,48	16	0,50	16	0,52	18
Penicillium funiculosum	1550 – 1560	0,05	5	0,01	1	0,02	2
	1275 – 1278	0,57	70	0,36	73	0,35	71
	831 – 835	0,57	25	0,39	26	0,40	27
Trichoderma viride	1550 - 1560	0,12	45	0,06	28	0,05	26
	1275 – 1278	0,48	42	0,38	54	0,39	56
	831 – 835	0,42	13	0,41	18	0,41	18

As the example of spectra the IR-spectra of surface layer of conidiums Aspergillus niger of various development times are presented in Figure 9. The IR-spectra of 15- and 30-days conidiums are almost completely coincide, that is explained by known factor of metabolic processes decay and reaching of cell quiescent state already after 13-14 days. So, for conidiums of Trichoderma viride type the concentration of amides in the surface layer of cell is decreased from 45% (the age is 3 days) down to 26-28% to the moment of 15 days. Weak rise of concentration of phospholipids (from 42 up o 56%) and glycane (from 13 up to 18%) components is observed. Probably synthesis of these substances is continued for more than 3 days.

When considering IR-spectra of surface layer of conidiums Penicillium funiculosum and A. terreus it is obvious that there are no significant changes in their spectra. Comparable analysis of spectral characteristics of walls of 3, 15 and 30-days conidiums of these types of microscopic fungi shows that surface layers of these cells are not differed from each other in chemical structure. Concentrations of amide, phospholipids and glycane compounds are practically not changed.

Analysis of spectra of surface layer of conidiums A. Niger of various ages shows that at the stage of growth the concentration of albumin components is increased (from 5 up to 58%) and phospholipids (from 71 down to 34%) and α-gkycanes (from 24 down to 8%) concentrations are decreased. Probably, albumin synthesis in the surface layer of A. niger is intensively proceeds for 15 days. In particular this type of conidiums was selected for investigation of their properties of adhesion to polymer materials.

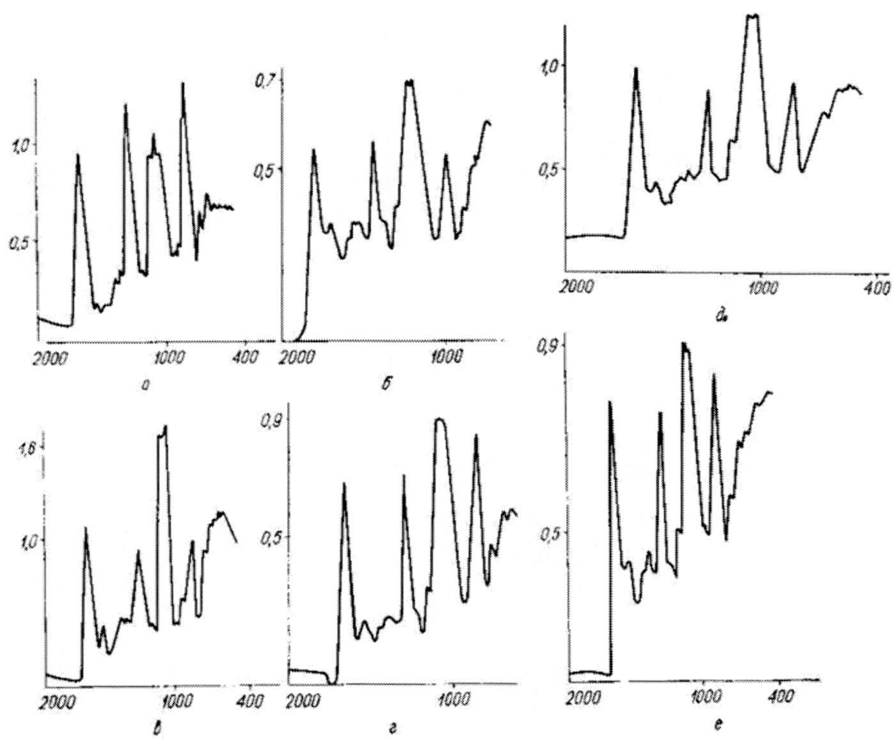

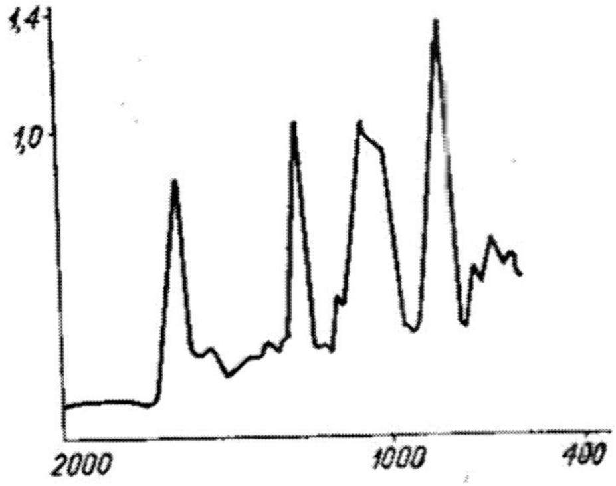

ж

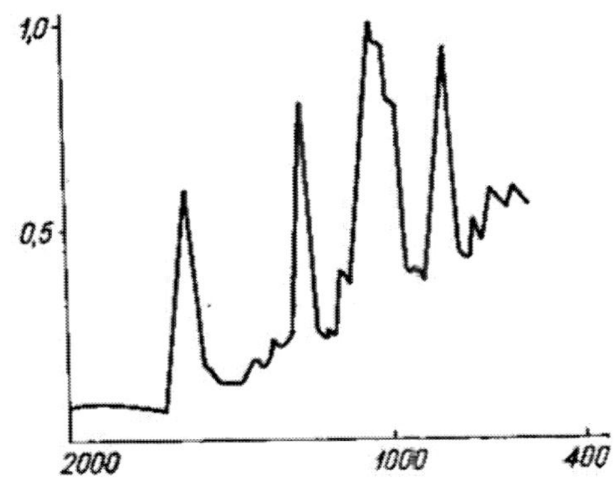

з

Figure 9. IR-spectra of the surface layer of conidiums: a – A. niger, 3 days; b – A. niger, 15 days, 30 days; c – A. Terreus, 3 – days; d – A. terreus, 15 days, 30 days; e – Tr. Viride, 3 days; f – Tr. Viride, 15 days, 30 days; j – P. Funiculosum, 3 days; h – P. Funiculosum, 15 days, 30 days.

Integral distribution of conidiums A. niger of various ages by adhesion forces to polyethylene at temperature 22°C and relative moisture of environment 98% is presented in Figure 10. As it is obvious from the Figure curves of 15 and 30 days conidiums are completely coincide. The force of adhesion of these conidiums is as follows:

$$F_{50}^{15} = F_{50}^{30} = (1,1 \pm 0,02)\ 10^{-2} \text{ dynes per cell},$$

where F_{50}^{15} and F_{50}^{30} – the forces of 50% detachment (γ = 50%) of 15- and 30-days conidiums correspondingly, that is in 3,4 times higher for adhesion force in the case of 3-days conidiums.

For 3-days conidiumes adhesion force is:

$$F_{50}^{3} = (3,2 \pm 0,2)\cdot 10^{-3} \text{ dyne/cell}.$$

when comparing data on adhesion of 15- and 30-days conidiums with IR-spectra of their surface layer we see that after 14-15 days of cultivation of fungi stabilization of al parameters, decay of metabolic processes and cell transition into quiescent state are observed.

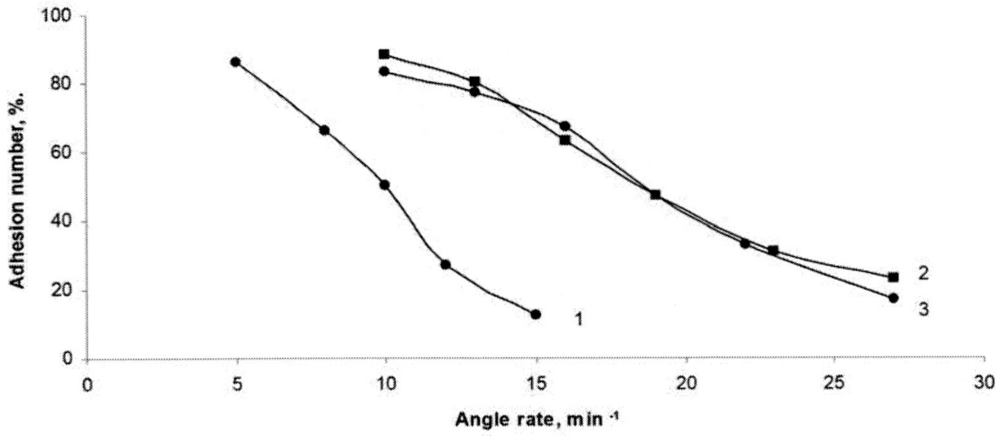

Figure 10. Distribution of adhesion number (γ) of conidiums A. niger in dependence on angle rate (ω) on polyethylene at T=22°C and φ = 98%. The time of conidiums ageing on the surface of polyethylene is 24 hours. 1 – s-days conidiums, 2 – 15-days conidiums, 3 – 30-days conidiums.

For conidiums with small sizes (A. terreus, P. Funiculosum) the concentrations of the main components of cell wall practically are not changed, whereas for conidiums with large sizes (T. Viride, A. niger) such changes are observed (Table 4). Obviously, for conidiums of small sizes the surface layer of cell is formed for the time not exceeding 3 days. For conidiums of larger sizes (for example A. niger) the process of surface layer formation is not stopped by the moment of 3 days and continued for longer time. At the first moment of adhesion the intensive evolving of albumin macromolecules occur [11]. For the case of A. niger conidiums the part of albumin component is increased in the course of cell growth and their adhesive interaction is also increased.

Thus, experimental material allows assuming that adhesion force of conidiums of microscopic fungi in the course of their growth is changed differently. Is depends on change of albumin components concentration in the surface layer of cell having mosaic structure. In those cases when these changes are significant (predominantly for conidiums of large sizes) the increase of force of adhesion interaction with surfaces occurs and if there are no clear changes (predominantly for conidiums of small sizes) the adhesion force is not changed.

References

[1] N.M. Emanuel, A.L. Buchachenko, Chemical physics of ageing and stabilization of polymers, Moscow: *Nauka* (1982) (in Russian).

[2] E.A. Ermilova, Theoretical and practical basis of microbiological destruction of chemical fibers, Moscow: *Nauka* (1991) (in Russian).

[3] P.S. Hocking, J. of Macromol. Sci., *Reviews in Chemistry and Physics*, 1, 35 (1992).

[4] M.R. Timmens, R.W. Lenz, *Trends in Polym. Sci.*, 1, 15 (1994).

[5] K.Z. Gumargalieva, G.E. Zaikov, Yu.V. Moiseev, *Uspekhi khimii*, 63, No.10, 905 (1994) (in Russian).

[6] K.Z. Gumargalieva, G.E. Zaikov, Yu.V. Moiseev, *Khimicheskaya fizika*, 14, No.10, 29 (1995) (in Russian).

[7] H.C. Flemming, *Polymer Degrad. And Stability.*, 59, No.1–3, 309 (1998).

[8] M.S. Fel'dman, S.I. Kirsh, V.M.Pozhidaev, in coll. *Biological basis of protection of indutrial materials from bio-damages*, N.Novgorod (1991) (in Russian).

[9] V.F. Smirnov, A.S. Semicheva, Conf. *Biological problems of ecological materials*, Penza (1995) (in Russian).

[10] F.B. Oppermann, S. Pickartz and A. Steinbuchel, *Polym. Degrad. And Stability.*, 59, No.1, 337 (1998).

[11] A.M. Gallardo – Moreno, M.L. Gonzblez, J.M. Bruque and C. Pijrez– Giraldo, 1-st Int. Meeting on Appl. Phys. Colloids and Surfaces A: *Physicochemical and Engineering Aspects.*, 249, No.1-3, 99 (2004).

[12] Microbiologic adhesion and aggregation, Ed. By Marshall K.C. Berlin etc.:Sprinser Verlag, (1984).

[13] V.A. Fomin, V.V. Guzeev, *Pkast. massy*, No.2, 42 (2001) (in Russian).

[14] L. Richard, W. Bizzoco, R. Bass, Thuy T. Vuong, James B. Vahl, Corona L. Hoang and Melina M. Diaz, *J. of Microbiol. Methods*, 55, No.3, 787 (2003).

[15] Bing – Mu Hsu and Chihpin Huang, Colloids and Surfaces A: *Physicochemical and Engineering Aspects*, 201, No.1, 201 (2002).

[16] M. Mercier – Bonin, K. Ouazzani, P. Schmitz and S. Lorthois, *J. of Colloid and Interface Sci.*, 271, No.2, 342 (2004).

[17] J. Schauersberger J., M. Amon, D. Aichinger and Apostoulos Georgopoulos, *J. of Cataract and Refractive Surgery*, 29, No.2, 361 (2003).

[18] A.D. Zimon, Adhesion of dust and powders, Moscow: *Khimiya* (1976) (in Russian).

[19] A.M. Raichur and S.P. Vijayalakshmi, *Fuel*, 82, No.2, 225 (2003).

[20] H. Onose, T. Miyazaki, S. Nomoto, *J. Dent. Res.*, 59, No.7, 1179 (1980).

[21] A.S. Gordon, S.M.Gerchakov, L.R Udey, *Cancd. J. Microbiol.*, 27, No.7, 698 (1981).

[22] P.J. Boyle, M. Walch, R. Mitchell, VIII Internat. Congr. Of Microbiology: Program and abstracts. Boston: *Internat. Union of Microbiol. Soc.*, 75 (1982) (in Russian).

USE AND ASSESSMENT OF RENEWABLE RESOURCES AND PLANT BY-PRODUCTS

In: Renewable Resources: Obtaining, Processing... ISBN 978-1-6060692-534-8
Editors: R. Kozlowski, G. Zaikov and F. Pudel

Chapter 12

POLYMERS AND CHEMICALS FROM THE AGRO-RESOURCES

R. Kozlowski and K. Bujnowicz

ABSTRACT

Recently, a strong need for using biomass to produce bio fuels is observed. Less attention has been given to use the biomass as a raw material for organic chemicals. Replacement of petroleum based chemicals with biomass based materials will play a very important role in sustaining the growth of the chemical industry. The 2004 report by the U.S. Department of Energy indicates considerable share of chemicals produced from biomass, which can be a starting material for obtaining many chemical products via traditional methods.

The U.S. Department of Agriculture showed that billion tons of lignocellulosic biomass after enzymatic and microbiological processing can be transformed into the sugars, lipids and polymers, for example: ferulic acid, extracted from corn fibers, is a precursor for numerous aromatic chemicals used in chemical industry.

Many chemicals are already produced by fermentation, including glutamic acid (~ 1,7 billion kg/year worldwide), citric acid (~ 1.6 billion kb/year) and lysine (~ 850 million kg/year).

In polymer applications derivatives of oils such as epoxides, polyols and dimerization products based on unsaturated fatty acids are used as plastic additives or components for composites or polyurethanes. Polyols based on vegetable oils and glycerol can be used for obtaining polyurethane polymers including elastomers, fibres, foams with the properties required for various applications.

POLYMERS FROM BIOMAS

In times of very high prices for crude oil on the world market, the research on obtaining different polymers from renewable raw materials gets a new importance. The products based on renewable raw materials are usually more environmental friendly compared to petrochemical substances, therefore, the use of natural resources for non-food applications grows every year.

The current need to use biomass for biofuel production is well known. Less attention has been paid to the biomass used for different useful organic compounds. Replacement of chemicals produced from petrochemical raw materials with those manufactured from biomass will play a key role in sustainable development of chemical industry. Currently, 13% of crude oil consumed in the USA is used for production of non-fuel chemicals.

The biomass is enzymatically and microbiologically processed to sugars, lipids and other organic compounds, including polymers. Chemicals produced by biological methods are an interesting history.

In 1945-1950, 60% of n-butanol and 10% of acetone was obtained from fermenting molasses and starch. Other products manufactured by fermentation in the second half of 20th century included acetic acid, citric acid, lactic acid and itaconic acid. Rising prices of sugar-bearing raw materials and reduction of petrochemical raw materials prices caused limitation in use of fermentation processes in production of raw materials. Currently, we can observe this trend to reverse.

The 2004 report of US Energy Department indicates a considerable share of biomass in production of chemicals, which are the initial products to obtain chemicals in traditional way. The U.S. Department of Agriculture showed that billion tons of lignocellulosic biomass after enzymatic and microbiological processing can be transformed into the sugars, lipids and polymers, for example: ferulic acid, extracted from corn fibers, is a precursor for numerous aromatic chemicals used in chemical industry.

Many chemicals are already produced by fermentation, including glutamic acid (~ 1,7 billion kg/year worldwide), citric acid (~ 1,6 billion kb/year) and lysine (~ 850 million kg/year).

In polymer applications, derivatives of oils such as epoxides, polyols and dimerization products based on unsaturated fatty acids are used as plastic additives or components for composites or polyurethanes.

Polyols based on vegetable oils and glycerol can be used for obtaining polyurethane polymers including elastomers, fibres, foams with the properties required for various applications.

In addition, such chemicals can be obtained from biologically manufactured monomers in combination with traditional chemical methods. Lactic acid obtained by fermentation can be chemically processed to methyl lactate, lactide and polylactic acid (PLA). This polymer is fully biodegradable and replaces PET. The efforts also are made to industrially process biologically lactic and hydroxypropionic acids to obtain methacrylic and acrylic acid. The world production of the latter is 1.6 billion kg/year. Another example can be studies conducted by Genencor and DuPont, who developed an effective fermentation system to obtain 1.3 propanediol, a key element to obtain poly-(propyleneterephthalate) (PPT, Sorona 3GT).

Application of enzymatic catalysis is also used to obtain monomers, e.g. application of commercially available catalyst lipases offers potential for processing sorbitol or glycerol to diacids. These methods reduce to considerable extent consumption of energy, necessary in polymerization of traditional chemicals.

Fermentation processes employing different microorganisms are used to obtain succinic acid which potentially can replace maleic anhydrite, currently produced from butane in volume of 1.8 billion kg/year.

Maleic anhydrite is a raw material used to obtain different polymers and industrial solvents. Another example of raw material produced with the help of microorganisms is shikimic acid which modified by *Escherichia coli* can be used for transformation of glucose to catechol and other aromatic alcohols. These products currently are manufactured using benzene.

A different metabolic route can be used to obtain *cis, cis*-muconic acid which chemically transformed to adipic acid is a component in production of Nylon 66. Polymers from biomass can also be obtained by microbiological synthesis resulting in Poly-3-hydroxyalkanoates (PHA). Corn fibers are high in ferulic acid which is a precursor of many aromatic substances used in chemical industry like vanillin or guaicol. Fast development of biofuels caused reduction of glycerol prices. Glycerol is a by-product in production of biofulels. Glycerol can be used as a cheap raw material. Companies like Dow Chemicals, Cargill i Archer Daniels Midland Corporation initiated projects for processing glycerin to propylene glycol. Dow Chemicals and Solvay plan setting factories which will use glycerol for production of epichlorohydrin. Other methods of processing fatty acids to polymers were developed by Cargill together with Kansas Polymer Research Center. Using chemical modification of double bonds in triglycerides they obtained polyols with hydroxyl and methyl groups. Polyols are precursors of many polyurethane materials. It is also worth to mention that biological processes do not require high pressures, temperatures which normally are used in typical chemical processes, hence there is a big chance for reduction of polymer manufacture costs.

POLYMERS FROM VEGETABLE OILS

From chemical point of view vegetable oils are triglycerides, i.e., esters of different fatty acids and glycerin. Fatty acids in oils are mostly unsaturated, causing the oils to be liquid, while fats are solid at room temperature. The most important unsaturated fatty acids in plant oils are:

- oleic acid (C18:1): $CH_3(CH_2)_7CH{=}CH(CH_2)_7COOH$
- linoleic acid (C18:2): $CH_3(CH_2)_4CH{=}CHCH_2CH{=}CH(CH_2)_7COOH$
- linolenic acid (C18:3): $CH_3CH_2CH{=}CHCH_2CH{=}CHCH_2CH{=}CH(CH_2)_7COOH$

Oleic acid is dominant in oils with lower unsaturation (olive, peanut, canola) while linoleic acid is predominant in corn, soybean and sunflower oils. Linolenic acid is present in linseed, soybean oils.

Most of triglycerides reactions is carried out by the carboxylic group (> 90%).

In polymer applications derivatives of oils such as epoxides, polyols and dimerization products based on unsaturated fatty acids, are used as plastic additives or components for composites or polyurethanes.

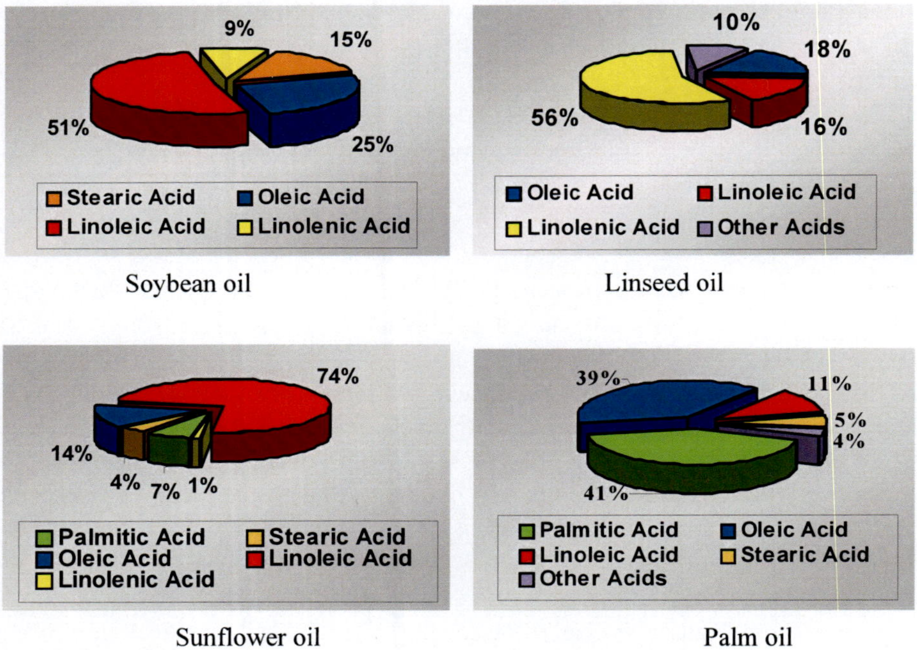

Soybean oil Linseed oil

Sunflower oil Palm oil

Figure 1. Fatty acids in vegetable oils.

One of the methods to obtain polymers from renewable resources is catalytic reaction of linseed oil with urea. It was found that in special conditions the linseed oil reacts quantitatively with urea.

$$ROCOCH_2CH(OCOR)CH_2OCOR + 2\ NH_2CONH_2 \xrightarrow{cat,Temp} NH_2OCOCH_2CH(OCOR)CH_2OCONH_2 + 2\ RCONH_2$$

or indirectly, in a two-stage reaction, in homogeneous system

$$C_4H_9OH + NH_2CONH_2 \xrightarrow{cat,Temp} C_4H_9OCONH_2 + NH_3$$

$$ROCOCH_2CH(OCOR)CH_2OCOR + 2\ C_4H_9OCONH_2 \xrightarrow{cat,Temp.} NH_2OCOCH_2CH(OCOR)CH_2OCONH_2 + 2\ RCOOC_4H_9\ ).$$

These studies allow for obtaining products capable of modifying properties of resins, changing for instance hydrophobic properties, elasticity, solubility, etc..

Reactions involving the alkyl chain or double bond represent less than 10% found mainly in polymer applications of epoxides, polyols, and dimerization products based on unsaturated fatty acids. These are used as plastic additives or components for composites or polymers like polyurethanes. Polyols based on vegetable oils can be used for obtaining polyurethane materials including elastomers and foams with the cell structure and properties required for various applications. An example of such natural product used on industrial scale in the field of polyurethane foams is castor oil. Only the oil, containing ricinoleic acid, has hydroxyl groups and can react with isocyanate groups. Other oils such as linseed or rapeseed and soybean, which do not contain the hydroxyl groups, can also be useful in polyurethane formulation but first, they need to be transformed. Double bonds of triglycerides are converted to hydroxyl groups. There is number of ways to prepare polyols from vegetable oils. Some include hydrolysis of esters bonds, others maintain the triglyceride structure. It is possible to obtain polyols through the epoxidation route and hydroformylation route.

The hydroformulated polyol has primary OH groups and extra CH_2 group. This configuration results in a lower cross-link density of polyurethane resin and gives a softer material. The hydroformulated product is very reactive. This kind of polyol can be used for preparing rigid polyurethane foams.

Epoxidized oils are also used for plastification of PVC. Many products stimulated studies on polymerisation of epoxidized oils and their application in coatings and as matrix resin for composites. Epoxidized vegetable oils have much less reactive internal epoxy groups. Acid catalysts, anhydrites and Lewis acids readily cause polymerisation of epoxy groups. Epoxidized soybean oil was cured with citric acid, phthalic anhydride and Lewis acids. Lewis acid initiators and citric acid gave rubbery materials of relatively low strength and low glass transition temperature. Phthalic anhydride gives glassy or rubbery materials depending on the molar ratio of epoxy to acid. Organic acids catalyse ring opening and ether formation. They also esterify, as a side reaction, the newly formed hydroxyl groups.

Other resins, modified acrylated or epoxidized soy oil, contain at least 50 wt% plant triglycerides and have mechanical properties comparable to synthetic resins such as vinyl esters, polyesters and epoxides. This resin could be used in composites with natural fibres. The matrix for the composites was prepared by mixing acrylated epoxidized soybean oil resin with styrene and divinylbenzene and subsequently subjected to free radical polymerisation curing during a resin transfer, then to molding process. This composite was envisaged as potential new material for high volume, low cost application.

Vegetable oil can be also a raw material for the cationic ring opening polymerisation of 2-oxazoline monomers. The monomers are synthesized by coupling soybean fatty acids with ethanolamine followed by ring closure in the presence of titanium catalyst. The resulting products are soy-based 2-oxazolines with different fatty acid side chains. Aqueous solutions of poly(2-oxazoline) block copolymers aggregate into micelles that could potentially be used in applications such as micellar catalysis and drug delivery.

Vegetable oil-based unsaturated polyesters are cured by a radical mechanism. It can be prepared by reacting an oil with maleic anhydride or epoxidized vegetable oil or with acrylic acid or maleic anhydride to introduce double bonds with dissolution of the resin in styrene. Other way to introduce reactive double bonds is to prepare alkyd resins under milder conditions. When triglycerides are reacted with glycerin, transesterification occurs giving monoglyceride and triglyceride esters of the fatty acids, witch may react with anhydrides to produce alkyds.

AGRO-FINE CHEMICALS

The possibility to use fibrous plants for obtaining very valuable agro fine chemicals are growing all the time. The relevant research covers lignans, cyclopeptides and waxes.

Obtaining these substances from natural raw materials is important for later use in cosmetic and pharmaceutical industry.

Lignans are secondary plant metabolites formulated by oxidizing dimerization of two phenyl-propane units. Flax seeds show the highest content of lignans among all foods. The main lignan found in flax seed (*Linum usitatissimum*) is diglucoside - sekoizolariciresinol. According to results of studies conducted so far, lingans show a number of curative properties such as anti oxidation, anti carcinogenic properties, anti virus, antibacterial and anti fungi; they also reduce risk of arteriosclerosis.

Research on cyclic peptides extracted from flax seed showed immunosuppressive action. These compounds show no strong toxic properties and could be an alternative for cyclosporine A, a commonly used immunosuppressant. A considerable number of cyclic peptides that could be isolated from numerous cultivars of flax encourages to start screening tests aiming at identification of new cyclic peptides – potential leading substances for immunosuppressants.

Epicular waxes are a mixture of very long chain aliphatic compounds, such as long chain alcohols, alkenes, acids, ketones, aldehydes, esters of fatty acids with chain alcohols (wax esters).

Our actions aim at finding new properties of fibers and fabrics, such as anti odor, UV barrier, antibacterial, anti fungus, self-cleaning, tonifiant, curative, etc. Obtaining new qualities in textiles increases their functionality.

SURFACTANTS

In recent years a trend towards surfactants based on renewable resources has become apparent. Factors driving the increasing use of renewable surfactants are sustainability and approaches to quantify the total environmental impact of these surfactants. In addition, the increasing education and pressure of the consumer has driven demand for renewables.

Cocomonoglyceride sulphate (CMGS) has been known for a long time but manufacturing methods have disadvantages. New method for obtaining Cocomonoglyceride sulphate is directly from coconut oil in a solvent-free two-stage process. The first stage is transesterification of coconut oil with glycerol and in the second step raw material is converted to CMGS by reaction with sulphur trioxide gas.

A major use for high-eruic oil is in the preparation of erucyl and behenyl alcohols . These are used in personal care products, laundry softeners and other household products.

LUBRICANTS

Fatty acid esters obtained from fatty acids and alcohols are used as biodegradable replacements for mineral oils. They would have a less negative environmental impact due to their inherent biodegradability and low-aquatic toxicity. Current developments refer to the use of tailor-made fatty acid esters with specific lubricant properties.

CONCLUSIONS

New methods for obtaining polymers can influence the growth in the economy of agriculture and bring progress in area of their application.Taking into account that European countries have a vast potential regarding cultivation of oil plants, positive results of this study may result in a new source of modern, sustainable, more ecological polymers.New polymers from renewable resources can be useful matrix component for composites reinforcement by natural fibres.New materials can meet flame retardancy and biodeterioration requirements.

REFERENCES

[1] Hill, K., (2000) Fats and oils as oleochemical raw materials, *Pure Appl. Chem., Vol. 72*, pp. 1255-1264.

[2] Petrovic, Z., Guo, A., Javini, I., Zhang, W., (2004) Plastics and Composites from Soybean Oil. In: Wallenberger, F. T., Weston, N.: *Natural Fibers, Plastics and Composites*, eds. Kluwer Academic, 2004.

[3] Bujnowicz, K., Kozlowski, R. Pielichowski, J., Wirpsza, Z., Polymers from Natural Fatty Acids. In: Pielichowski, K., Modern Polymeric Materials for Environmental Applications, Vol.1, 1st International Seminar, Krakow 2004.

[4] Kozlowski, R., Pielichowski, J., Wirpsza, Z., Bujnowicz, K., Przepiera, A., Polymers from Linseed Oil, *The Proceedings of the 3rd Global Workshop "Bast Fibrous Plants for Healthy Life"*, Banja Luka, Bosnia and Hercegowina, October 24-28, 2004.

[5] Kozlowski, R., Pielichowski, J., Wirpsza, Z., Bujnowicz, K, Polymers from Natural Fatty Acids, *The Proceedings of the 10th Internationanal Conference for Renewable Resources and Plant Biotechnology*, Magdeburg, Germany, June 7-8, 2004.

[6] Kozłowski, R., Pielichowski, J., Prociak, A., Marek, M., Linseed Oil Based Polyols for Polyurethane Foams, *8th International Conference on Frontiers of Polymers and Advanced Materials (ICFPAM)*, Cancun, 2005.

[7] Daute, P., Gruetzmacher, R., Hoefer, R., Westfechtel, A., *Saponification resistant polyols for polyurethane applications based on oleochemical raw material, Fett Wiss. Technol., 1993; 95*: p. 91-94.

[8] Guo A., Cho Y.J., Petrovic Z.S., *Structure and Properties of Halogenated and non-halogenated soy-based polyols, J. Polym. Sci. Part A: Polym. Chem.* 2000; 38, p. 3900-3910.

[9] Petrovic Z.S., Zhang W., Miller R., Javni I., *ANTEC 2002*, San Francisco: SPE, 2002.

[10] Williams G. I., Wool R.P., *Composites from Natural Fibres and Soy Oil Resins., Aplied Composite Materials 2000*: 7, p. 421-432.

[11] Warth H., Muelhaupt R., Hoffmann B., Lawson S., *Polyester Networks Based Upon Epoxidized and Maleinated Natural Oils, Angew. Makromol. Chem. 1997*: 249, p. 79-92.

[12] Wool R., Kusefoglu S., Palmese G., Khot S., Zhao R., *U.S. Patent 6,121,398.*

[13] A. Behler A., Hensen H., Vier J., *Fett/Lipid 98*,309, 1996: b) Behler A., Hensen H., Vier J. *Henkel Referate* 33, 7, 1997.

[14] Piazza G. J., Foglia T. A. ,. *Rapeseed oil for oleochemical usage, Eur. J. Lipid Sci. Technol.* 103:450-454. 2001.

[15] Wildes S., Tomac S., *The use of soy oil for industrial products and applications, Lipid Technology*, 11/1999, p. 57-61.

[16] Hoogenboom R., Wiesbrock F., Schubert U. S., *Microwave-assisted cationic ring-opening polymerization of a soy-based 2-oxazoline monomer, The 230th ACS National Meeting*, Washington, USA, August 28 – September 1, 2005.

[17] Kaniewski R., Banach J., Kuzminski R., *Mozliwosc wykorzystania konopi wloknistych – Cannabis sativa L. w biologicznych preparatach ochrony roslin, Biuletyn Informacyjny Polskiej Izby Lnu i Konopi, „Len i konopie" 09/2007*.pp. 28-34.

[18] Kaniewski R., Konczewicz W., *Banach J.. Eteryczne olejki konopne. Biuletyn informacyjny Polskiej Izby Lnu i Konopi nr 6*. Poznan, January, pp. 52-57 (2006).

[19] Kozlowski R. et al., *Potential of medical application of 'agro fine chemicals' from fibrous plants for special medical textiles, poster at FiberMed06*, Tampere, Finland,06.2006.

[20] Kozlowski R., Bobkiewicz-Kozlowska T., Forjasz-Grus R., Kozlowska J., Grygorowicz Z. *Beneficial Effect of Linseed as a Source of Pharmaceutical and Dietetic Products. Proceedings of the 60th Flax Institute Of the United States,* (2004), Fargo, ND, pp.72-79.

[21] Kozlowska J., Grygorowicz Z., Biskupski M., Kuczynski A. *Zastosowanie oleju lnianego w przemysle farmaceutycznym, suplementach diety i kosmetykach. Len i konopie*. 2006, 6: 58-63.

In: Renewable Resources: Obtaining, Processing… ISBN 978-1-6060692-534-8
Editors: R. Kozlowski, G. Zaikov and F. Pudel

Chapter 13

ANTHOCYANINS: THEIR ROLE, PROCESSING AND APPLICATION

Anna Bakowska-Barczak*[*,1,3], *Jae-Jong Lee*[2], *Soo-Hee Hong*[2] and *Paul P. Kolodziejczyk*[1]

[1]Olds College School of Innovation, Olds College, Olds, AB, T4H 1R6, Canada;
[2]Department of Biochemistry, Hallym University, Chuncheon, 200-702 Ganwon-do, Korea (South);
[3]Fruits, Vegetables and Grain Technology Department, University of Environmental and Life Sciences, Norwida 25, 50-375 Wroclaw, Poland

ABSTRACT

Anthocyanins, which are used as a food coloring, are widely distributed in human diets, suggesting that human's ingest large amounts of anthocyanins from plant-based products. The antioxidant, anti-inflammatory and anti-diabetic *in vitro* activity of anthocyanins isolated from Canadian berries and purple corn has been evaluated. In addition, the *in vivo* ant-diabetic potential of purple corn extract has been analyzed. Canadian berries, saskatoon and black currant, showed good antioxidant activity, 3.65 and 4.25 mM/100g, respectively. The black currant anthocyanin extract showed COX-I and COX-II inhibition activity comparable to ibuprofen and indomethacin at 38.96 and 63.72 μg/ml, respectively. Saskatoon berry anthocyanins inhibited COX-I activity (IC50 = 68.0 μg/ml), and were relatively week against the COX-II isozyme (IC50 = 102 μg/ml). The purple corn anthocyanins displayed minimal COX and LOX activity (COX-I IC50 = 132.07 μg/ml and COX-II IC50 = 118.57 μg/ml, LOX-I IC50 = 155.2 μg/ml). Canadian berries were also active against lipoxygenase (black currant anthocyanins IC50 = 54.21 μg/ml and saskatoon berry anthocyanins IC50 = 75.96 μg/ml). Although, purple corn extract showed week antioxidant and anti-inflammatory activities *in vitro*, it showed excellent anti-diabetic activity both *in vitro* and *in vivo*. Blood glucose level in db/db mice fed with a diet containing purple corn extract was significantly lower than in control animals. Purple corn polyphenols supplemented mice reversed the decrease of pancreatic beta cells more efficiently than glimepiride, a known insulin secretion agent.

[*] corresponding author: abakowska@oldscollege.ca

INTRODUCTION

Anthocyanins are the largest water-soluble pigments in the plant kingdom (Figure 1). They are widely distributed in the human diet through crops, fruits, beans, vegetables, and red wine (Mazza and Miniatti 1993), suggest that human ingest significant amount of anthocyanins from plant-based daily diets. In general, anthocyanin pigments are stable under acidic conditions, but are unstable in neutral or basic conditions. Therefore, anthocyanins have not been recognized as a physiological functional food factor (Brouillard 1988). However, in recent years, numerous studies have shown that anthocyanins display a wide range of biological activities including antioxidant, antimicrobial and anti-carcinogenic activities; improvement of vision; induction of apoptosis; and neuroprotective effects (Kalt *et al.* 2008, Prior *et al.* 2008, Tsuda 2008, Tsuda *et al.* 2003). Recently, attention has been focused on anti-inflammatory activities of anthocyanins (Seeram *et al.* 2001, Schauss *et al.* 2006, Wang *et al.* 1999). Probably two of the most important pathways to inflammation are mediated by two enzymes lipoxygenase (LOX) and cyclooxygenase (COX). These enzymes catalyze the oxidation of arachidonic acid to leukotrienes and prostaglandins, respectively, which play important roles in the initiation and progression of the inflammatory response.

Cyclooxygenase (COX) is a rate-limiting enzyme for the synthesis of dienoic eicosanoids such as prostaglandin (PG) E2. COX exists in two isoforms (Funk *et al.* 1991; Hempel *et al.* 1994). COX-I is expressed constitutively in many types of cells and is responsible for the production of prostaglandins under physiological conditions.

COX-II is induced by pro-inflammatory stimuli, including mitogens, cytokines and bacterial lipopolysaccharide (LPS) in macrophages (Hempel *et al.* 1994) and epithelial cells (Kelley *et al.* 1997; Mitchell *et al.* 1994).

Accumulated data indicate that COX-II is involved in many inflammatory processes and induced in various carcinomas, suggesting that COX-II plays a key role in inflammation and tumorigenesis (Hla *et al.* 1993; Mestre *et al.* 1999). Interestingly, some anti-oxidants with chemopreventive effects inhibit the expression of COX-II by interfering with the signaling mechanisms that regulate the COX-II gene (Chinery *et al.* 1998).

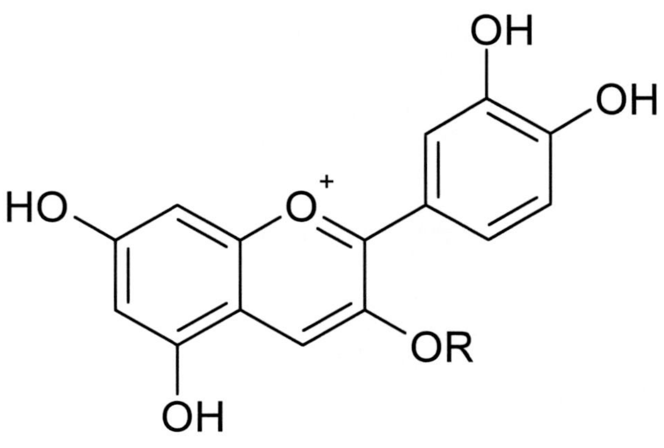

R- sugar moiety.

Figure 1. Chemical structure of anthocyanins.

Thus, the COX-II gene has been used as a biomarker to screen the cancer chemopreventive effects from phytochemicals and the identification of a COX-II inhibitor is considered to be a promising approach to protect against inflammation and tumorigenesis.

Lipoxygenase (LOX) possess regiospecificity during interaction with substrates and on this basis have been designed as arachidonate 5-, 8-, 12-, 15- lipoxygenase (5-LOX, 8-LOX, 12-LOX, 15-LOX) (Funk 2001). The four distinct enzymes insert oxygen at carbon 5, 8, 12, 15 of arachidonate acid. The primary products are 5S-, 8S-, 12S-, or 15S-hydroperoxyeicosatetraenoic acid (HPETE), which can be further reduced by gluthatione peroxidase to the hydroxyl forms (5-, 8-, 12-, 15-HETE) (Funk 2001). The variety of lipoxygenases and their products indicates a great complexity of the LOX enzymatic pathway. Accumulating evidence suggest that the 5-LOX pathway has a profound influence on the development and progression of human cancers (Steele *et al.* 1999). Inhibition of the 5-LOX pathway has a chemopreventive effect in animal lung carcinogenesis, and blocks the oxidation of several potent carcinogens (Rioux and Castonguay 1998, Moody *et al.* 1998). 8-LOX product, 8(S)-HETE plays a role in keratinocyte differentiation 12(S)-HETE, the major metabolite of the 12-LOX pathway, has been shown to correlate with metastatic potential and stimulate the expression of integrin receptors leading to increased tumor cell adhesion (Tang and Honn 1994, Tang *et al.* 1993). Additionally, 12-LOX human prostate carcinoma stimulates angiogenesis and tumor growth (Nie *et al.* 1999). 15-LOX and its products have been linked to cell maturation and differentiation (Kuhn and Chan 1997).

The effect of plant flavonoids on mammalian cells, and inflammation, heart disease and cancer has been reviewed by Middleton *et al.* (2000).

A recent study (Acquaviva *et al.* 2003) provided evidence that anthocyanins have capacity to inhibit DNA cleavage, scavenge free radicals and inhibit another enzyme: xanthine oxidase. These diverse protective effects of anthocyanins, have contributed towards a growing interest in the application of these flavonoids as dietary supplements and cosmetic agents.

More recently, Jayaprakasam *et al.* (2005) reported insulin secretion ability of anthocyanins and anthocyanidins in a culture assay. Their results suggest that Cornelian cherry (*Cornus oficinalis*) anthocyanins are useful for the prevention of type II diabetes. Type II diabetes is frequently linked to dietary habits and it is essentially preventable. This disease is deemed to be a result of a lifestyle associated with poor dietary habits, lack of exercise, and other factors.

As "fast food diets" displaced the more balanced diets, cases of diabetes type II began to increase sharply. Glycemic response is the pathway by which insulin reacts with glucose. Many of the foods we eat today are high in carbohydrates that are broken down to glucose and rapidly absorbed from the intestine into the bloodstream. This rapid absorption of glucose leads to a high glycemic response or glycemic index (GI) and a consequent rapid secretion of insulin from the pancreas. Increased insulin levels are thought to be a key factor in the development of several diseases, including non-insulin dependent diabetes, cardiovascular disease, metabolic syndrome, and insulin resistance.

Diabetes is responsible for the major share of the $117 billion in annual costs attributed to obesity according to a recent report issued by the U.S. Surgeon General's office. Type II diabetes is one condition that is worsened by further weight gain and is actually improved by weight loss.

A number of polyphenol compounds (including anthocyanins) have shown considerable biological activity such as scavenging free radicals, assisting in healthy weight loss, and lowering blood glucose levels, in culture assays.

However, *in vivo* studies and clinical evaluation of these compounds must be carried out to further validate the *in vitro* results.

Accordingly, the present study was designed to examine the role of Canadian berry, and purple corn anthocyanins in regulating blood glucose level *in vitro* and *in vivo* (with db/db mice) studies. Additionally, the anti-inflammatory properties of anthocyanin extracts were evaluated.

Materials and Methods

Cyclooxygenase I and II, and lipoxyganese were purchased from Sigma Chemicals (Oakville, Canada). Tris, Hematin, phenol, phosphatidylcholine, ibuprofen, indomethacin from Sigma Chemicals, while arachidonic acid was purchased from Nu-Chek Prep Inc. (Elysian, MN, USA). Cyanidin-3-glucoside, cyanidin-3,5-diglucoside, and cyanidin-3-rutinoside standards were obtained from Indofine (Hillsborough, NJ, USA); cyanidin-3-arabinoside was purchased from Extrasynthese (Geney Cedex, France).

Fruit Samples

Saskatoon berries cv. Northline and black currant berries cv. Ben Alder were obtained from a commercial grower (Pearson Berry Farm, AB, Canada). Purple corn grains were donated by local growers.

Extraction of Anthocyanins from Berries

Anthocyanin extracts were prepared according to Bakowska-Barczak et al. (2007). Briefly, frozen fruits (20g) were homogenized with 70ml of 80% aqueous methanol (0.1% formic acid) in a food processor and sonicated for 20min. The extracts were then centrifuged (1000xg, 30min, 4°C) and the supernatant removed. The samples were extracted once more with 70ml of 80% aqueous methanol (0.1% formic acid) using the same procedure and the supernatants were combined. The supernatants were then evaporated at 40°C under vacuum to remove solvents. The residue was dissolved in water (0.1% formic acid) and applied to a column of nonionic polymeric absorbent (Amberlite XAD-16, Rohm and Haas, Philadelphia, USA). After washing with water (0.1% formic acid) the polyphenol fraction was collected by elution with methanol (0.1% formic acid) and then evaporated to dryness using a rotary evaporator. The residue was dissolved with methanol (0.1% formic acid) and anthocyanins were separated with ethyl acetate.

Anthocyanins Extraction from Purple Corn

Anthocyanins in ground grain materials were extracted according to Abdel-Aal et al. (2006) with modifications. Ten grams of the ground material was extracted twice by mixing with 100ml of methanol acidified with 1.0N HCl (85:15 v/v) and sonicating for 30min. The crude extracts were centrifuged (1000xg, 30min, 4°C) and then refrigerated for 1 day to precipitate large molecules. Methanol from the combined crude extracts was evaporated at 40°C under vacuum. The concentrated extract was applied to a Amberlite XAD-16 resin and purified in the same manner as the berry extracts.

HPLC-ESI-MS/MS Analysis of Individual Anthocyanins

The HPLC-ESI-MS/MS analyses were performed using a 1100 Series Agilent Technologies LC/MSD system equipped with a diode array detector (DAD) coupled to a mass spectrometer (quadrupole analyzer) equipped with an ESI (electrospray ionization) interface (Agilent Technologies, Mississagua, ON, Canada). The reverse-phase separation was performed on a Vydac RP C18 column 250x4.6mm (Grace Vydac, Hespedia, CA, USA). The compounds were separated with gradient elution using 4.5% aqueous formic acid (A) and 80% acetonitrile in solution A (B) as eluents. The elution system was as follows: 0-7min 15% B, 7.0-15min 20% B, 15-16min 100% B, 16-24min 0.0% B. MS parameters were as follows: capillary voltage, 4000V; drying gas temperature, 350°C; gas flow (N_2) 12 l/min; nebulizer pressure, 60 psi; smart parameter setting (SPS); compound stability, 70%. The instrument was operated in positive ion mode scanning from *m/z* 100 to 1500 at a scan rate of 2.0 s/cycle.

Antioxidant Activity – *In Vitro* Assay

The antioxidant activity of the extracts was analyzed by Trolox Equivalent Antioxidant Capacity with ABTS radical according to Re *et al.* (1999). Briefly, to the spectrophotometric cuvet containing 30μl of extract, 3ml of $ABTS^{\cdot+}$ radical was added. Absorbance at 734nm was measured after 6 minutes of $ABTS^{\cdot+}$ addition. Determination of antioxidant activity was carried out in triplicate. The percentage inhibition of absorbance at 734nm was expressed as Trolox equivalent per 100g of fresh plant material.

Anti-Inflammatory Activity – *In Vitro* Assay

COX and LOX enzyme inhibitory assays were performed in an oxygen chamber by monitoring the initial rate of oxygen uptake using a YSI oxygen electrode attached to a YSI model 5300A biological oxygen monitor (Yellow Springs Instruments, Inc., Yellow Springs, OH) at 37°C. The assays were conducted according to a previously reported procedure (Wang et al. 1999) with modification. Each assay mixture (3ml) contained 2.97ml of 0.1M Tris HCl (pH 7.0), 10 μl of ethanol or sample, 1mM phenol, 1μM hematin, 10mM sodium chloride (only for LOX assay) and 10 μl of COX-I (0.1mg protein/ml), COX-II (0.2mg protein/ml) or

LOX (3300U/ml) enzyme. The reaction was initiated by addition of arachidonic acid (10 μl of a 10 μM final concentration). The assay was conducted in duplicate, and standard deviation was calculated for $n = 2$. The data was recorded using an Omega Engineering dataloger and software (Omega Engineering Inc., Stamford, CT). The IC50, the concentration of anthocyanins or polyphenols at which the initial oxygen consumption rate decreased by 50% was used to express the enzyme inhibition activity.

Anti-Diabetic Activity – *In Vitro* Assay

A hamster pancreatic beta cell line, HIT-T15 was used in the assay. After pre-incubating for 30 min at 37°C in KRB buffer, cells were stimulated for 60 min with 11.1mM glucose in KRB buffer and 1, 10 and 100 μg/ml of anthocyanin extract. Cells were collected by centrifugation for 5min at 3000*g* and the supernatant was used at -80°C for insulin assay. Insulin was determined by ELISA insulin assay kit (Shibayagi Co., Japan) using rat insulin as a standard, according to the manufacturer's instructions. Protein content was measured by the BCA protein assay reagent kit using albumin as a standard (Pierce). Values of secreted insulin were normalized to protein content. The average and standard deviations of three repeated experiments were calculated and plotted.

Anti-Diabetic Activity – *In Vivo* Assay

Specific-pathogen-free 6 week old female C57BL/KsJ-db/db mice (type 2 non-insulin dependent diabetes mellitus model) were obtained and used for anti-diabetic analysis of purple corn polyphenols (PCP). PCP or glimepiride (Glim) (a positive control, an insulin secretion agent) was administered through oral gavage for 7 weeks (10 mg/kg/day). Each group including the untreated control had 5 db/db mice.

(1) Measurement of Physiological and Biochemical Parameters

Body weight was measured every week. Blood glucose was measured by a portable glucose meter using OneTouch Ultra in blood samples taken from the inpraorbital vein 15 hrs after fasting every week.

(2) Measurement of HbA1c Level

HbA1c level was measured by an ELISA with anti-HbAic antibody in the blood sample taken in the 1st week and the 7th week.

(3) Oral Glucose Tolerance Test (OGTT)

After 7-week treatment, mice were administered orally with glucose (0.5 g/kg body weight) after overnight fasting. Blood samples were taken at various time points (0–120 min) from the vein, and blood glucose level was measured by a portable glucose meter using OneTouch Ultra.

(4) Immunohistochemistry

The animal was deeply anesthetized with phenobarbital sodium (10 mg/100 g body weight i.p.) and perfused through the heart with 0.9% saline and 4% paraformaldehyde in 0.1 mol/L PBS (pH 7.4). The pancreas was removed and postfixed in 4% paraformaldehyde for 4–6 h and embedded in paraffin, then sectioned into 5-μm thick slices. The sections were incubated overnight for 14 h at 4 °C with primary antibodies (guinea pig anti-human insulin antibody) diluted in PBS containing 1% bovine serum albumin (BSA). The sections were incubated with biotinylated secondary antibody (biotinylated goat anti-guinea pig IgG antibody1:1000) for 30 min at room temperature. Sections were then treated with horseradish peroxidase-conjugated streptavidin. Photographs of insulin secreting beta cells were taken under the microscope.

Toxicity *in Vitro* Assay

A hamster pancreatic beta cell line, HIT-T15 was cultured in RPMI 1640 medium containing 10% FBS, 2m l-glutamine, 100 IU of penicillin-G and 100 μg/ml of streptomycin at 37°C in a humidified atmosphere of 5% CO_2. The subculture cells were seeded at a density of 1.0–5.0×10^5cells/ml in multiwall plates or culture dishes. When they reached more than 80% confluence, the cells were used for experiments. Throughout the cell growth period the culture media were replaced every 2 days. To evaluate the effect of anthocyanins on cell viability the HIT-T15 cells (1×10^5) were incubated in 1, 10 and 100 μg/ml of anthocyanins in 24-well plates. After 24h of incubation, the cell viability was determined by the MTT reduction assay, according to the manufacturer's instructions.

CONCLUSIONS

HPLC-ESI-MS/MS analysis of individual anthocyanins. The HPLC analysis of anthocyanins showed that saskatoon berries and black currant fruits contained four main anthocyanin compounds (Figure 2). The anthocyanin composition in the saskatoon berry was reported before by Mazza (1986) and Bakowska-Barczak *et al.* (2007) and confirmed in the present study. The four primary anthocyanins in black currant fruits have been identified before by Masumoto *et al.* (2001), and Slimestad and Solheim (2002). The HPLC analysis of purple corn extract showed that it contained 13 main anthocyanins, from which 11 has been identified by HPLC-ESI-MS/MS (Table 1). The quantities of anthocyanins in 1 gram of powder are presented in Table 1. The anthocyanin composition in purple corn has been investigated before by Aoki *et al.* (2002). The authors identified 8 main anthocyanins in purple corn and their identification correlated with our present data. Through the HPLC quantification, it was found that concentrations of anthocyanins in black currant and saskatoon berry extracts were 299.2 and 306.8 mg/g, respectively. The purple corn extract contained 20.5 mg/g of anthocyanins. The anthocyanins composition and concentration in the extracts is presented in Table 1.

Antioxidant activity - in vitro assay. The TEAC assay with ABTS radical was used in the evaluation of antioxidant activity of berries and purple corn. The black currants and saskatoon berry showed high *in vitro* antioxidant activity (Figure 3).

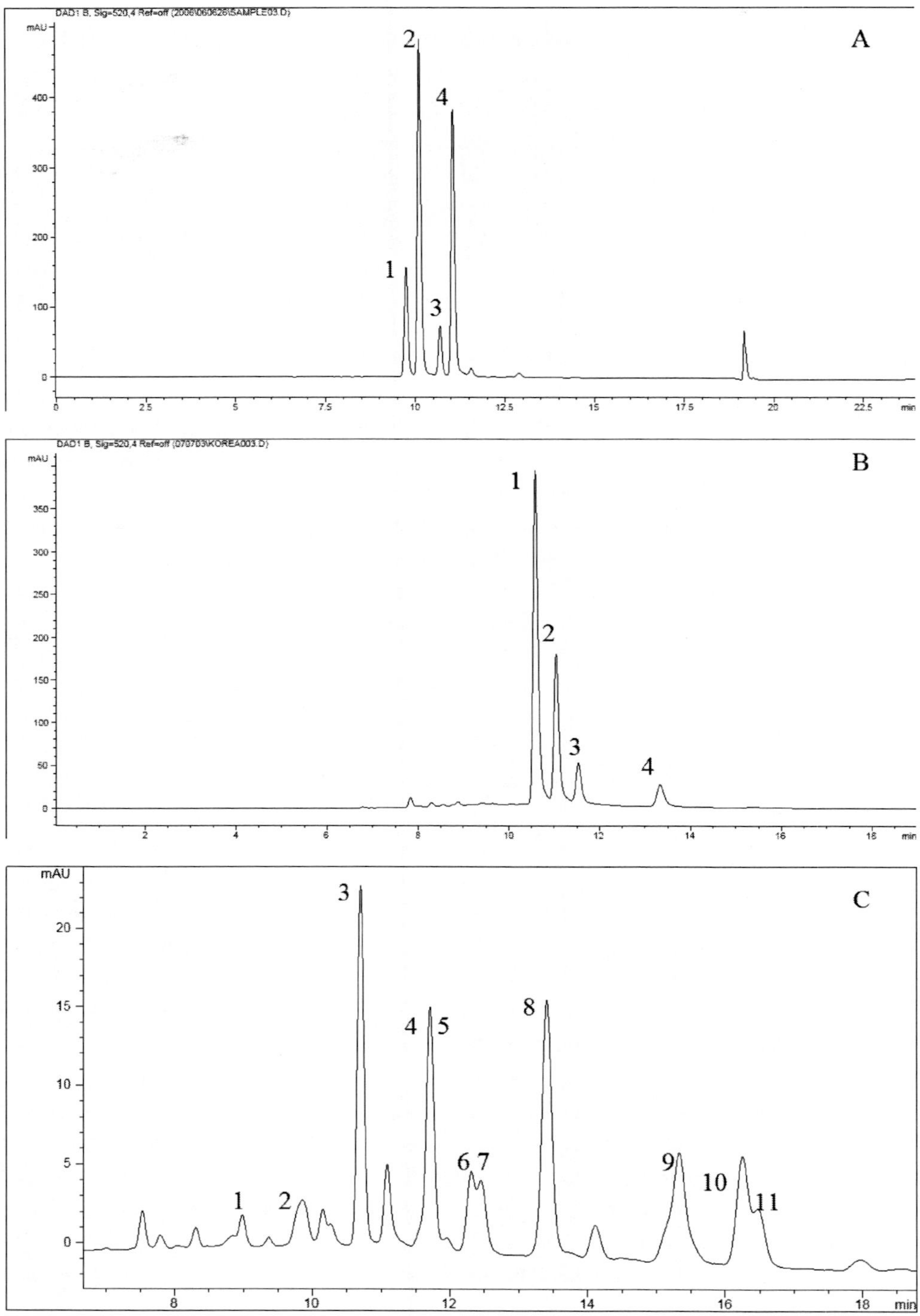

Figure 2. HPLC chromatogram (520nm) of black currant (A), saskatoon berry (B), and purple corn (C) anthocyanins.

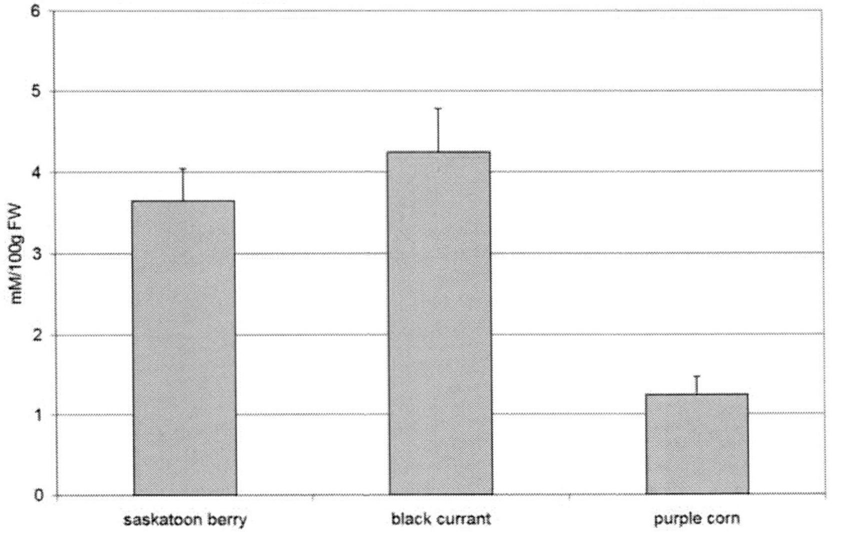

Figure 3. Antioxidant activity of saskatoon berry, black currant and purple corn.

Table 1. Composition and concentration of black currant, saskatoon berry, and purple corn anthocyanins

Peak nr	Anthocyanin	Concentration (mg/g)
Black currant		
1	delphinidin-3-glucoside*	63.7
2	delphinidin-3-rutinoside*	103.6
3	cyanidin-3-glucoside	31.4
4	cyanidin-3-rutinoside	87.6
	others*	12.9
	Total	299.2
Saskatoon berry		
1	cyanidin-3-galactoside	175.1
2	cyanidin-3-glucoside	39.7
3	cyanidin-3-arabinoside	33.5
4	cyanidin-3-xyloside	58.5
	Total	306.8
Purple corn		
1	cyanidin-3.5-diglucoside	0.35
2	cyanidin-3-galactoside	0.78
3	cyanidin-3-glucoside	2.65
4	cyanidin-3-arabinoside	2.22
5	pelargonidin-3-glucoside*	1.97
6	peonidin-3-glucoside	0.67
7	cyanidin-3-(6-malonyl)galactoside*	0.66
8	cyanidin-3-(6-malonyl)glucoside*	2.46
9	pelargonidin-3-(6-malonyl)glucoside*	1.74
10	cyanidin-3-(3,6-dimalonyl)glucoside*	1.95
11	peonidin-3-(6-malonyl) glucoside*	0.84
	others*	4.17
	Total	20.46

* Concentration calculated as cyanidin-3-glucoside.

The low antioxidant activity *in vitro* of purple corn can be correlated with the presence of acylated anthocyanins in the corn. According to Rice-Evans *et al.* (1996) the antioxidant activity is correlated with the glycosylation in anthocyanin. The highest antioxidant potential is displayed by the aglycon of the anthocyanin (anthocyanidin) than by the glycoside. The presence of acid substituents in the anthocyanin structure would cause a further decrease in the antioxidant potential of anthocyanins.

Anti-inflammatory activity - in vitro assay. The anti-inflammatory assay was done using cyclooxyganase I and II (COX) isozymes and lipoxyganese (LOX) enzyme. These enzymes are used to measure the anti-inflammatory effects of natural products (Jang and Pezzuto 1997, Seeram *et al.* 2001, Schewe *et al.* 2002, Skrzypczak-Jankun *et al.* 2003) and the assay is based on the ability of the enzymes to convert arachidonic acid to prostglandins (COX) or leukotriens (LOX), which evoke the physiological response of inflammation (Meade *et al.* 1993). The saskatoon berry, black currant, and purple corn anthocyanin extracts were assayed. Positive controls used were ibuprofen and indomethacin, which showed 50% COX-I and COX-II inhibition activity at 36.9 and 41.9 μg/ml (ibuprofen) and 26.8 and 29.3 μg/ml (indomethacin), respectively (Figure 4). The black currant anthocyanin extract showed COX-I and COX-II inhibition activity comparable to ibuprofen and indomethacin at 38.96 and 63.72 μg/ml, respectively. Anthocyanins from saskatoon berries were active against the COX-I enzyme (IC50 = 68.0 μg/ml), and relatively week against the COX-II isozyme (IC50 = 102 μg/ml). The purple corn polyphenols displayed minimal COX activity (COX-I IC50 = 132.07 μg/ml and COX-II IC50 = 118.57 μg/ml).

Anthocyanins from black currant and saskatoon berry were active against the lipoxygenase (LOX) enzyme (Figure 4). The black currant anthocyanins showed the best lipoxygenase inhibition of all extracts tested (IC50 = 54.21 μg/ml).

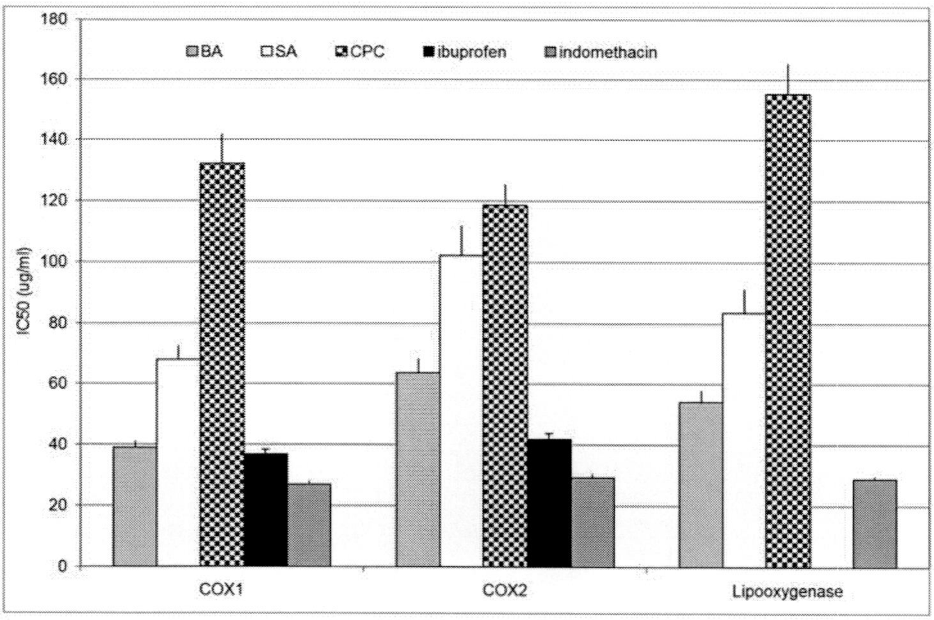

Figure 4. The effect of saskatoon berry anthocyanins (SA), black currant anthocyanins (BA), purple corn polyphenols (PCP) and commercial drugs on COX-1, COX-2, and Lipoxygenase activity.

Saskatoon berry anthocyanins showed good inhibition activity against LOX (IC50 = 75.96 μg/ml), while purple corn polyphenols showed minimal LOX inhibition (IC50 = 155.20 μg/ml).

In recent years, numerous studies have shown that anthocyanins display an anti-inflammatory activity. Seeram *et al.* (2001) reported that anthocyanins from raspberries and blackberries showed the best inhibition of COX-I enzyme as compared to the other fruits tested. At the same time tart cherries displayed the best COX-II inhibition activity. The tart cherry anthocyanins anti-inflammatory activity was also investigated by Wang *et al.* (1999). The authors showed good cyclooxygenase inhibitory activity of the tart cherry anthocyanins mixture as compared to ibuprofen and aspirin. Recently, Schauss *et al.* (2006) studied COX inhibition effect of acai freeze-dried powder. The authors reported mild COX-I and COX-II inhibitory activities; COX-I IC50 = 97.4 μg of polyphenols per 1 ml and COX-II IC50 = 175.0 μg of polyphenols per 1 ml.

The data obtained in the present research revealed that black currant anthocyanins extract showed considerable cyclooxygenases inhibitory activity. The anti-inflammatory activity of black currant extract might be related to the high amount of delphinidins in this extract; where as saskatoon berry and purple corn extracts have no delphinidins. This correlated with the observation by Hou *et al.* (2004) who reported superior inhibition of COX-II enzyme by bilberry anthocyanin extract (rich in delphinidins) and pure delphinidin.

Anti-diabetic activity – in vitro assay. The anti-diabetic activity of Canadian berry and purple corn extracts *in vitro* assay was conducted with hamster pancreatic beta cell, HIT-T15. The results obtained showed very strong insulin-releasing activity of purple corn polyphenols at 100μg/ml concentration (Figure 5). The black currant and saskatoon berry anthocyanins showed no insulin secretion activity.

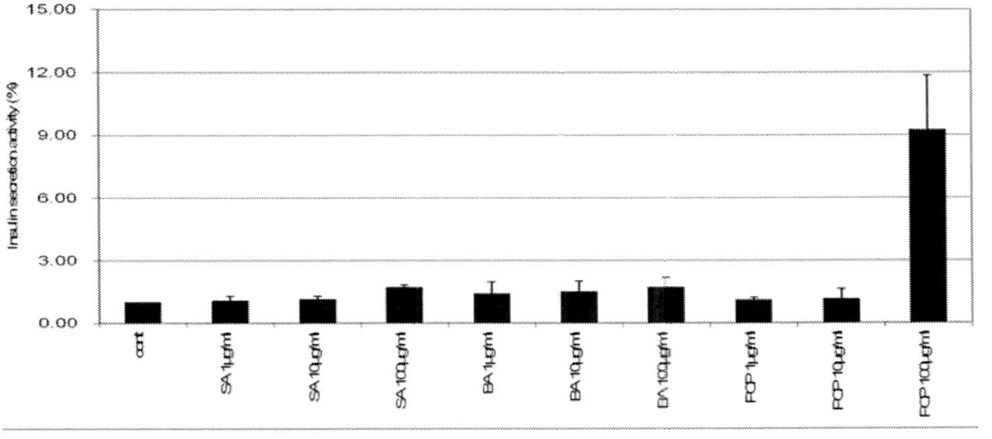

Figure 5. The influence of saskatoon berry anthocyanins (SA), black currant anthocyanins (BA) and purple corn polyphenols (PCP) on insulin secretion activity.

Anti-Diabetic Activity – *In Vivo* Assay

(1) Measurement of physiological and biochemical parameters. The body weight of the control, Glim and PCP groups did not differ throughout the experimental period (Figure 6).

The blood glucose level was 30% and 15.5% lower than in control in PCP and Glim groups, respectively (Figure 6). The purple corn polyphenols lowered the glucose level in mice blood more effectively than glimepiride.

(2) Measurement of HbA1c level. HbA1c represents glycated hemoglobin and shows the status of long term blood glucose level (about 2 months). After 7 weeks the AbA1c level was 28%, 16% and 15% higher in the control group than in the Glim and PCP groups, respectively (Figure 7). The results indicated that glycated hemoglobin level was regulated at a moderate level in the Glim and PCP treated mice.

(3) Oral glucose tolerance test (OGTT). After a 7-week treatment, mice were administered orally with glucose (0.5 g/kg body weight) after overnight fasting. Blood samples were taken from the vein at 0, 30, 60, 120 min after glucose intake, and blood glucose level was measured. The results showed short term glycolytic activity of blood glucose (Figure 8). Oral glucose tolerance activity was high in PCP treated mice, and very low in Glim treated and untreated mice.

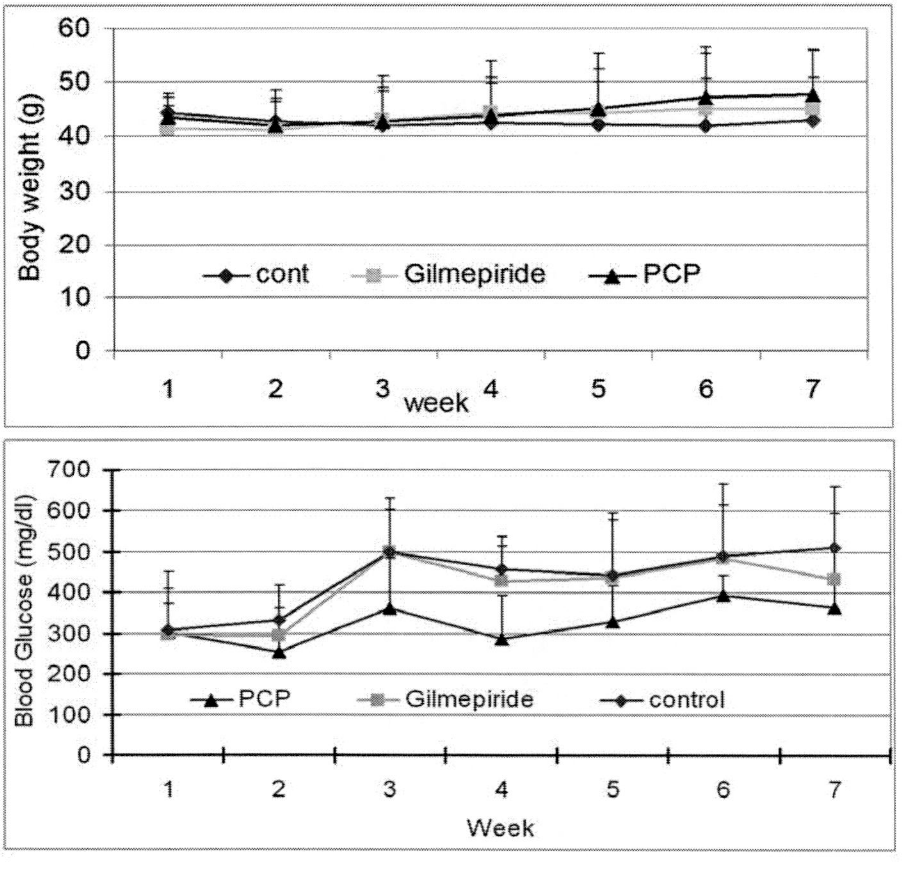

Figure 6. The influence of purple corn polyphenols (PCP) and glimepiride (positive control) on changes in body weight and blood glucose level after 7 weeks of mice feeding.

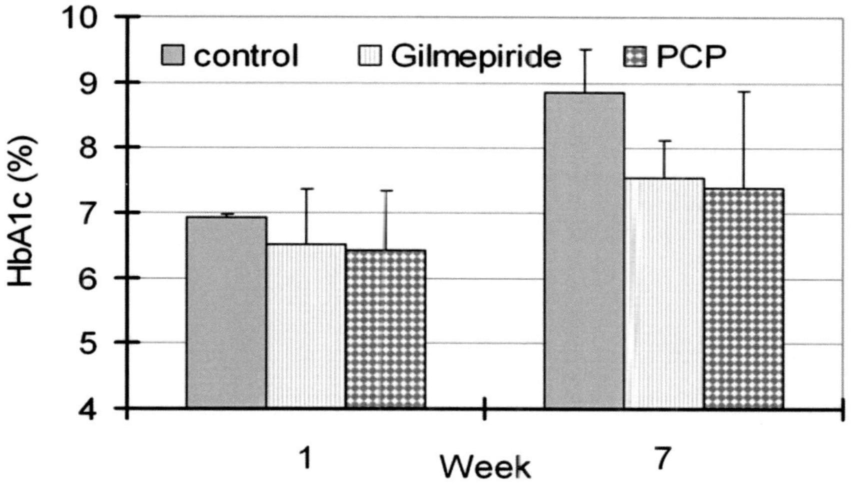

Figure 7. The influence of purple corn polyphenols (PCP) and glimepiride (positive control) on changes in long term blood glucose level after 7 weeks of mice feeding.

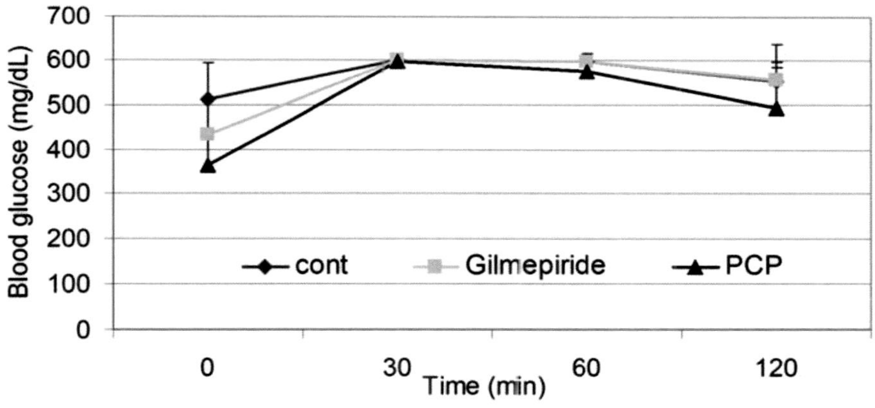

Figure 8. Effect of purple corn polyphenols (PCP) and glimepiride (positive control) in diet on changes in blood glucose level after glucose intake.

(4) Immunohistochemistry. The control db/db mice exhibited islet boundary definition loss and degeneration, whereas db/db mice supplemented with PCP and glimepiride preserved islet architecture (Figure 9).

Previous studies (Kaneto *et al.* 2001, Marshak *et al.* 1999) have shown that high blood glucose causes the deterioration of pancreatic beta cells due to oxidative stress. Therefore, antioxidants can have beneficial effects on pancreatic beta cells by neutralizing the oxidative stress. Normal beta cells compensate for insulin resistance by increasing glucose-stimulated insulin secretion or beta cell mass (Kaneto *et al.* 2001). In the current study, PCP supplemented mice reversed the decrease of pancreatic beta cells more efficiently than glimepiride, a known insulin secretion agent. A similar effect of polyphenols on the pancreas of diabetic mice was observed by Jung *et al.* (2006) who studied the antihyperglycemic properties of caffeic acid. Previously, Balasubashini *et al.* (2004) observed the ability of ferulic acid to slow the age-dependent insulin decline by a reduction of pancreatic beta cell mass. More recently, Jung *et al.* (2004) showed stimulation of insulin secretion in the pancreatic beta cells by hesperidin and naringin citrus bioflavonoids.

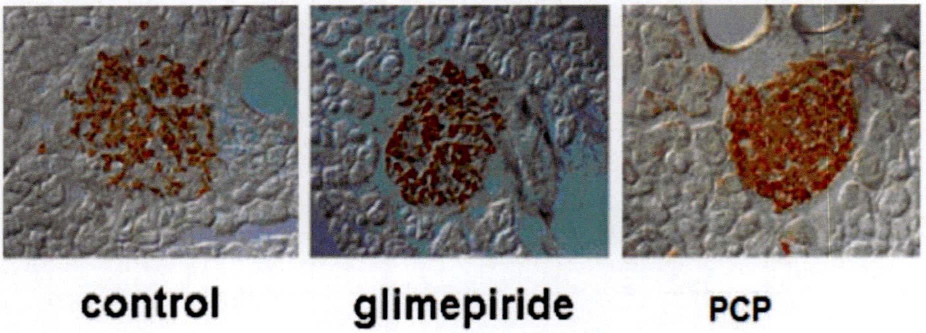

Figure 9. Immunohostochemical photograph of pancreas in C57BL/KsJ-db/db mice fed with diets supplemented with purple corn polyphenols (PCP) or glimepiride (positive control).

Toxicity - in vitro assay. A hamster pancreatic beta cell line, HIT-T15 was used for cell viability measurement. The results obtained showed that Canadian berry and purple corn extracts at the concentrations used have no growth-inhibiting effect on HIT-T15 cells, indicating no toxicity of these extracts. Interestingly, purple corn polyphenols extract showed viability-enhancing activity at 100 μg/ml (Figure 10).

Conclusion. In this study, anthocyanin extracts obtained from saskatoon berries and black currants grown in Western Canada have been shown to be powerful antioxidants. The black currant anthocyanin extract was also shown to be a potential COX-I and COX-II inhibitor. The purple corn polyphenols reduced blood glucose level and reversed the decrease of pancreatic beta cells more efficiently than glimepiride, a known insulin secretion agent. Dietary intake of purple corn polyphenol-rich food might be beneficial in preventing the onset of type 2 diabetes mellitus.

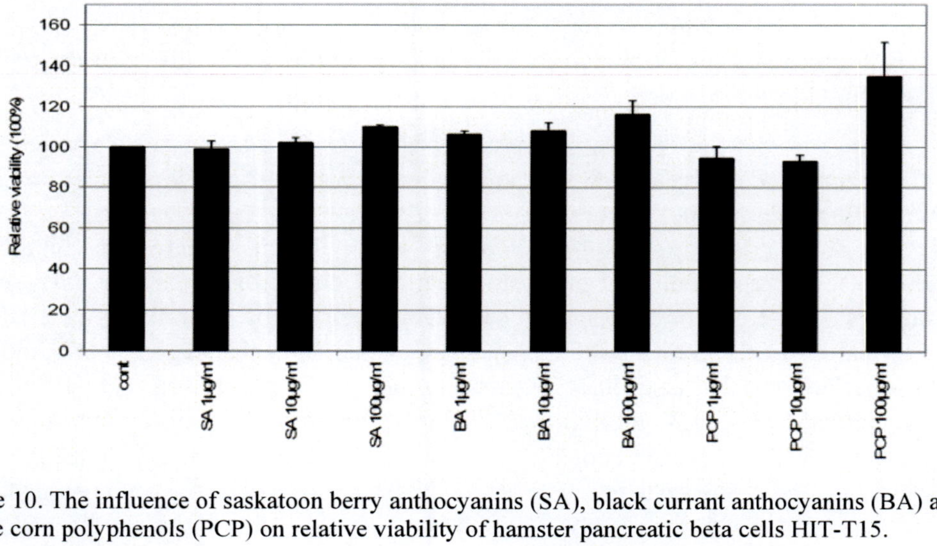

Figure 10. The influence of saskatoon berry anthocyanins (SA), black currant anthocyanins (BA) and purple corn polyphenols (PCP) on relative viability of hamster pancreatic beta cells HIT-T15.

ACKNOWLEDGMENTS

This work was supported by Alberta Innovation and Science and Korean Research Authority within the "Gangwon – Alberta Research Collaboration Program" and the Alberta Ingenuity Fund.

REFERENCES

Acquaviva, R; Russo, A; Galvano, G; Barcellona, M; Li Volti, G; Vanella, A. Cyanidin and cyanidin 3-O- β-D-glucoside as DNA cleavage protectors and antioxidants. *Cell Biology and Toxicology,* 2003, 19, 243-252.

Abdel-Aal, E-SM; Young, JC; Rabalski, I. Anthocyanin composition in black, pink, purple, and red cereal grains. *J. Agric. Food Chem.*, 2006, 54, 4696-4704.

Aoki, H; Kuze, N; Kato, Y; Gen, S-E. Anthocyanins isolated from purple corn. *Foods and Food Ingredients J. Japan*, 2002, 199, 41-45.

Bakowska-Barczak, AM; Marianchuk, M; Kolodziejczyk, P. A survey of bioactive components in Western Canadian berries. *Can. J. Physiol. Pharmacol.*, 2007, 85, 1139-1152.

Balasubashini, MS; Rukkumani, R; Viswanathan, P; Menon, VP. Ferulic acid alleviates lipid peroxidation in diabetic rats. *Phytother. Res.*, 2004, 18, 310-314.

Brouillard, R. Flavonoids and flower color. In *The flavonoids*; Harborne J.B., Ed.; Chapman and Hall, London, U.K., 1988, 525-538.

Chinery, R; Beauchamp, RD; Shyr, Y; Kirkland SC; Coffey, RJ; Morrow, JD. Antioxidants reduce cyclooxygenase-2 expression, prostaglandin production, and proliferation in colorectal cancer cells. *Cancer Res.,* 1998, 58, 2323–2327.

Funk, CD. Prostaglandins and leukotrienes: advances in eicosanoid biology. *Science*, 2001, 294, 1871-1875.

Funk, CD; Funk, LB; Kennedy, ME; Pong, AS; Fitzgerald, GA. Human platelet/erythroleukemia cell prostaglandin G/H synthase: cDNA cloning, expression, and gene chromosomal assignment. *FASEB J.* 1991, 5, 2304–2312.

Hempel, SL; Monick, MM; Hunninghake, GW. Lipopolysaccharide induces prostaglandin H synthase-2 protein and mRNA in human alveolar macrophages and blood monocytes. *J. Clin. Invest.* 1994, 93, 391–396.

Hla, T; Ristimaki, A; Appleby, S; Barriocanal, JG. Cyclooxygenase gene expression in inflammation and angiogenesis. *Ann.. NY Acad. Sci.* 1993, 696, 197–204.

Hou, D-X; Fujii, M; Terahara, N; Yoshimoto, M. Molecular mechanisms behind the chemopreventive effects of anthocyanidins. *J. Biomed. Biotech.* 2004, 5, 321-325.

Jang, M-S; Pezzuto, JM. Assessment of cyclooxygenase inhibitors using in vitro assay systems. *Methods in Cell Sci.*, 1997, 19, 25-31.

Jayaprakasam, B; Vareed, SK; Olson, LK; Nair, MG. Insulin secretion by bioactive anthocyanins and anthocyanidins present in fruits. *J. Agric. Food Chem.* 2005, 53, 28-31.

Jung, UJ; Lee, MK; Jeong, KS; Choi, MS. The hypoglycemic effects of hesperidin and naringin are partly mediated by hepatic glucose-regulating enzymes in C57BL/KsJ-db/db mice. *J. Nutr.,* 2004, 134, 2499-2503.

Jung, UJ; Lee, MK; Park, YB; Jeon, SM; Choi, MS. Antihyperglycemic and antioxidative properties of caffeic acid in *db/db* mice. *J. Pharm. Exp. Ther.,* 2006, 318, 476-483.

Kalt, W; Blumberg, JB; McDonald, JE; Vinqvist-Tymchuk, MR; Fillmore, SAE; Graf, BA; O'Leary, JM; Milbury, PE. Identification of Anthocyanins in the liver, eye, and brain of blueberry-fed pigs. *J. Agric. Food Chem.,* 2008, 56, 705-712.

Kaneto, H; Xu, G; Song, KH; Suxuma, K; Bonner-Weir, S; Sharma, A; Weir, GC. Activation of the hexosamine pathway leads to deterioration of pancreatic β-cell function through the induction of oxidative stress. *J. Biol. Chem.,* 2001, 276, 31099-31104.

Kelley, DJ; Mestre, JR; Subbaramaiah, K; Sacks, PG; Schantz, SP; Tanabe, T. Benzo-α-pyrene up-regulates cyclooxygenase-2 gene expression in oral epithelial cells. *Carcinogenesis*, 1997, 18, 795–799.

Kuhn, H; Chan, L. The role of 15-lipoxygenase in atherogenesis: pro- and antiatherogenic actions. *Curr. Opin. Lipidol.*, 1997, 8, 111-117.

Mahyar-Roemer, M; Roemer, K. p21 Wafl/Cipl can protect human colon carcinoma cells against p53-dependent and p53-independent apoptosis induced by natural chemopreventive and therapeutic agents. *Oncogene*, 2001, 20, 3387–3398.

Marshak, S; Leibowitz, G; Bertuzzi, F; Socci, C; Kaiser, N; Gross, DJ; Cerasi, E; Melloul, D. Impaired β-cell functions induced by chronic exposure of cultured pancreatic islets to high glucose. *Diabetes*, 1999, 48, 1230-1236.

Matsumoto, H; Hanamura, S; Kawakami, T; Sato, Y; Hirayama, M. Preparative-scale isolation of four anthocyanin components of black currant (*Ribes nigrum* L.) fruits. *J. Agric. Food Chem.*, 2001, 49, 1541-1545.

Mazza, G. 1986. Anthocyanins and other phenolic compounds of saskatoon berries (*Amelanchier alnifolia* Nutt.). *J. Food Sci*. 51: 1260–1264.

Mazza, G and Miniatti, E. (1993). *Anthocyanins in fruit, vegetables and grains*. CRC Press, Boca Ranton, Florida, USA.

Meade, EA; Smith, WL; Dewitt, DL. Differential inhibition of prostaglandin endoperoxide synthase (cyclooxygenase) isozymes by aspirin and other non-steroidal anti-inflammatory drugs. *J. Biol. Chem.*, 1993, 268, 6610-6614.

Mestre, JR; Chan, G; Zhang, F; Yang, EK; Sacks, PG; Boyle, JO. Inhibition of cyclooxygenase-2 expression. An approach to preventing head and neck cancer. *Ann N Y Acad Sci* 1999; 889:62–71.

Middleton, Jr E; Kandaswami, C; Theoharides, TC. The effects of plant flavonoids on mammalian cells: implication for inflammation, heart disease, and cancer. *Pharmacol. Rev.,* 2000; 52: 673-751.

Mitchell, JA; Belvisi, MG; Akarasereenont, P; Robbins, RA; Kwon, OJ; Croxtall, J. Induction of cyclooxygenase-2 by cytokines in human pulmonary epithelial cells: regulation by dexamethasone. *Br. J. Pharmacol.*, 1994;113:1008–14.

Moody, TW; Leyton, J; Martinez, A; Hong, S; Malkinson, A; Mulshine, JL. Lipoxygenase inhibitors prevent lung carcinogenesis and inhibit ana-small cell lung cancer growth. *Exp. Lung Res*., 1998, 24, 617-628.

Nie, D; Hillman, GG; Geddes, T; Tang, K; Pierson, C; Gringon, DJ; Honn, KV. Platelet-type 12-lipoxygenase regulates angiogenesis in human prostate cancer. *Urology*, 1999, 469, 623-630.

Okano, Y; Obayashi, K; Yahagi, S; Kurihara, K; Kaburagi, S; Kurata, Y; Masaki, H. Improvement of wrinkles by an all-trans-retinoic acid derivative, D-δ-tocopheryl retinoate. *J. Derma. Sci.*, 2006, 2, 65-74.

Prior, RL; Wu, X; Gu, L; Hager, TJ; Hager, A; Howard, LR. Whole berries versus berry anthocyanins: interactions with dietary fat levels in the C57BL/6J mouse model of obesity. *J. Agric. Food Chem.*, 2008, 56, 847-863.

Radhakrishnan, S; Nguyen, LT; Ciric, B; Flies, D; Van Keulen, VP; Tamada, K; Chen, L; Rodriguez, M; Pease, LR. Immunotherapeutic Potential of B7-DC (PD-L2) Cross-Linking Antibody In Conferring Antitumor Immunity. *Cancer Research*, 2004, 64, 4965-4972.

Re, R; Pellegrini, N; Proteggente, A; Pannala, A; Yang, M; Rice-Evans, C. Antioxidant activity applying an improved ABTS radical cation decolorization assay. *Free Radic. Biol. Med.*, 1999, 26, 1231-1237.

Rice-Evans, CA; Miller, NJ; Paganga, G. Structure-antioxidant activity relationships of flavonoids and phenolic acids. *Free Rad. Biol. Med.*, 1996, 20, 933-965.

Rioux, N; Castonguay, A. Inhibitors of liupoxygenase: a new class of cancer chemopreventive agents. *Carcinogenesis*, 1998, 19, 1393-1400.

Schauss, AG; Wu, X; Prior, RL; Ou, B; Huang, D; Owens, J; Agarwal, A; Jensen, GS; Hart, AN; Shanbrom, E. Antioxidant capacity and other bioactivities of the freeze-dried Amazonian palm berry, *Euterpe oleraceae* Mart. (Acai). *J. Agric. Food Chem.,* 2006, 54, 8604-8610.

Schewe, T; Kuh, H; Sies, H. Flavonoids of cocoa inhibit recombinat human 5-lipoxygenase. *J. Nutr.*, 2002, 132, 1825-1829.

Seeram, NP; Momin, RA; Nair, MG; Bourquin, LD. Cyclooxygenase inhibitory and antioxidant cyanidin glycosides in cherries and berries. *Phytomedicine*, 2001, 8, 362-369.

Skrzypczak-Jankun, E; Zhou, K; Jankun, J. Inhibition of lipoxygenase by (-)-epigallocatechin gallate: X-ray analysis at 2.1Å reveals degradation of EGCG and shows soybean LOX-3 complex with EGC instead. *Int. J. Mol. Med.*, 2003, 12, 415-422.

Slimestad, R; Solheim, H. Anthocyanins from black currants (*Ribes nigrum* L.). *J. Agric. Food Chem.*, 2002, 50, 3228-3231.

Steele, VE; Holmes, CA; Hawk, ET; Kopelovich, L; Lubet, RA; Crowell, JA; Sigman, CC; Kelloff, GJ. Lipoxygenase inhibitors as potential cancer chemopreventives. Cancer Epidemiol. *Biomarkers Prev.*, 1999, 8, 467-483.

Tang, DG; Grossi, IM; Chen, YQ; Diglio, CA; Honn, KV. 12(S)-HETE promotes tumor-cell adhesion by increasing surface expression of alpha V beta 3 integrins on endothelial cells. *Int. J. Cancer.*, 1993, 54, 102-111.

Tang, DG; Honn, KV. 12-Lipoxygenase, 12(S)-HETE, and cancer metastasis. *Ann. NY. Acad. Sci.*, 1998, 744, 199-215.

Wang, H; Nair, MG; Strasburg, GM; Chang, YC; Booren, AM. Antioxidant and antiinflammatory activities of anthocyanins and their aglycon, cyanidin, from tart cherries. *J. Nat. Pro.*, 1999, 62, 294-296.

In: Renewable Resources: Obtaining, Processing... ISBN 978-1-6060692-534-8
Editors: R. Kozlowski, G. Zaikov and F. Pudel

Chapter 14

BIODEGRADATION AND MEDICAL APPLICATION OF MICROBIAL POLY(3-HYDROXYBUTYRATE)

***A. P. Bonartsev*[*1,3]*, A. L. Iordanskii*[2]*, G. A. Bonartseva*[1] *and G. E. Zaikov*[2]**

[1]A. N. Bach's Institute of Biochemistry, Russian Academy of Sciences, Leninskii prosp. 33, 119071 Moscow, Russia

[2]General Institute of Chemical Physics, Russian Academy of Sciences, Kosygin str. 4, 119991 Moscow, Russia

[3]Faculty of Biology, Moscow State University, Leninskie gory 1-12, 119992 Moscow, Russia

ABSTRACT

This review is designed to be a comprehensive source for biodegradable polymer : poly(3-hydroxybutyrate) research, including fundamental structure/property relationships and biodegradation kinetics for samples of different geometry. In addition to presenting the scientific framework for the advances in PHB research, this review focuses on applications of PHB in biomedicine and environment with a discussion on commercial applications and health/safety concerns for biodegradable materials.

Keywords: microbial poly(3-hydroxybutyrate), nonenzymatic hydrolysis, enzymatic degradation, biodegradation in soil, degradation in animal tissues, biocompatibility, applications.

Over last decade an intense development of biomedical application of microbial poly((R)-3-hydroxybutyrate) (PHB) in producing of biodegradable polymer implants and controlled drug release systems needs for comprehensive understanding of the PHB biodegradation process [1-3]. Examination of PHB degradation process is also necessary for

[*] E-mail: bonar@inbi.ras.ru bonar@inbi.ras.ru

development of novel friendly environment polymer packaging [4-6]. It is generally accepted that biodegradation of PHB both in living systems and in environment occurs via enzymatic and non-enzymatic processes that take place simultaneously under natural conditions. It is, therefore, important to understand both processes [1,7]. Opposite to PGA and PLGA, PHB is considered to be moderately resistant to degradation in vitro as well as to biodegradation in animal body. The rates of degradation are influenced by the characteristics of the polymer, such as chemical composition, crystallinity, morphology and molecular weight [8,9]. In spite of that PHB application in vitro and in vivo has been intensively investigated, the most of the available data are often incomplete and even contradictory. The presence of conflicting data can be partially explained by the fact that biotechnologically produced PHB with standardized properties is relatively rare and is not readily available due to a wide variety of PHB biosynthesis sources and different PHB manufacturing processes.

Above contradictoriness can be explained also by excess applied trend in PHB degradation research. At most of the papers observed in this review PHB degradation process has been investigated in the narrow framework of development of specific medical device on the base of PHB. Depending on applied biomedical purposes biodegradation of PHB was investigated under different geometry: films and plates with various thickness [10-13], cylinders [13-16], monofilament threads [17-18] and microspheres [19]. At these experiments PHB was used from various sources, with different molecular weight and crystallinity. Besides, different technologies of PHB devices manufacture affect such important characteristics as polymer porosity and surface structure [11,12]. Reports regarding the complex theoretical research of mechanisms of hydrolysis, enzymatic degradation and biodegradation in vivo of PHB processes are relatively rare [10-11,13,20-22] that attaches great value and importance to these investigations. Nevertheless, the effect of thickness, size and geometry of PHB device, molecular weight and crystallinity of PHB on the mechanism of PHB hydrolysis and biodegradation was not yet well clarified.

1.1. Nonenzymatic Hydrolysis of PHB *In Vitro*

Examination of hydrolytic degradation of natural poly((R)-3-hydroxybutyrate) in vitro is a very important step for understanding of PHB biodegradation. There are several very profound and careful examinations of PHB hydrolysis that was carried out 10-15 years [20-23]. Hydrolytic degradation of PHB was usually examined under standard experimental conditions simulating internal body fluid: in buffered solutions with pH=7.4 at 37°C but at some cases the higher temperature (55°C, 70°C and more) and other values of pH (from 2 to 11) were selected.

The classical experiment for examination of PHB hydrolysis in comparison with hydrolysis of other widespread biopolymer, polylactic acid (PLA), was carried out by Koyama N. and Doi Y. [20]. They selected films (10 x 10 mm size, 50 μm thickness, 5 mg initial mass) from PHB (M_n = 300 kDa, M_w = 650 kDa) and polylactic acid (M_n = 9 kDa, M_w = 21 kDa) prepared by solvent casting and aged for 3 weeks to reach equilibrium crystallinity. It was shown that hydrolytic degradation of natural PHB is very slow process. The mass of PHB film remained unchanged at 37°C in 10 mM phosphate buffer (pH=7.4) over a period of 150 days, while the mass of the PLA film rapidly decreased with time and reached 17% of the initial mass after 140 days. The rate of decrease in the M_n of the PHB was also much slower than the rate of decrease in the M_n of PLA. The M_n of the PHB decreased approximately to 65% of the initial PHB M_n after 150 days, while the M_n of the PLA decreased to 20% (2 kDa) of the initial PLA M_n at the same time point. As PLA used at this research was with low molecular weight we should compare this data with the data of investigation of hydrolysis of PLA with the same molecular weight as observed PHB. We examined rates of in vitro mass loss of polymer films with the same thickness (40 μm) from PLA and PHB with the same molecular weight (M_w = 450 kDa). It was shown that the mass of PLA film decreased to 87%, whereas the mass of PHB film remained unchanged at 37°C in 25 mM phosphate buffer (pH=7.4) over a period of 84 days [24-25].

The cleavage of polyester chains is known to be catalysed by the carboxyl end groups, and its rate is proportional to the concentrations of water and ester bonds that may be constant during the hydrolysis, owing to the presence of a large excess of water molecules and of ester bonds of polymer chains. Thus, the kinetics of nonenzymatic hydrolysis can be expressed by the following equation [26-27]:

$$\ln M_n = \ln M_n^0 - kt \tag{1}$$

where M_n and M_n^0 are the number-average molecular weights of a polymer component at time t and zero, respectively.

The average number of bond cleavage per original polymer molecule, N, is given by equation 2:

$$N = (M_n^0/M_n) - 1 = k_d P_n^0 t, \tag{2}$$

where k_d is the rate constant of hydrolytic degradation, and P_n^0 is the number-average degree of poymerization at time zero. Thus, if the chain scission is completely random, the value of N is linearly dependent on time.

The molecular weight decrease with time is the distinguishing feature of mechanism in nonenzymatic hydrolysis condition in contrast to enzymatic hydrolysis condition of PHB when M_n values remained almost unchanged. It was supposed also that water-soluble oligomers of PHB may accelerate chain scissions of PHB homopolymer [20]. In contrast, Freier T. et al. [11] showed that PHB hydrolysis was not accelerated by the addition of pre-degraded PHB: the rate of mass and M_w loss of blends (70/30) from high-molecular PHB (M_w = 641 kDa) and low-molecular PHB (M_w = 3 kDa) was the same with degradation rate of pure high-molecular PHB. Meanwhile, the addition of amorphous atactic PHB (atPHB) (Mw = 10 kDa) to blend with high-molecular PHB caused significant acceleration of PHB hydrolysis: the mass loss of PHB/atPHB blends was 7% in comparison with 0% mass loss of pure PHB, the decrease of M_w was 88% in comparison with 48% M_w decrease of pure PHB [11,28]. We have showed that the rate of hydrolysis of PHB films depends on M_w of PHB. The films from PHB of high molecular weight (450 and 1000 kDa) degraded slowly as it was described above whereas films from PHB of low molecular weight (150 and 300 kDa) lost weight relatively gradually and more rapidly [24-25].

To enhance the hydrolysis of PHB a higher temperature was selected for degradation experiments: 55°C, 70°C and more [20]. It was showed by the same research team that the weights of films (12 mm diameter, 55 μm thick) from PHB (M_n = 768 and 22 kDa, M_w =

1460 and 75 kDa) were unchanged at 55°C in 10 mM phosphate buffer (pH=7.4) over a period of 58 days. The M_n value decreased from 768 to 245 kDa for 48 days. The film thickness increased from 65 to 75 µm for 48 days, suggesting that water permeated the polymer matrix during the hydrolytic degradation. Examination of the surface and cross-section of PHB films before and after hydrolysis showed that surface after 48 days of hydrolysis was apparently unchanged, while the cross-section of the film exhibited a more porous structure (pore size < 0.5 µm). It was shown also that the rate of hydrolytic degradation is not dependent upon the crystallinity of PHB film. The observed data indicates that the nonenzymatic hydrolysis of PHB in the aqueous media proceeds via a random bulk hydrolysis of ester bonds in the polymer chain films and occurs throughout the whole film, since water permeates the polymer matrix [20-21]. Moreover, as the molecular weight distribution was unimodal over the whole degradation time which, together with the observed first-order kinetics, indicates a random chain scission both in the crystalline and the amorphous regions of PHB [11,29]. For synthetic amorphous atactic PHB it was shown that the hydrolysis of PHB is the two-step process. First, the random chain scission proceeds The scission accompanies by a molecular weight decrease. Then, at a molecular weight of about 10000, mass loss begins [23].

The analysis of literature data shows a great spread in values of rate of PHB hydrolytic degradation in vitro. It can be explained by different thickness of PHB films or geometry of PHB devices used for experiment as well as by different sources, purity degree and molecular weight of PHB (Table 1). At 37°C and pH=7.4 the weight loss of PHB (unknown M_w) films (500 µm thick) was 3% after 40 days incubation [32], 0% after 52 weeks (364 days) and after 2 years (730 days) incubation (640 kDa PHB, 100 µm films) [11-12], 0% after 150 days incubation (650 kDa PHB, 50 µm film) [20], 7.5% after 50 days incubation (279 kDa PHB, unknown thickness of films) [31],, 0% after 3 months (84 days) incubation (450 kDa PHB, 40 µm films), 12% after 3 months (84 days) incubation (150 kDa PHB, 40 µm films) [24-25], 0% after 180 days incubation of monofilament threads (30 µm in diameter) from PHB (470 kDa) [17-18]. The molecular weight of PHB dropped to 36% of the initial values after 2 years (730 days) of storage in buffer solution [12], to 87% of the initial values after 98 days [32], to 58% of the initial values after 84 days [24-25] (Table 1).

At acidic or alkaline aqueous media PHB degrades more rapidly: 0% after 20 weeks (140 days) incubation in 0.01 NaOH (pH=11) (200 kDa PHB, 100 µm films) with surface changing [33], 0% after 180 days incubation of PHB threads in phosphate buffer (pH=5.2 and 5.9) [18], complete PHB films biodegradation after 19 days (pH=13) and 28 days (pH=10) [31]. It was demonstrated that after 20 weeks of exposure to NaOH solution, the surfaces of PHB samples became rougher, along with an increased density of whole formation on their surfaces. From these results, it can be surmised that the non-enzymatic degradation of PHAs progresses on their surfaces before noticeable weight loss occurs (Figure 1) [33].

It was shown also that treatment of PHB film with 1M NaOH caused a reduce in pore size on film surface from 1-5 µm to around 1 µm that indicates a partially surface degradation of PHB in alkaline media [34-35]. At higher temperature no weight loss of PHB films and threads was observed after 98 and 182 days incubation in phosphate buffer (pH=7.2) at 55°C and 70°C, respectively [17], 12% and 39% of PHB (450 and 150 kDa, respectively) films after 84 days incubation at 70°C [35,40], 50% and 25% after 150 days incubation of microspheres (250-850 µm diameter) from PHB (50 kDa and 600 kDa, respectively) [36].

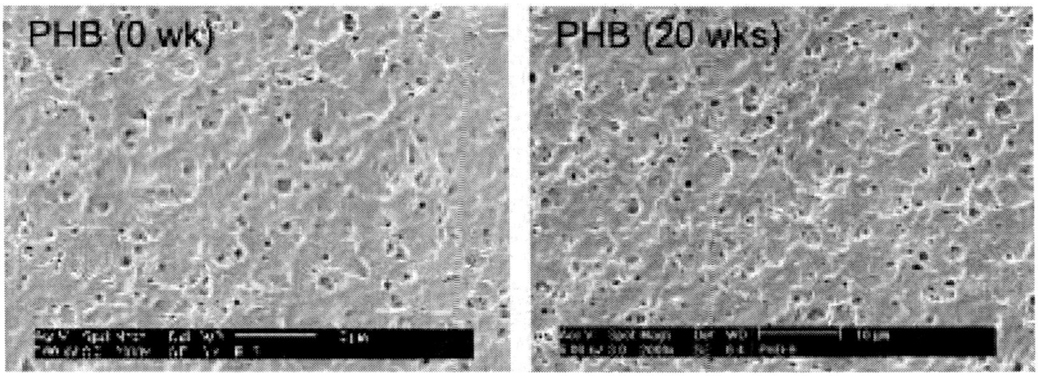

Figure 1. Scanning electron microscopy photographs of PHB films both before (initial sample, panel on the left) and after 20 weeks (panel on the right) of non-enzymatic hydrolysis in 0.01 N NaOH solution (scale bars, 10 μm) [33].

Table 1. Nonenzymatic hydrolysis of PHB in vitro

Type of device	Initial M_w of PHB, kDa	Size/ Thickness, μm	Conditions	Rela-tive mass loss of PHB, %	Rela-tive decrease of PHB M_w, %	Time, days	Links
film	650	50	37°C, pH=7.4	0	35	150	20
film	640	100	37°C, pH=7.4	0	64	730	12
film	640	100	37°C, pH=7.4	0	45	364	11
film	450	40	37°C, pH=7.4	0	42	84	24-25
film	150	40	37°C, pH=7.4	12	63	84	24-25
film	279	-	37°C, pH=7.4	7.5	-	50	31
plate	-	500	37°C, pH=7.4	3	-	40	30
plate	380	1000	37°C, pH=7.4	0	-	28	37
plate	380	2000	37°C, pH=7.4	0	8	98	32
thread	470	30	37°C, pH=7.0	0	-	180	18
thread	-	-	37°C, pH=7.2	0	-	182	17
micro-spheres	50	250-850	37°C, pH=7.4	0	0	150	36
thread	470	30	37°C, pH=5.2	0	-	180	17
film	279	-	37°C, pH=10	100	-	28	31
film	279	-	37°C, pH=13	100	-	19	31
film	650	50	55°C,pH=7.4	0	68	150	20
plate	380	2000	55°C, pH=7.4	0	61	98	32
film	640	100	70°C, pH=7.4	-	55	28	11
film	150	40	70°C, pH=7.4	39	96	84	24-25
film	450	40	70°C, pH=7.4	12	92	84	24-25
micro-spheres	50	250-850	85°C, pH=7.4	50	68	150	36
micro-spheres	600	250-850	85°C, pH=7.4	25	-	150	36

During degradation of PHB monofilament threads, films and plates in vitro the change of mechanical properties was observed under different conditions [17,37]. It was shown that a number of mechanical indices of threads became worse: load at break lost 36%, strain at break lost 33%, Young's modulus didn't change, tensile strength lost 42% after 182 days incubation in phosphate buffer (pH=7.2) at 70°C. But at 37°C the changes were more complicated: at first load at break increased from 440 g to 510 g (16%) at 90th day and then decreased to the initial value at 182nd day, strain at break increased rapidly from 60 to 70% (in 17%) at 20th day and then gradually increased to 75% (in 25%) at 182nd day, Young's modulus didn't change [17]. For PHB films it was demonstrated a gradual 32% decrease in Young's modulus and 77% fall in tensile strength during 120 days incubation in phosphate buffer (pH=7.4) at 37°C [37]. For PHB plates more complicated changes were observed: at first tensile strength dropped in 13% for 1st day and then increased to the initial value at 28th day, Young's modulus dropped in 32% for 1st day and then remain unchanged up to 28th day, stiffness decreased sharply also in 40% for 1st day and then remain unchanged up to 28th day [38].

1.2. ENZYMATIC DEGRADATION OF PHB *IN VITRO*

The examination of enzymatic degradation of PHB in vitro is the following important step for understanding of PHB operation in animal tissues and in environment. The most papers observed degradation of PHB by depolymerases of its own bacterial producers. The degradation of PHB in vitro by depolymerase was thoroughly examined and mechanism of enzymatic PHB degradation was perfectly clarified by Doi Y. [20-21]. At these early works it was shown that 68-85% and 58% mass loss of PHB (Mw = 650-768 and 22 kDa, respectively) films (50-65 μm thick) occurred for 20 h under incubation at 37°C in phosphate solution (pH=7.4) with depolymerase (1.5-3 μg/ml) isolated from *A. faecalis*. The rate (k_e) of enzymatic degradation of films from PHB (M_n = 768 and 22 kDa) was 0.17 and 0.15 mg/h, respectively. The thickness of polymer films dropped from 65 to 22 μm (32% of initial thickness) during incubation. The scanning electron microscopy examination showed that the surface of the PHB film after enzymatic degradation was apparently blemished by the action of PHB depolymerase, while no change was observed inside the film. Moreover, the molecular weight of PHB remained almost unchanged after enzymatic hydrolysis: the M_n of PHB decreased from 768 to 669 kDa or unchanged (22 kDa) [20-21].

The extensive literature data on enzymatic degradation of PHB by specific PHB depolymerases was collected in detail in review of Sudesh K., Abe H. and Doi Y. [39]. We would like to summarize some the most important data. But at first it is necessary to note that PHB depolymerase is very specific enzyme and the hydrolysis of polymer by depolymerase is the unique process. But in animal tissues and even in environment the enzymatic degradation of PHB is occurred mainly by nonspecific esterases [19,40]. Thus, in the frameworks of this review, it is necessary to observe the fundamental mechanisms of PHB enzymatic degradation.

The rate of enzymatic erosion of PHB by depolymerase is strongly dependent on the concentration of the enzyme. The enzymatic degradation of solid PHB polymer is heterogeneous reaction involving two steps, namely, adsorption and hydrolysis. The first step is adsorption of the enzyme onto the surface of the PHB material by the binding domain of the PHB depolymerase, and the second step is hydrolysis of polyester chains by the active site of the enzyme. The rate of enzymatic erosion for chemosynthetic PHB samples containing both monomeric units of (R)- and (S)-3-hydrohybutyrate is strongly dependent on both the stereocomposition and on the tacticity of the sample as well as on substrate specificity of PHB depolymerase. The water-soluble products of random hydrolysis of PHB by enzyme showed a mixture of monomers and oligomers of (R)-3-hydrohybutirate. The rate of enzymatic hydrolysis for melt-crystallized PHB films by PHB depolymerase decreased with an increase in the crystallinity of the PHB film, while the rate of enzymatic degradation for PHB chains in an amorphous state was approximately 20 times higher than the rate for PHB chains in a crystalline state. It was suggested that the PHB depolymerase predominantly hydrolyzes polymer chains in the amorphous phase and then, subsequently, erodes the crystalline phase. The surface of the PHB film after enzymatic degradation was apparently blemished by the action of PHB depolymerase, while no change was observed inside the film. Thus, depolymerase hydrolyses of the polyester chains in the surface layer of the film and polymer erosion proceeds in surface layers, while dissolution, the enzymatic degradation of PHB are affected by many factors as monomer composition, molecular weight and degree of crystallinity [39].

At the next step it is necessary to observe enzymatic degradation of PHB under the conditions that modeled the animal tissues and body fluids containing nonspecific esterases. In vitro degradation of PHB films in the presence of various lipases as nonspecific esterases was carried out in buffered solutions containing lipases [41-42], in digestive juices (for example, pancreatin) [11], biological media (serum, blood etc.) [18] and crude tissue extracts containing a mixture of enzymes [19] to examine the mechanism of nonspecific enzymatic degradation process. It was noted that a Ser..His..Asp triad constitutes the active center of the catalytic domain of both PHB depolymerase [43] and lipases [44]. The serine is part of the pentapeptide Gly X1-Ser-X2-Gly, which has been located in all known PHB depolymerases as well as in lipases, esterases and serine proteases [43].

On the one hand, it was shown that PHB was not degraded for 100 days with a quantity of lipases isolated from different bacteria and fungi [41,42]. On the other hand, the progressive PHB degradation by lipases was shown [24-25,34-35]. PHB enzymatic biodegradation was studied also in biological media: it was shown that with pancreatin addition no additional mass loss of PHB was observed in comparison with simple hydrolysis [11], the PHB degradation process in serum and blood was demonstrated to be similar to hydrolysis process in buffered solution [24-25], whereas progressive mass loss of PHB sutures was observed in serum and blood: 16% and 25%, respectively, after 180 days incubation [18], crude extracts from liver, muscle, kidney, heart and brain showed the activity to degrade the PHB: from 2% to 18% mass loss of PHB microspheres after 17 h incubation at pH 7.5 and 9.5 [19]. The degradation rate in solution with pancreatin addition, obtained from the decrease in M_w of pure PHB, was accelerated about threefold: 34% decrease in M_w after incubation for 84 days in pancreatin (10 mg/ml in Sorensen buffer) vs. 11% decrease in Mw after incubation in phosphate buffer [11].The same data was obtained for PHB biodegradation in buffered solutions with porcine lipase addition: 72% decrease in M_w of PHB (450 kDa) after incubation for 84 days with lipase (20 U/mg, 10 mg/ml in Tris-buffer) vs. 39% decrease in Mw after incubation in phosphate buffer [24-25]. This observation is in contrast to enzymatic degradation by PHB depolymerases which was reported to proceed on the surface of the polymer film with an almost unchanged molecular weight [20-21]. It has been proposed that for depolymerases the relative size of the enzyme compared with the void space in solvent cast films is the limiting factor for diffusion into the polymer matrix [Jesudason J.J. et al., 1993] whereas lipases can penetrate into the polymer matrix through pores in PHB film [34-35]. It was shown that lipase (0.1 g/l in buffer) treatment for 24 h caused significant morphological change in PHB film surface: transferring from native PHB film with many pores ranging from 1 to 5 μm in size into a pore free surface without producing a quantity of hydroxyl groups on the film surface. It was supposed that the pores had a fairly large surface exposed to lipase, thus it was degraded more easily (Figure 2) [34-35]. It indicates also that lipase can partially penetrate into pores of PHB film but the enzymatic degradation proceeds mainly on the surface of the coarse polymer film achievable for lipase. Two additional effects reported for depolymerases could be of importance. It was concluded that segmental mobility in amorphous phase and polymer hydrophobicity play an important role in enzymatic PHB degradation by nonspecific esterases [11]. Significant impairment of the tensile strength and other mechanical properties were observed during enzymatic biodegradation of PHB threads in serum and blood. It was shown that load at break lost 29%, Young's modulus lost 20%, and tensile strength didn't change after 180 days of threads incubation, the mechanical properties changed gradually [18].

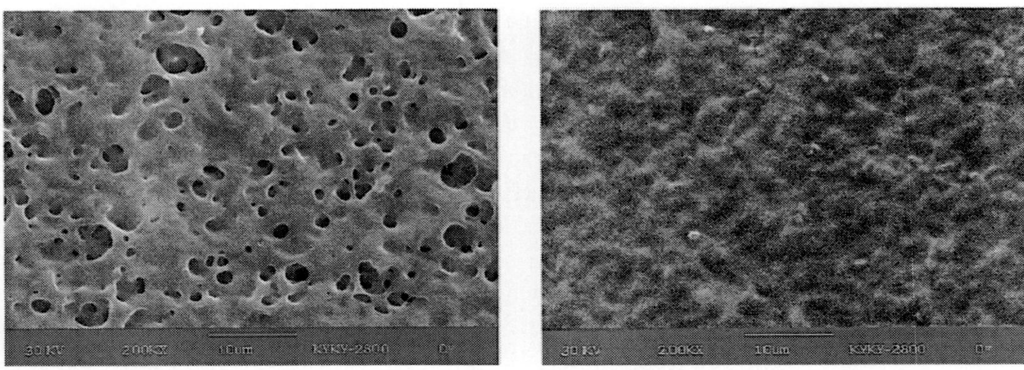

Figure 2. Scanning electron microscopy photographs documented the surface structure of PHB polymer films: (a) PHB film; (b) PHB film treated with lipase (0.1 g/l at 30°C and pH=7.0 for 24 h) [35].

1.3. Biodegradation of PHB by Soil Microorganisms

Polymers exposed to the environment are degraded through their hydrolysis, mechanical, thermal, oxidative, and photochemical destruction, and biodegradation [4,32,45,46]. One of the valuable properties of PHB is its biodegradability, which can be evaluated using various field and laboratory tests. Requirements for the biodegradability of PHB may vary in accordance with its applications. The most attractive property of PHB with respect to ecology is that it can be completely degraded by microorganisms finally to CO_2 and H_2O. This property of PHB allows to manufacture biodegradable polymer objects for various applications (figure 3) [2].

The degradation of PHB and its composites in natural ecosystems, such as soil, compost, and bodies of water, was described in a number of publications [2,32,45,46]. Maergaert et al. isolated from soil more than 300 microbial strains capable of degrading PHB in vitro [45].

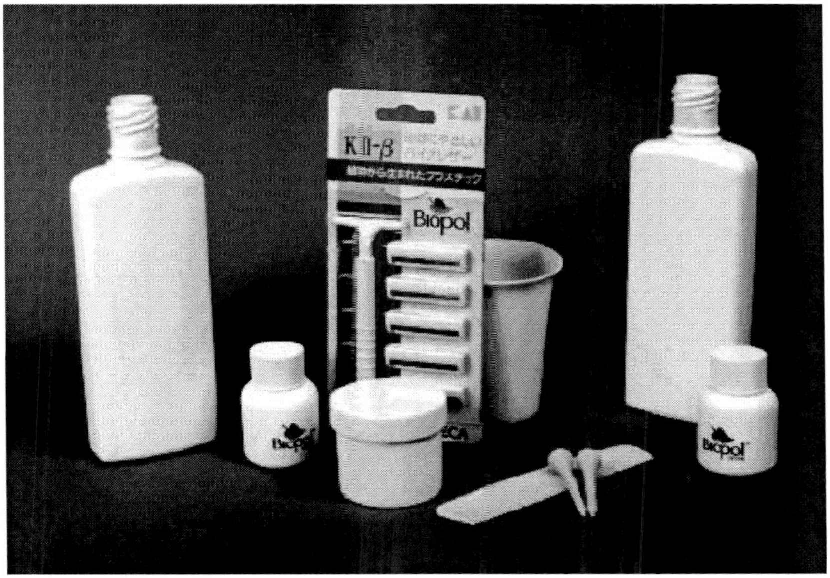

Figure 3. Moulded PHB objects for various applications. In soil burial or composting experiments, such objects biodegrade in about three months [2].

The bacteria detected on the degraded PHB films were dominated by the genera *Pseudomonas*, *Bacillus, Azospirillum, Mycobacterium,* and *Streptomyces etc.* The samples of PHB have been tested for fungicidity and resistance to fungi by estimating the growth rate of test fungi from the genera *Aspergillus, Aureobasidium, Chaetomium, Paecilomyces, Penicillum, Trichoderma* under optimal growth conditions. PHB film did not exhibit neither fungicide properties, nor the resistance to fungal damage, and served as a good substrate for fungal growth [47].

It was studied biodegradability of PHB films under aerobic, microaerobic and anaerobic condition in the presence and absence of nitrate by microbial populations of soil, sludge from anaerobic and nitrifying/denitrifying reactors, and sediment of a sludge deposit site, as well as to obtain active denitrifying enrichment culture degrading PHB (figure 4) [48]. Changes in molecular mass, crystallinity, and mechanical properties of PHB have been studied. A correlation between the PHB degradation degree and the molecular weight of degraded PHB was demonstrated. The most degraded PHB exhibited the highest values of the crystallinity index. As it has been shown by Spyros *et al.,* PHAs contain amorphous and crystalline regions, of which the former are much more susceptible to microbial attack [49]. If so, the microbial degradation of PHB must be associated with a decrease in its molecular weight and an increase in its crystallinity, which was really observed in the experiments. Moreover, microbial degradation of the amorphous regions of PHB films made them more rigid. However, further degradation of the amorphous regions made the structure of the polymer much looser [48].

PHB biodegradation in the enriched culture obtained from soil on the medium used to cultivate denitrifying bacteria (Gil'tai medium) has been also studied. The dominant bacterial species, *Pseudomonas fluorescens* and *Pseudomonas stutzeri*, have been identified in this enrichment culture. Under denitrifying conditions, PHB films were completely degraded for seven days.

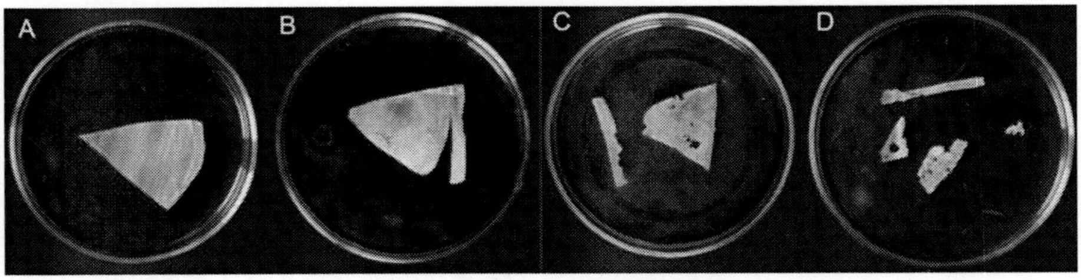

Figure 4. Undegraded PHB film (A) and PHB films with different degrees of degradation after 2 months incubation in soil suspension: anaerobic conditions without nitrate (B), microaerobic conditions without nitrate (C), and microaerobic conditions with nitrate (D) [24,48].

Both the film weight and M_w of PHB decreased with time. In contrast to the data of Doi et al. [21] who found that M_w of PHB remained unchanged upon enzymatic biodegradation in an aquatic solution of PHB- depolymerase from *Alcaligenes faecalis,* in our experiments, the average viscosity molecular weight of the higher- and lower-molecular polymers decreased gradually from 1540 to 580 kDa and from 890 to 612 kDa, respectively. The "exo"-type cleavage of the polymer chain, i. e. a successive removal of the terminal groups, is known to occur at a higher rate than the "endo"-type cleavage, i. e., a random breakage of the polymer chain at the enzyme-binding sites. Thus, the former type of polymer degradation is primarily responsible for changes in its average molecular weight. However the "endo"-type attack plays the important role at the initiation of biodegradation, because at the beginning, a few polymer chains are oriented so that their ends are accessible to the effect of the enzyme [50]. Biodegradation of the lower-molecular polymer, which contains a higher number of terminal groups, is more active, probably, because the "exo"-type degradation is more active in lower than in higher molecular polymer [48,51].

1.4. Biodegradation of PHB *In Vivo* in Animal Tissues

The first scientific works on biodegradation of PHB in vivo in animal tissues were carried out 15-20 years ago by Miller N.D. et al. and Saito T. et al [17,19]. They are high-qualitative researches that disclosed many important characteristics of this process. As it was noted above the both enzymatic and non-enzymatic processes of biodegradation of PHB in vivo can occur simultaneously under normal conditions. But it doesn't mean that polymer biodegradation in vivo is a simple combination of non-enzymatic hydrolysis and enzymatic degradation. Moreover, in vivo the biodegradation (decrease of molecular weight and mass loss) of PHB) is a controversial subject in the literature. As it was noted above for in vitro PHB hydrolysis, the main reason for the controversy, is the use of samples made by various processing technologies and the incomparability of different implantation and animal models. The most of researches on PHB biodegradation was carried out with use of prototypes of various medical devices on the base of PHB: solid films and plates [10,13,24,52], porous patches [11,12], nonwoven patches consisted of fibers [53-57], screws [24], cylinders as nerve guidance channels and conduits [13,15-16], monofilament sutures [17-18], microspheres [19,58]. In vivo biodegradation researches were carried out on various laboratory animals: rats [11,16-19,24], mice [13,58], rabbits [10,52,59], minipigs [12], cats [15], calves [53], sheep [54-56], patients [57]. It is obviously that these animals differ in level of metabolism very much: for example, only weight of these animals differs from 10-20 g (mice) to 50 kg (calves). The implantation of devices from PHB was carried out through different ways: subcutaneously [10,13,17,18,24,59], intraperitoneally on a bowel [11], subperiostally on the osseus skull [12,52], nerve wrap-around [14-16], intramuscularly [58-59], into the pericardium [54-57], into the atrium [53] and intravenously [19]. The terms of implantation were also different: 2.5 h, 24 h, 13 days, 2 months [19]; 7, 14, 30 days [16], 2, 7, 14, 21, 28, 55, 90, 182 days [17]; 1, 3, 6 months [10,13,14]; 3, 6, 12 months [53]; 6, 12 months [15]; 6, 24 months [57]; 3, 6, 9, 12, 18, 24 months [56] (Table 2).

The most entire study of PHB in vivo biodegradation was fulfilled by Gogolewski S. et al. and Qu X.-H. et al. [10,13]. It was shown that PHB lost about 1.6% (injection-molded film, 1.2 mm thick, M_w of PHB = 130 kDa) [13] and 6% (solvent-casting film, 40 µm thick, M_w = 534 kDa) [10] of initial weight after 6 months of implantation. But the observed small weight loss was partially due to the leaching out of low molecular weight fractions and impurities present initially in the implants. The M_w of PHB decreased from 130 kDa to 74 kDa (57% of initial M_w) [13] and from 534 kDa to 216 kDa (40% of initial Mw) [10] after 6 months of implantation. The polydispersity of PHB polymers narrowed following implantation. PHB showed a constant increase in crystallinity (from 60.7 to 64.8%) up to 6 months [13] or an increase (from 65.0 to 67.9%) after 1 month and a fall again (to 64.5%) after 6 month of implantation [10] which suggests the degradation process had not affected the crystalline regions. This data is in accordance with data of PHB hydrolysis [20] and enzymatic PHB degradation by lipases in vitro [11] where M_w decrease was observed. The initial biodegradation of amorphous regions of PHB in vivo is similar to PHB degradation by depolymerase [39].

Thus, the observed biodegradation of PHB showed coexistence of two different degradation mechanisms in hydrolysis in the polymer: enzymatically or non-enzymatically catalyzed degradation. Although non-enzymatical catalysis occurred randomly in homopolymer, indicated by M_w loss rate in PHB, at some point in a time, a critical molecular weight is reached whereupon enzyme-catalyzed hydrolysis accelerated degradation at the surface because easier enzyme/polymer interaction becomes possible. However considering the low weight loss of PHB, the critical molecular weight appropriate for enzymes predominantly does not reach, yet resulting low molecular weight and crystallinity in PHB could provide some sites for the hydrolysis of enzymes to accelerate the degradation of PHB [10,13]. Additional data revealing the mechanism of PHB biodegradation in animal tissues was obtained by Kramp B. et al. in long-term implantation experiments. A very slow, clinically not recordable degradation of films and plates was observed during 20 month (much more than in experiments mentioned above). A drop in the PHB weight loss evidently took place between the 20th and 25th month. Only initial signs of degradation were to be found on the surface of the implant until 20 months after implantation but no more test body could be detected after 25 months [52]. The complete biodegradation in vivo in the wide range from 3 to 30 months of PHB was shown by other researches [53,55-57,60], whereas almost no weight loss and surface changes of PHB during 6 months of biodegradation in vivo was shown [13,17]. Residual fragments of PHB implants were found after 30 months of the patches implantation [54,56]. A reduction of PHB patch size in 27% was shown in patients after 24 months after surgical procedure on pericardial closure with the patch [57]. Significantly more rapid biodegradation in vivo was shown by other researches [10,15,18,37,53].

Table 2. Biodegradation of PHB in vivo (data for comparison)

Type of device	Thickness/ diameter, μm	Animal	Site of implantation/ surgical procedure	Relative mass loss of PHB, %	Relative loss of PHB molecular weight, %	Time, months	Links
film (injection-molded)	1200	mouse	subcutaneously	1.6	43	6	13
film (solvent-casting)	150-200	rabbit	subcutaneously (dorsal side)	6	60	6	10
film (solvent-casting)	50	rat	subcutaneously (ventral side)	100	100	3	24,25
porous PHB/ atactic PHB patch	100	rat	intraperitoneally to repair a bowel defect	>90*	62	6.5	11
porous PHB/ atactic PHB patch	250	minipigs	with contact to bone and dura mater to cover rhinobasal skull defects	>50*	65	6.5	12
Films and plates	100-1000	rabbit	subperiostally on the osseus skull or respectively on cut trough zygomatic arches	100	-	25	52
Films and plates	100 and 500	rabbit	subperiostally on the osseus skull or respectively on cut trough zygomatic arches	<10*	-	20	52
Plates and screws	500 and 1500	rabbit	subperiostally on the osseus skull or respectively trough osseus skull	0	-	12	25
cylinder (nerve conduits)	150 (of wall)	rat	nerve wrap-around to bridge an irreducible nerve gap	0	-	1	16
cylinder (nerve conduits)	150 (of wall)	cat	nerve wrap-around to bridge an irreducible nerve gap	>25*	-	12	15
mono-filament suture	-	rat	subcutaneously (dorsal side)	0	-	6	17
mono-filament suture	30	rat	subcutaneously (fold of neck)	30	-	6	18

Type of device	Thickness/ diameter, μm	Animal	Site of implantation/ surgical procedure	Relative mass loss of PHB, %	Relative loss of PHB molecular weight, %	Time, months	Links
thin films and ground particles	-	rabbit	subcutaneously and intramuscularly in the legs	>30*	-	2	59
nonwoven patch (consisted of fibers)	200-600 (of patch) 2-20 (of fibers)	sheep	on the wall of pericardium to close artificial defect and prevent pericardial adhesions	>90*	-	24	55
nonwoven patch (consisted of fibers)	200-600 (of patch) 2-20 (of fibers)	sheep	as transannular patches on the wall of right ventricular outflow tract and pulmonary artery	>99*	-	12	56
nonwoven patch (consisted of fibers)	200-600 (of patch) 2-20 (of fibers)	calve	on the septal of right atrium to close artificial septal defect	>99*	-	12	53
nonwoven patch (consisted of fibers)	200-600 (of patch) 2-20 (of fibers)	patient	on the wall of pericardium to close artificial defect and prevent pericardial adhesions	27	-	24	57
microspheres	0.5-0.8	rat	intravenously	8*	-	2	19
microspheres	100-300	mice	intramuscularly in the legs	0*	-	2	58
rivet-shaped plate	2300	rabbit	intraosseously, into the lateral condyle of femur	<10*	-	6	37

* - indirect data.

It was shown that 30% mass loss of PHB sutures occurred gradually during 180 days of in vivo biodegradation with minor changes in the microstructure on the surface and in volume of sutures [18]. It was shown that PHB nonwoven patches (made to close atrial septal defect in calves) was slowly degraded by polynucleated macrophages, and 12 months postoperatively no PHB device was identifiable but only small particles of polymer were still seen. The absorption time of PHB patches was long enough to permit regeneration of a normal tissue [53]. The PHB sheets progressive biodegradation was demonstrated qualitatively at 2, 6 and 12 months after implantation as weakening of the implant surface, tearing/cracking of the implant, fragmentation and a decrease in the volume of polymer material [15,37,59]. The complete biodegradation of PHB (M_w = 150-1000 kDa) thin films (10-50 μm) for 3-6 months was shown and degradation process was described. The process of PHB biodegradation consists of several phases. At initial phase PHB films was covered by fibrous capsule. At second phase capsulated PHB films very slowly lost weight with simultaneous increase of crystallinity and decrease of M_w and mechanical properties of PHB. At third phase PHB films were rapidly disintegrated and then completely degraded. At 4th phase empty fibrous capsule resolved (figure 5) [24-25]. Interesting data were obtained for biodegradation in vivo of PHB microspheres (0.5-0.8 μm in diameter). It was demonstrated indirectly that PHB loss about 8% of weight of microspheres accumulated in liver after 2 month of intravenous injection. It was demonstrated also a presence of several types PHB degrading enzymes in the animal tissues extracts [19].

Some researches studied a biodegradation of PHB threads with a tendency of analysis of its mechanical properties in vivo [17-18]. It was shown that at first load at break index decreased rapidly from 440 g to 390 g (12%) at 15th day and then gradually increased to the initial value at 90th and remain almost unchanged up to 182nd day [17] or gradually decreased in 27% during 180 days [18], strain at break decreased rapidly from 60 to 50% (in 17% of initial value) at 10th day and then gradually increased to 70% (in 17% of initial value) at 182nd day [17] or didn't change significantly during 180 days [18].

It was demonstrated that the primary reason of PHB biodegradation in vivo was a lysosomal and phagocytic activity of polynucleated macrophages and giant cells of foreign body reaction. The activity of tissue macrophages and nonspecific enzymes of body liquids made a main contribution to significantly more rapid rate of PHB biodegradation in vivo in comparison with rate of PHB hydrolysis in vitro.

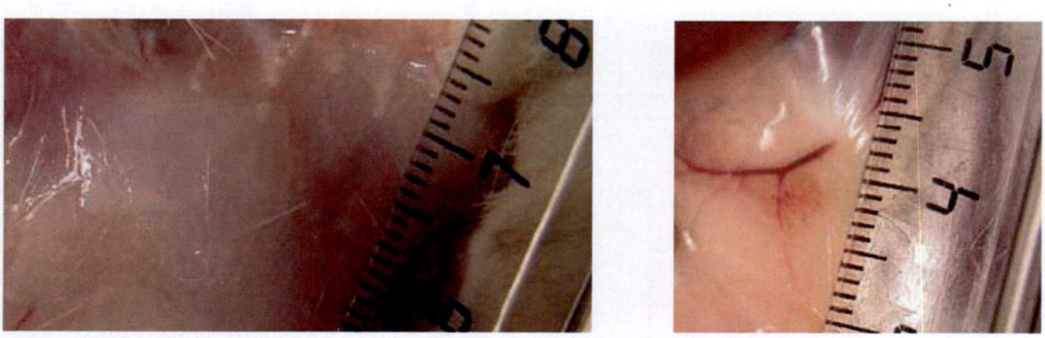

Figure 5. Biodegradation of PHB films in vivo. Connective-tissue capsule with PHB thin films (outlined with broken line) 2 weeks (98% residual weight of the film) (left photograph) and 3 months (0% residual weight of the film) (left photograph) after subcutaneous implantation [24,25].

The PHB material was encapsulated by degrading macrophages. Presence of PHB stimulated uniform macrophage infiltration, which is important for not only the degradation process but also the restoration of functional tissue. The long absorption time produced a foreign-body reaction, which was restricted to macrophages forming a peripolymer layer [18,53,56,59].

Very important data that clarifies the tissue response that contributes to biodegradation of PHB was obtained by Lobler M. It was demonstrated an significant increase of expression of two specific lipases after 7 and 14 days of PHB contact with animal tissues. Moreover, liver specific genes were induced with similar results. It is striking that pancreatic enzymes are induced in the gastric wall after contact with biomaterials [40]. Saito T. et al. suggested the presence of at least two types of degradative enzymes in rat tissues: liver serine esterases with the maximum of activity in alkaline media (pH=9.5) and kidney esterases with the maximum of activity in neutral media [19].

The mechanism of PHB biodegradation by macrophages was demonstrated at cultured macrophages incubated with particles of low-molecular weight PHB [61]. It was shown that macrophages and, to a lesser level, fibroblasts have the ability to take up (phagocytize) PHB particles (1-10 μm). At high concentrations of PHB particles (>10 μg/mL) the phagocytosis is accompanied by toxic effects and alteration of the functional status of the macrophages but not the fibroblasts.

This process is accompanied by cell damage and cell death. The elevated production of nitric oxide (NO) and tumor necrosis factor alfa (TNF-α) by activated macrophages was observed. It was suggested that the cell damage and cell death may be due to phagocytosis of large amounts of PHB particles; after phagocytosis, polymer particles may fill up the cells, and cause cell damage and cell death. It was demonstrated also that phagocytized PHB particles disappeared in time due to an active PHB biodegradation process (figure 6) [61].

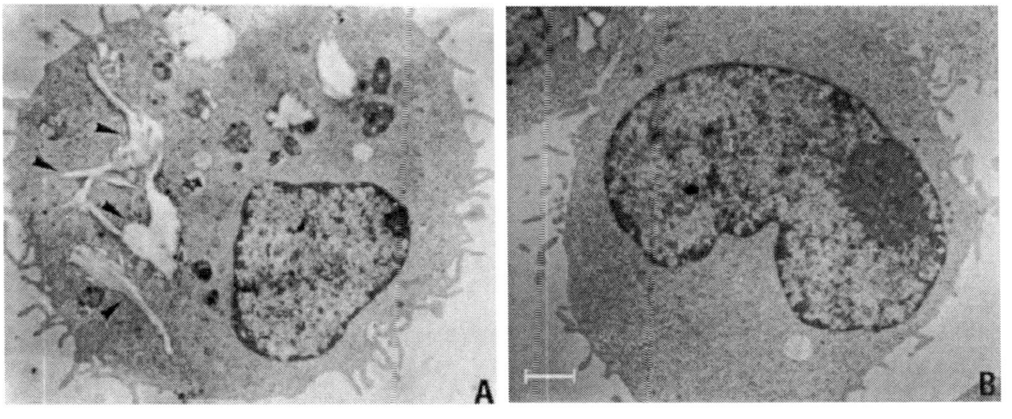

Figure 6. Phagocytosis of microparticles of PHB in macrophages. TEM analysis of cultured macrophages in the presence (A) or absence (B) of 2 μg PHB microparticles/mL for 24 h. Bar in B represents 1 μm, for A and B.

2. Application of PHB

2.1. Medical Devices on the Base of PHB and PHB *In Vivo* Biocompatibility

The perspective area of PHB application is development of implanted medical devices for dental, cranio-maxillofacial, orthopaedic, cardiovascular, hernioplastic and skin surgery. A number of potential medical devices on the base of PHB: bioresorbable surgical sutures [17-18,62-63], biodegradable screws and plates for cartilage and bone fixation [24,52], biodegradable membranes for periodontal treatment, surgical meshes with PHB coating for hernioplastic surgery [24], wound coverings [64], patches for repair of a bowel, pericardial and osseous defects [11-12,53-57], nerve guidance channels and conduits [15-16] etc. was developed (figure 7).

The tissue reaction in vivo to implanted PHB films and medical devices was studied. In most cases a good biocompatibility of PHB was demonstrated. In general, no acute inflammation, abscess formation, or tissue necrosis was observed in tissue surrounding of the implanted PHB materials. In addition, no tissue reactivity or cellular mobilization occurred in areas remote from the implantation site [10,13,24,58]. On the one hand, it was shown that PHB elicited similar mild tissue response as PLA did [13], but on the other hand the use of implants consisting of polylactic acid, polyglicolic acid and their copolymers is not without a number of sequelae related with the chronic inflammatory reactions in tissue [65-69].

Subcutaneous implantation of PHB films for 1 month has shown that the samples were surrounded by a well-developed, homogeneous fibrous capsule of 80-100 μm in thickness. The vascularized capsule consists primarily of connective tissue cells (mainly, round, immature fibroblasts) aligned parallel to the implant surface. A mild inflammatory reaction was manifested by the presence of mononuclear macrophages, foreign body cells, and lymphocytes. Three months after implantation, the fibrous capsule has thickened to 180-200 μm due to the increase in the amount of connective tissue cells and a few collagen fiber deposits. A substantial decrease in inflammatory cells was observed after 3 months, tissues at the interface of the polymer were densely organized to form bundles. After 6 months of implantation, the number of inflammatory cells had decreased and the fibrous capsule, now thinned to about 80-100 μm, consisted mainly of collagen fibers, and a significantly reduced amount of connective tissue cells.

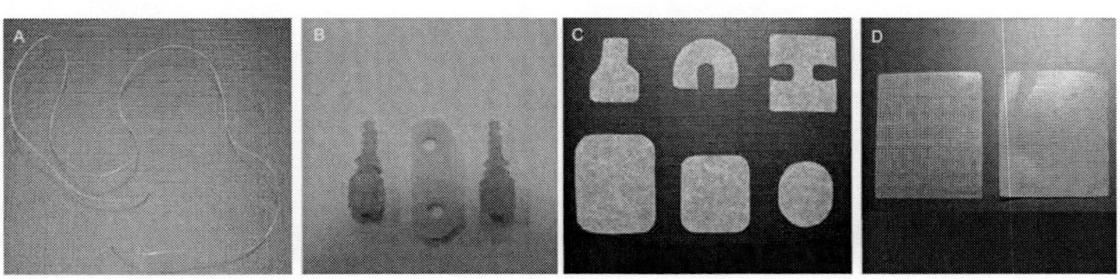

Figure 7. Medical devices on the base of PHB. (A) bioresorbable surgical suture; (B) biodegradable screws and plate for cartilage and bone fixation; (C) biodegradable membranes for periodontal treatment; (D) surgical meshes with PHB coating for hernioplastic surgery, pure (left) and loaded with antiplatelet drug, dipyridamole (right) [24].

A little inflammatory cells effusion was observed in the tissue adherent to the implants after 3 and 6 months of implantation [10,13]. The biocompatibility of PHB has been demonstrated in vivo under subcutaneous implantation of PHB films. Tissue reaction to films from PHB of different molecular weight (300, 450, 1000 kDa) implanted subcutaneously was relatively mild and didn't change from tissue reaction to control glass plate [24].

At implantation of PHB with contact to bone the overall tissue response was favorable with a high rate of early healing and new bone formation with some indication of an osteogenic characteristic for PHB compared with other thermoplastics, such as polyethylene. Initially there was a mixture of soft tissue, containing active fibroblasts, and rather loosely woven osteonal bone seen within 100 μm of the interface. There was no evidence of a giant cell response within the soft tissue in the early stages of implantation. With time this tissue became more orientated in the direction parallel to the implant interface. The dependence of the bone growth on the polymer interface is demonstrated by the new bone growing away from the interface rather than towards it after implantation of 3 months. By 6 months post-implantation the implant is closely encased in new bone of normal appearance with no interposed fibrous tissue. Thus, PHB-based materials produce superior bone healing [37].

Regeneration of a neointima and a neomedia, comparable to native arterial tissue, was observed at 3–24 months after implantation of PHB nonwoven patches as transannular patches into the right ventricular outflow tract and pulmonary artery. In the control group, a neointimal layer was present but no neomedia comparable to native arterial tissue. Three layers were identified in the regenerated tissue: neointima with a endothelium-like lining, neomedia with smooth muscle cells, collagenous and elastic tissue, and a layer with polynucleated macrophages surrounding istets of PHB, capillaries and collagen tissue. Lymphocytes were rare. It was concluded that PHB nonwoven patches can be used as a scaffold for tissue regeneration in low-pressure systems. The regenerated vessel had structural and biochemical qualities in common with the native pulmonary artery [56]. Biodegradable PHB patches implanted in atrial septal defects promoted formation of regenerated tissue that macroscopically and microscopically resembled native atrial septal wall. The regenerated tissue was found to be composed of three layers: monolayer with endothelium-like cells, a layer with fibroblasts and some smooth-muscle cells, collagenous tissue and capillaries, and a third layer with phagocytizing cells isolating and degrading PHB. The neointima contained a complete endothelium-like layer resembling the native endothelial cells. The patch material was encapsulated by degrading macrophages. There was a strict border between the collagenous and the phagocytizing layer. Presence of PHB seems to stimulate uniform macrophage infiltration, which was found to be important for the degradation process and the restoration of functional tissue. Lymphocytic infiltration as foreign-body reaction, which is common after replacement of vessel wall with commercial woven Dacron patch, was wholly absent when PHB. It was suggested that the absorption time of PHB patches was long enough to permit regeneration of a tissue with sufficient strength to prevent development of shunts in the atrial septal position [53]. The prevention of postoperative pericardial adhesions by closure of the pericardium with absorbable PHB patch was demonstrated. The regeneration of mesothelial layer after implantation of PHB pericardial patch was observed. The complete regeneration of mesothelium, with morphology and biochemical activity similar to findings in native mesothelium, may explain the reduction of postoperative pericardial adhesions after operations with insertion of absorbable PHB patches [55]. The regeneration of normal filament structure of restored tissues was observed by immunohistochemical methods after

PHB devices implantation [54]. The immunohistochemical demonstration of cytokeratine, an intermediate filament, which is constituent of epithelial and mesodermal cells, agreed with observations on intact mesothelium. Heparan sulfate proteoglycan, a marker of basement membrane, was also identified [54].

PHB patches for the gastrointestinal tract were tested using animal model. Patches made from PHB sutured and PHB membranes were implanted to close experimental defects of stomach and bowel wall. The complete regeneration of tissues of stomach and bowel wall was observed at 6 months after patch implantation without strong inflammatory response and fibrosis [11,70].

Recently an application of biodegradable nerve guidance channels (conduits) for nerve repair procedures and nerve regeneration after spinal cord injury was demonstrated. Polymer tubular structures from PHB can be modulated for this purpose. Successful nerve regeneration through a guidance channel was observed as early as after 1 month. Virtually all implanted conduits contained regenerated tissue cables centrally located within the channel lumen and composed of numerous myelinated axons and Schwann cells. The inflammatory reaction had not interfered with the nerve regeneration process. Progressive angiogenesis was present at the nerve ends and through the walls of the conduit. The results demonstrate good-quality nerve regeneration in PHB guidance channels [16,71].

Biocompatibility of PHB was evaluated by implanting microspheres from PHB (M_w = 450 kDa) into the femoral muscle of rats. The spheres were surrounded by one or two layers of spindle cells, and infiltration of inflammatory cells and mononuclear cells into these layers was recognized at 1 week after implantation. After 4 weeks, the number of inflammatory cells had decreased and the layers of spindle cells had thickened. No inflammatory cells were seen at 8 weeks, and the spheres were encapsulated by spindle cells. The toxicity of PHB microspheres was evaluated by weight change and survival times in L1210 tumor-bearing mice. No differences were observed in the weight change or survival time compared with those of control. These results suggest that inflammation accompanying microsphere implantation is temporary as well as toxicity to normal tissues is minimal [58].

The levels of tissue factors, inflammatory cytokines, and metabolites of arachidonic acid were evaluated. Growth factors derived from endothelium and from macrophages were found. These factors most probably stimulate both growth and regeneration occurring when different biodegradable materials were used as grafts [40,53,55,70]. The positive reaction for thrombomodulin, a multifunctional protein with anticoagulant properties, was found in both mesothelial and endothelial cells after pericardial PHB patch implantation. Prostacycline production level, which was found to have cytoprotective effect on the pericardium and prevent adhesion formation, in the regenerated tissue was similar to that in native pericardium [53,55]. The PHB patch seems to be highly biocompatible, since no signs of inflammation were observed macroscopically and also the level of inflammation associated cytokine mRNA did not change dramatically, although a transient increase of interleukin-1β and interleukin-6 mRNA through days 1-7 after PHB patch implantation was detected. In contrast, tumor necrosis factor-α mRNA was hardly detectable throughout the implantation period, which agrees well with a observed moderate fibrotic response [40,70].

2.2. PHB as Tissue Engineering Material and PHB *In Vitro* Biocompatibility

Biopolymer PHB is promising material in tissue engineering due to high biocompatibility in vitro. Cell cultures of various origins including murine and human fibroblasts [12,34,72-74], human mesenchymal stem cells [75], rabbit bone marrow cells (osteoblasts) [30,73,76], human osteogenic sarcoma cells [77], human epithelial cells [74,77], human endothelial cells [78-79], rabbit articular cartilage chondrocytes [80-81] and rabbit smooth muscle cells [82] in direct contact with PHB when cultured on polymer films and scaffolds exhibited satisfactory levels of cell adhesion, viability and proliferation. Moreover, it was shown that fibroblasts, endothelium cells, and isolated hepatocytes cultured on PHB films exhibited high levels of cell adhesion and growth (figure 8) [83].

It was shown also that cultured cells produced collagen II and glycosaminoglycan, the specific structural biopolymers formed the extracellular matrix [77,80,81]. A good viability and proliferation level of macrophages and fibroblasts cell lines was obtained under culturing in presence of particles from short-chain low-molecular PHB [61]. However it was shown that cell growth on the PHB films was relatively poor: the viable cell number ranged from 1×10^3 to 2×10^5 [34,73,81]. An impaired interaction between PHB matrix and cytoskeleton of cultured cells was also demonstrated [77]. It was reported that a number of polymer properties including chemical composition, surface morphology, surface chemistry, surface energy and hydrophobicity play important roles in regulating cell viability and growth [84]. The investigation showed that this biomaterial can be used to make scaffolds for in vitro proliferous cells [34,76,80].

The most widespread methods to manufacture the PHB scaffolds for tissue engineering by means of improvement of cell adhesion and growth on polymer surface are change of PHB surface properties and microstructure by salt-leaching methods and enzymatic/chemical/physical treatment of polymer surface [34,76,80,85]. Adhesion to polymer substrates is one of the key issues in tissue engineering, because adhesive interactions control cell physiology. One of the most effective techniques to improve adhesion and growth of cells on PHB films is treatment of polymer surface with enzymes, alkali or low pressure plasma [34,85].

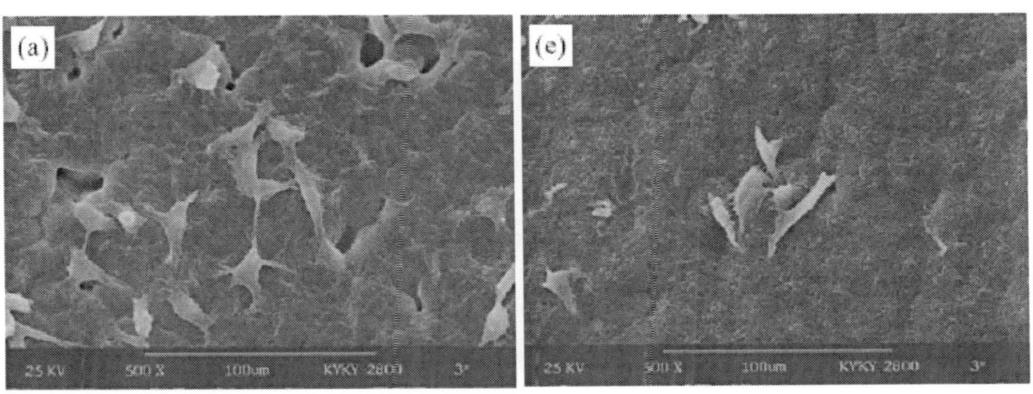

Figure 8. Scanning electron microscopy image of 2 days growth of fibroblast cells on films made of (a) PHB; (e) PLA; (500 x). Cell density of fibroblasts grown on PHB film is significantly higher vs. cell density of fibroblasts grown on PLA film [73].

Lipase treatment increases the viable cell number on the PHB film from 100 to 200 times compared to the untreated PHB film. NaOH treatment on PHB film also indicated an increase of 25 times on the viable cell number compared with the untreated PHB film [34]. It was shown that treatment of PHB film surface with low pressure ammonia plasma improved growth of human fibroblasts and epithelial cells of respiratory mucosa due to increased hydrophylicity (but with no change of microstructure) of polymer surface [74]. It was suggested that the improved hydrophilicity of the films after PHB treatment with lipases, alkali and plasma allowed cells in its suspension to easily attach on the polymer films compared to that on the untreated ones. The influence of hydrophilicity of biomaterial surface on cell adhesion was demonstrated earlier [86].

But a microstructure of PHB film surface can be also responsible for cell adhesion and cell growth [87-89]. Therefore, noticed above modification of polymer film surface after enzymatic and chemical treatment (in particular, reduced pore size and a surface smoothing) is expected to play an important role for enhanced cell growth on the polymer films [34]. Different cells prefer different surface. For example, osteoblasts preferred rougher surfaces with appropriate size of pores [87,88] while fibroblast prefer smoother surface, yet epithelial cells only attached to the smoothest surface [89]. This appropriate roughness affects cell attachment as it provides the right space for osteoblast growth, or supplies solid anchors for filapodia. A scaffold with appropriate size of pores provided better surface properties for anchoring type II collagen filaments and for their penetration into internal layers of the scaffolds implanted with chondrocytes. This could be illuminated by the interaction of extracellular matrix proteins with the material surface. The right surface properties may also promote cell attachment and proliferation by providing more spaces for better gas/nutrients exchange or more serum protein adsorption. [30,76,80]. Additionally Sevastianov et al. found that PHB films in contact with blood did not activate the hemostasis system at the level of cell response, but they did activate the coagulation system and the complement reaction [90].

The high biocompatibility of PHB may be due to several reasons. First of all, PHB is a natural biopolymer involved in important physiological functions both prokaryotes and eukaryotes. PHB from bacterial origin has property of stereospecificity that is inherent to biomolecules of all living things and consists only from residues of D(-)-3-hydrohybutyric acid [91]. Low molecular weight PHB (up to 150 resides of 3-hydrohybutyric acid), complexed to other macromolecules (cPHB), was found to be a ubiquitous constituent of both prokaryotic and eukaryotic organisms of nearly all phyla [92-96]. Complexed cPHB was found in a wide variety of tissues and organs of mammals (including human): blood, kidney, vessels, nerves, vessels, eye, brain, as well as in organelles, membrane proteins, lipoproteins, and plaques. cPHB concentration ranged from 3-4 μg/g wet tissue weight in nerves and brain to 12 μg/g in blood plasma [97,98]. In humans, total plasma cPHB ranged from 0.60 to 18.2 mg/l, with a mean of 3.5 mg/l. [98]. It was shown that cPHB is a functional part of ion channels of erythrocyte plasma membrane and hepatocyte mitochondria membrane [99,100]. The singular ability of cPHB to dissolve salts and facilitate their transfer across hydrophobic barriers defines a potential physiological niche for cPHB in cell metabolism [94]. However a mechanism of PHB synthesis in eukaryotic organisms is not well clarified that requires additional studies. Nevertheless, it could be suggested that cPHB is one of products of symbiotic interaction between animals and gut microorganisms. It was shown, for example, that E.coli is able to synthesize low molecular weight PHB and cPHB plays various physiological roles in bacteria cell [96,101].

Intermediate product of PHB biodegradation, D(-)-3-hydroxybutyric acid is also a normal constituent of blood at concentrations between 0.3 and 1.3 mM and contains in all animal tissues [102,103]. As it was noted above PHB has a rather low degradation rate in the body in comparison to, e.g., poly(lactic-co-glycolic) acids, that prevent from increase of 3-hydroxybutyric acid concentration in surrounding tissues [10,13], whereas polylactic acid release, following local pH decrease in implantation area and acidic chronic irritation of surrounding tissues is a serious problem in application of medical devices on the base of poly(lactic-co-glycolic) acids [104,105]. Moreover, chronic inflammatory response to polylactic and polyglycolic acids that was observed in a number of cases may be induced by immune response to water-soluble oligomers that released during degradation of synthetic polymers [105-107].

2.3. Novel Drug Dosage Forms on the Base of PHB

An improvement of medical devices on the base of biopolymers by encapsulating different drugs opens up the wide prospects in applications of these new devices with pharmacological activity in medicine. The design of injection systems for sustained drug delivery in the forms of microparticles (microspheres, microcapsules) prepared on the base of biodegradable polymers is extremely challenging in the modern pharmacology. The fixation of pharmacologically active component with the biopolymer and following slow drug release from the microparticles provides an optimal level of drug concentration in local target organ during long-term period (up to several months), that provides effective pharmaceutical action. At curative dose the prolonged delivery of drugs from the systems into organism permits to eliminate the shortcomings in peroral, injectable, aerosol, and the other traditional methods of drug administration. Among those shortcomings hypertoxicity, instability, pulsative character of rate delivery, ineffective expenditure of drugs should be pointed out. Alternatively, applications of therapeutical polymer systems provide orderly and purposefully the deliverance for an optimal dose of agent that is very important at therapy of acute or chronic diseases [108]. An ideal biodegradable microsphere formulation would consist of a free-flowing powder of uniform-sized microspheres less than 125 μm in diameter and with a high drug loading. In addition, the drug must be released in its active form with an optimized release profile. The manufacturing method should produce such microspheres in a process that is reproducible, scalable, and benign to some often delicate drugs, with a high encapsulation efficiency [109,110].

PHB as biodegradable and biocompatible is a promising material for producing of polymer systems for controlled drug release. A number of drugs with various pharmacological activities were used for development of polymer controlled release systems on the base of PHA, mainly on the base of poly(3-hydroxybutyrate-co-3-hydroxyvalerate) and poly(3-hydroxybutyrate-co-4- hydroxybutyrate) copolymers: model drugs (2,7-dichlorofluorescein [111], dextran-FITC [112], methyl red [113,114], 7-hydroxethyltheophylline [115,116]), antibiotics and antibacterial drugs (rifampicin [117,118], tetracycline [119], cefoperazone and gentamicin [120], sulperazone and duocid [121-124], sulbactam and cefoperazone [125]), anticancer drugs (5-fluorouracil [126], 2',3'-diacyl-5-fluoro-2'-deoxyuridine [58]), anti-inflammatory drug (indomethacin [127]), analgesics (tramadol [128], vasodilator and antithrombotic drugs (dipyridamole [24,127,129], nitric oxide donor [130,131]). The biocompatibility and pharmacological activity of some of these systems was studied [24,58,117,123-125,128,131]. But only a few drugs were used for production of drug controlled release systems on the base of PHB homopolymer: 7-hydroxethyltheophylline, methyl red, 2',3'-diacyl-5-fluoro-2'-deoxyuridine, rifampicin, tramadol, indomethacin and dipyridamole [58,113-118,127-131].

The first drug sustained delivery system on the base of PHB was developed by Korsatko W. et al., who observed a rapid release of encapsulated drug, 7-hydroxethyltheophylline, from tablets of PHB (M_w = 2000 kDa), as well as weight losses of PHB tablets containing the drug after subcutaneous implantation. It was suggested that PHB with molecular weight greater than 100 kDa was undesirable for long-term medication dosage [115].

Pouton C.W. and Akhtar S. describing the release of low molecular drugs from PHB matrices reported that the latter have the tendention of enhanced water penetration and pore formation [132]. The entrapment and release of model drug, methyl red, from melt-crystallized PHB matrices was found to be a function of polymer crystallization kinetics and morphology whereas overall degree of crystallinity was shown to cause no effect on drug release kinetics. Methyl red released from PHB films for more than 7 days with initial phase of rapid release ("burst effect") and second phase with relatively uniform release. Release profiles of PHB films crystallized at 110°C exhibited a greater burst effect when compared to those crystallized at 60°C. This was explained by better trapping of drug within polymeric spherulites with the more rapid rates of PHB crystallization at 110°C [113,114].

Kawaguchi T. et al showed that chemical properties of drug and polymer molecular weight had a great impact on drug delivery kinetics from PHB matrix. Microspheres (100-300 μm in diameter) from PHB of different molecular weight (65, 135, and 450 kDa) were loaded with prodrugs of 5-fluoro-2'deoxyuridine (FUdR) synthesized by esterification with aliphatic acids (propionate, butyrate, and pentanoate). Prodrugs have different physicochemical properties, in particular, solubility in water (from 70 mg/ml for FUdR to 0.1 mg/ml for butyryl-FUdR). The release rates from the spheres depended on both the lipophilicity of the prodrug and the molecular weight of the polymer. Regardless of the polymer, the relative release rates were propionyle-FUdR > butyryl- FUdR > pentanoyl-FUdR. The release of butyryl- FUdR and pentanoyl-FUdR from the spheres consisting of low-molecular-weight polymer (M_w = 65 kDa) was faster than that from the spheres of higher molecular weight (M_w = 135 or 450 kDa). The effect of drug content on the release rate was also studied. The higher the drug content in the PHB microspheres, the faster was the drug release. The release of FUdR continued for more than 5 days [58].

Kassab A.C. developed a well-managed technique for manufacture of PHB microspheres loaded with drugs. Microspheres were obtained within a size of 5-100 μm using a solvent evaporation method by changing the initial polymer/solvent ratio, emulsifier concentration, stirring rate, and initial drug concentration. Very high drug loading of up to 408 g rifampicin/g PHB were achieved. Drug release rates were rapid: the maximal duration of rifampicin delivery was 5 days. Both the size and drug content of PHB microspheres were found to be effective in controlling the drug release from polymer microspheres [118].

The sustained release of analgesic drug, tramadol, from PHB microspheres was demonstrated by Salman M.A. et al. It was shown that 58% of the tramadol (the initial drug content in PHB matrix = 18%) was released from the microspheres (7.5 μm in diameter) in the first 24 h. Drug release decreased with time. From 2 to 7 days the drug release was with zero-order rate. The entire amount of tramadol was released after 7 days [128].

The kinetics of different drug release from PHB films and microspheres was studied by our team [24,127]. It was found that the release occurs via two mechanisms, diffusion and degradation, operating simultaneously. Vasodilator and antithrombotic drug, dipyridamole, and anti-inflammatory drug, indomethacin, diffusion processes determine the rate of the release at the early stages of the contact of the system with the environment (the first 6-8 days). The coefficient of the release diffusion of a drug depends on its nature, the thickness of the PHB films containing the drug, the weight ratio of dipyridamole and indomethacin in polymer, and the molecular weight of PHB. Thus, it is possible to regulate the rate of drug release by changing of molecular weight of PHB, for example. A number of other drugs have been also used for development polymeric systems of sustained drug delivery: antibiotics (rifampicin, metronidazole, ciprofloxacin, levofloxacin), anti-inflammatory drugs (flurbiprofen, dexamethasone, prednisolone), and antitumor drugs (paclitaxel) [127]. The biodegradable microspheres on the base of PHB designed for controlled release of dipyridamole were kinetically studied. The profiles of release from the microspheres with different diameters 4, 9, 63, and 92 μm present the progression of nonlinear and linear stages. Diffusion kinetic equation describing both linear (PHB hydrolysis) and nonlinear (diffusion) stages of the dipyridamole release profiles from the spherical subjects has been written down as the sum of two terms: desorption from the homogeneous sphere in accordance with diffusion mechanism and the zero-order release. In contrast to the diffusivity dependence on microsphere size, the constant characteristics of linearity are scarcely affected by the diameter of PHB microparticles. The view of the kinetic profiles as well as the low rate of dipyridamole release are in satisfactory agreement with kinetics of weight loss measured in vitro for the PHB films and observed qualitatively for PHB microspheres. Taking into account kinetic results, it was supposed that the degradation of both films and PHB microspheres is responsible for the linear stage of dipyridamole release profiles. Thus, a good method for production of systems with sustained drug release was demonstrated. The sustained invariable drug release is an essential property of injectable therapeutic polymer systems that allows to keep constant the adjusted drug dosing. PHB films and microspheres with sustained uniform drug release for more that 1 month were developed (figure 9) [24,127, 129].

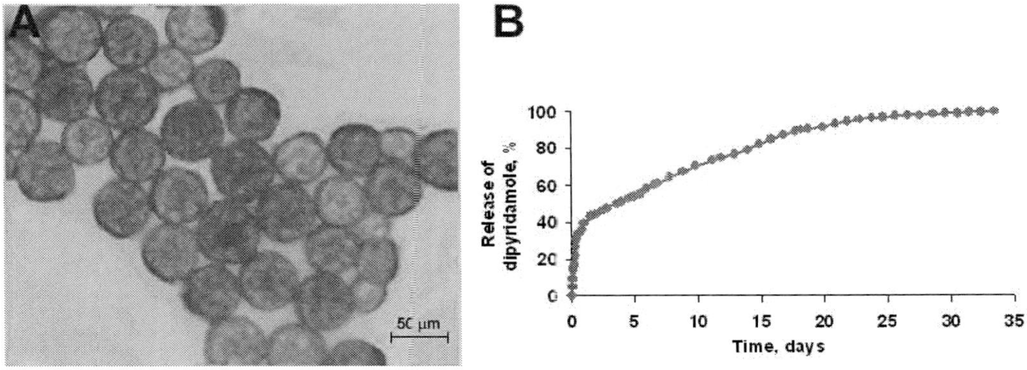

Figure 9. PHB microspheres for sustained delivery of drugs. (A) PHB microspheres (average diameter = 63 μm, PHB Mw = 1000 kDa) loaded with dipyridamole (10% w/w); (B) Sustained delivery of dipyridamole from PHB microspheres for more than 1 month [24,131].

The biocompatibility and pharmacological activity of advanced drug delivery systems on the base of PHB was studied [24,58,117,128]. It was shown that implanted PHB films loaded with dipyridamole and indomethacin caused the mild tissue reaction. The inflammation accompanying implantation of PHB matrices is temporary and additionally toxicity relative to normal tissues is minimal [24]. No signs of toxicity were observed after administration of PHB microspheres loaded with analgesic, tramadol, [128]. A single intraperitoneal injection of PHB (M_w = 450 kDa) microspheres containing anticancer prodrugs, butyryl-FUdR and pentanoyl-FUdR, resulted in high antitumor effects against P388 leukemia in mice over a period of five days [58].

Embolization with PHB microspheres in vivo at dogs as test animals has been studied by Kasab et al. Renal angiograms obtained before and after embolization and also the histopathological observations showed the feasibility of using these microspheres as an alternative chemoembolization agent [117]. Epidural analgesic effects of tramadol released from PHB microspheres were observed for 21 h, whereas an equal dose of free tramadol was effective for less than 5 h. It was suggested that controlled release of tramadol from PHB microspheres in vivo is possible, and pain relief during epidural analgesia is prolonged by this drug formulation compared with free tramadol [128].

The observed data indicate the wide prospects in applications of drug-loaded medical devices and microspheres on the base of PHB as implantable and injectable therapeutic systems in medicine for treatment of various diseases: cancer, cardio-vascular diseases, tuberculosis, osteomyelitis, arthritis etc. Besides application of PHB for producing of medical devices and systems of sustained drug delivery, PHB can be used for production of systems for controlled release of activators or inhibitors of enzymes.

The use of these systems allows the development of the physiological models for prolonged local activation or inhibition of enzymes in vivo. PHB is a perspective tool in design of novel physiological models due to minimal adverse inflammatory tissue reaction to PHB matrices implantation or PHB microspheres administration. A system of sustained nitric oxide (NO) donor delivery on basis of PHB was developed.

This system can be used for study of prolonged NO action on normal tissues of blood vessels in vivo. The development of in vivo model of prolonged NO local action on vascular tissues is a difficult problem, because NO donors deliver NO at most only for a few minutes. We have developed a model of prolonged local NO action on appropriate artery on basis of PHB loaded with a new effective NO donor, FPTO [133]. It has been shown that FPTO-loaded PHB cylinders can release FPTO (and consequently NO) for up to 1 month with relatively constant rate. FPTO-loaded PHB cylinders with sustained FPTO delivery were implanted around left carotid artery of Wistar rats, pure PHB cylinders were implanted around right carotid artery as control. At 1st, 4th and 10th days after implantation arteries and cylinders have been isolated. The elevated levels of the main metabolic products of NO, nitrites and nitrates, in arterial tissues were observed that indicates the possibility of application of this system for production of physiological model of NO prolonged action on arterial tissues in vivo [24,130,131].

ACKNOWLEDGEMENTS

This review was supported by grant no. 06-04-49339 and 08-03-00929 from the Russian Foundation for Basic Research (RFBR) and by grant no. 07-II-10 from Foundation for Assistance to Small Innovative Enterprises (FASIE).

REFERENCES

[1] Chen G.Q., Wu Q. The application of polyhydroxyalkanoates as tissue engineering materials. *Biomaterials*, 2005, 26(33):6565-6578.

[2] Lenz R.W., Marchessault R.H. Bacterial Polyesters: Biosynthesis, Biodegradable Plastics and Biotechnology. *Biomacromolecules,* 2005, 6(1): 1-8.

[3] Anderson A.J. and Dawes E.A. Occurrence, metabolism, metabolic role, and industrial uses of bacterial polyhydroxyalkanoates. *Microbiological Reviews*, 1990, Vol. 54, No. 4, p. 450-472.

[4] Jendrossek D., Handrick R. Microbial degradation of polyhydroxyalkanoates. *Annu. Rev. Microbiol.* 2002; 56:403-432.

[5] Kim, D.Y. and Y.H. Rhee. 2003. Biodegradation of microbial and synthetic polyesters by fungi. *Appl. Microbiol. Biotechnol.* 61, 300-308.

[6] Steinbuchel, A. and Lutke-Eversloh, T. (2003) Metabolic engineering and pathway construction for biotechnological production of relevant polyhydroxyalkanoates in microorganisms. *Biochem. Eng. J.* 16, 81–96.

[7] Marois Y., Zhang Z., Vert M., Deng X., Lenz R., Guidoin R. Hydrolytic and enzymatic incubation of polyhydroxyoctanoate (PHO): a short-term in vitro study of a degradable bacterial polyester. *J. Biomater. Sci. Polym.* Ed., 1999, 10, 483-499.

[8] Abe H., Doi Y. Side-chain effect of second monomer units on crystalline morphology, thermal properties, and enzymatic degradability for random copolyesters of (R)-3-hydroxybutyric acid with (R)-3-hydroxyalkanoic acids. *Biomacromolecules*. 2002, 3(1): 133-138.

[9] Renstad, R., S. Karlsson, and A.C. Albertsson. 1999. The influence of processing induced differences in molecular structure on the biological and non-biological degradation of poly(3-hydroxybutyrate-co-3-hydroxyvalerate), P(3-HB-co-3-HV). *Polym. Degrad. Stab.* 63, 201-211.

[10] Qu X.H., Wu Q., Zhang K.Y., Chen G.Q. In vivo studies of poly(3-hydroxybutyrate-co-3-hydroxyhexanoate) based polymers: biodegradation and tissue reactions. *Biomaterials,* 2006, 27(19):3540-3548 (c).

[11] Freier T., Kunze C., Nischan C., Kramer S., Sternberg K., Sass M., Hopt U.T., Schmitz K.P. In vitro and in vivo degradation studies for development of a biodegradable patch based on poly(3-hydroxybutyrate). *Biomaterials*. 2002, 23(13):2649-2657.

[12] Kunze C, Edgar Bernd H, Androsch R, Nischan C, Freier T, Kramer S, Kramp B, Schmitz KP. In vitro and in vivo studies on blends of isotactic and atactic poly (3-hydroxybutyrate) for development of a dura substitute material. *Biomaterials*. 2006 Jan; 27(2):192-201 (a).

[13] Gogolewski S., Jovanovic M., Perren S.M., Dillon J.G., Hughes M.K. Tissue response and in vivo degradation of selected polyhydroxyacids: polylactides (PLA), poly(3-hydroxybutyrate) (PHB), and poly(3-hydroxybutyrate-co-3-hydroxyvalerate) (PHB/VA). *J. Biomed. Mater. Res.* 1993, 27(9):1135-1148.

[14] Borkenhagen M., Stoll R.C., Neuenschwander P., Suter U.W., Aebischer P. In vivo performance of a new biodegradable polyester urethane system used as a nerve guidance channel. *Biomaterials* 1998, 19(23): 2155-2165.

[15] Hazari A., Johansson-Ruden G., Junemo-Bostrom K.,. Ljungberg C., Terenghi G., Green C., Wiberg M. A new resorbable wrap-around implant as an alternative nerve repair technique. *Journal of Hand Surgery* (British and European Volume, 1999) 24B: 3: 291–295 (a).

[16] Hazari A., Wiberg M., Johansson-Rudén G., Green C., Terenghi G. A resorbable nerve conduit as an alternative to nerve autograft. *British Journal of Plastic Surgery* (1999), 52, 653–657 (b).

[17] Miller N.D, Williams D.F. On the biodegradation of poly-beta-hydroxybutyrate (PHB) homopolymer and poly-beta-hydroxybutyrate-hydroxyvalerate copolymers. *Biomaterials.* 1987 Mar;8(2):129-137.

[18] Shishatskaya E.I., Volova T.G., Gordeev S.A., Puzyr A.P. Degradation of P(3HB) and P(3HB-co-3HV) in biological media. *J. Biomater. Sci. Polym. Ed.* 2005;16(5):643-657.

[19] Saito T., Tomita K., Juni K., Ooba K. In vivo and in vitro degradation of poly(3-hydroxybutyrate) in rat. *Biomaterials*. 1991, 12(3):309-312.

[20] Koyama N. and Doi Y. Morphology and biodegradability of a binary blend of poly((R)-3-hydroxybutyric acid) and poly((R,S)-lactic acid). *Can. J. Microbiol.*, 1995, 41(Suppl. 1): 316-322.

[21] Doi Y., Kanesawa Y., Kunioka M., Saito T. Biodegradation of microbial copolyesters: poly(3-hydroxybutyrate-co-3-hydroxyvalerate) and poly(3-hydroxybutyrate-co-4-hydroxybutyrate). *Macromolecules*, 1990a, 23:26-31.

[22] Holland S.J., Jolly A.M., Yasin M., Tighe B.J. Polymers for biodegradable medical devices. II. Hydroxybutyrate-hydroxyvalerate copolymers: hydrolytic degradation studies. *Biomaterials*, 1987, 8(4):289-295.

[23] Kurcok P., Kowalczuk M., Adamus G., Jedlinrski Z., Lenz R.W. Degradability of poly (b-hydroxybutyrate)s. Correlation with chemical microstucture. *JMS-Pure Appl. Chem.* 1995, A32:875–880.

[24] Bonartsev A.P., Myshkina V.L., Nikolaeva D.A., Furina E.K., Makhina T.A., Livshits V.A., Boskhomdzhiev A.P., Ivanov E.A., Iordanskii A.L., Bonartseva G.A. Biosynthesis, biodegradation, and application of poly(3-hydroxybutyrate) and its copolymers - natural polyesters produced by diazotrophic bacteria. *Communicating Current Research and Educational Topics and Trends in Applied Microbiology*, Ed: A. Méndez-Vilas, Formatex, Spain, 2007, V.1, p. 295-307.

[25] Bonartsev A.P. private communication.

[26] Cha Y., Pitt C.G. The biodegradability of polyester blends. *Biomaterials* 1990, 11(2):108-112.

[27] Schliecker G., Schmidt C., Fuchs S., Wombacher R., Kissel T. Hydrolytic degradation of poly(lactide-co-glycolide) films: effect of oligomers on degradation rate and crystallinity. *Int. J. Pharm.* 2003, 266(1-2): 39-49.

[28] Scandola M, Focarete ML, Adamus G, Sikorska W, Baranowska I, Swierczek S, Gnatowski M, Kowalczuk M, Jedlinr ski Z. Polymer blends of natural poly(3-hydroxybutyrate-co-hydroxyvalerate) and a synthetic atactic poly(3-hydroxybutyrate). Characterization and biodegradation studies. *Macromolecules* 1997; 30:2568–2574.

[29] Doi Y, Kanesawa Y, Kawaguchi Y, Kunioka M. Hydrolytic degradation of microbial poly(hydroxyalkanoates). *Makrom. Chem. Rapid. Commun*. 1989;10:227–230.

[30] Wang Y.-W., Yang F., Wu Q., Cheng Y.C., Yu P.H., Chen J., Chen G.-Q. Evaluation of three-dimensional scaffolds made of blends of hydroxyapatite and poly(3-hydroxybutyrate-co-3-hydroxyhexanoate) for bone reconstruction. *Biomaterials*. 2005, 26(8), 899–904 (a).

[31] Muhamad I.I., Joon L.K., Noor M.A.M. Comparing the degradation of poly-β-(hydroxybutyrate), poly-β–(hydroxybutyrate-co-valerate)(PHBV) and PHBV/Cellulose triacetate blend. *Malaysian Polymer Journal*, 2006, 1:39-46.

[32] Mergaert J., Webb A., Anderson C., Wouters A., Swings J. Microbial degradation of poly(3-hydroxybutyrate) and poly(3-hydroxybutyrate-co-3-hydroxyvalerate) in soils. *Applied and environmental microbiology*, 1993, 59(10):3233-3238.

[33] Choi G.G., Kim H.W., Rhee Y.H. Enzymatic and non-enzymatic degradation of poly(3-hydroxybutyrate-co-3-hydroxyvalerate) copolyesters produced by Alcaligenes sp. MT-16. *The Journal of Microbiology*, December 2004, 42(4):346-352.

[34] Yang X., Zhao K., Chen G.Q. Effect of surface treatment on the biocompatibility of microbial polyhydroxyalkanoates. *Biomaterials*. 2002; 23(5):1391-1397.

[35] Zhao K., Yang X., Chen G.Q., Chen J.C. Effect of lipase treatment on the biocompatibility of microbial polyhydroxyalkanoates. *J. Material Science: materials in medicine*. 2002; 13:849-854.

[36] Wang H.T., Palmer H., Linhardt R.J., Flanagan D.R., Schmitt E. Degradation of poly(ester) microspheres. *Biomaterials*. 1990; 11(9):679-685.

[37] Doyle C., Tanner E.T. and Bonfield W. In vitro and in vivo evaluation of polyhydroxybutyrate and of polyhydroxybutyrate reinforced with hydroxyapatite. *Biomaterials* 1991, 12:841-847.

[38] Coskun S., Korkusuz F. and Hasirci V. Hydroxyapatite reinforced poly(3-hydroxybutyrate) and poly(3-hydroxybutyrate-co-3-hydroxyvalerate) based degradable composite bone plate. *J. Biomater. Sci. Polymer Edn*, 2005, Vol. 16, No. 12, pp. 1485–1502.

[39] Sudesh K., Abe H., Doi Y. Synthesis, structure and properties of polyhydroxyalkanoates: biological polyesters. *Prog. Polym. Sci*. 2000, 25: 1503-1555.

[40] Lobler M., Sass M., Kunze C., Schmitz K.P., Hopt U.T. Biomaterial patches sutured onto the rat stomach induce a set of genes encoding pancreatic enzymes. *Biomaterials*, 2002, 23:577–583.

[41] Tokiwa Y, Suzuki T, Takeda K. Hydrolysis of polyesters by Rhizopus arrhizus lipase. *Agric. Biol. Chem*.1986; 50:1323–1325.

[42] Hoshino A. and Isono Y. Degradation of aliphatic polyester films by commercially available lipases with special reference to rapid and complete degradation of poly(L-lactide) film by lipase PL derived from Alcaligenes sp. *Biodegradation* 13: 141–147, 2002.

[43] Jendrossek D., Schirmer A., Schlegel H.G. Biodegradation of polyhydroxyalkanoic acids. *Appl. Microbiol. Biotechnol*. 1996; 46:451–463.

[44] Winkler F.K., D'Arcy A., Hunziker W. Structure of human pancreatic lipase. *Nature* 1990; 343: 771–774.

[45] Mergaert J., Anderson C., Wouters A., Swings J., Kersters K. Biodegradation of polyhydroxyalkanoates. *FEMS Microbiol. Rev*. 1992, 9(2-4): 317-321.

[46] Tokiwa Y., Calabia B.P. Degradation of microbial polyesters. *Biotechnol. Lett*. 2004, 26(15):1181-1189.

[47] Mokeeva V., Chekunova L., Myshkina V., Nikolaeva D., Gerasin V., Bonartseva G. Biodestruction of poly(3-hydroxybutyrate) by microscopic fungi: tests of polymer on resistance to fungi and fungicidal properties. *Mikologia and Fitopatologia* 2002, 36(5): 59-63.

[48] Bonartseva G.A., Myshkina V.L., Nikolaeva D.A., Kevbrina M.V., Kallistova A.Y., Gerasin V.A., Iordanskii A.L., Nozhevnikova A.N. Aerobic and anaerobic microbial degradation of poly-beta-hydroxybutyrate produced by Azotobacter chroococcum. *Appl. Biochem. Biotechnol*. 2003, 109(1-3): 285-301.

[49] Spyros A., Kimmich R., Briese B.H., Jendrossek D. 1H NMR Imaging Study of Enzymatic Degradation in Poly(3-hydroxybutyrate) and Poly(3-hydroxybutyrate-co-3-hydroxyvalerate). Evidence for Preferential Degradation of the Amorphous Phase by PHB Depolymerase B from Pseudomonas lemoignei. *Macromolecules,* 1997, 30(26): 8218-8225.

[50] Hocking P.J., Marchessault R.H., Timmins M.R., Lenz R.W., Fuller R.C., Enzymatic Degradation of Single Crystals of Bacterial and Synthetic Poly(-hydroxybutyrate). *Macromolecules*, 1996, 29(7), 2472-2478.

[51] Bonartseva G.A., Myshkina V.L., Nikolaeva D.A., Rebrov A.V., Gerasin V.A., Makhina T.K. [The biodegradation of poly-beta-hydroxybutyrate (PHB) by a model soil community: the effect of cultivation conditions on the degradation rate and the physicochemical characteristics of PHB] *Mikrobiologiia* 2002, 71(2):258-263. Russian.

[52] Kramp B., Bernd H.E., Schumacher W.A., Blynow M., Schmidt W., Kunze C., Behrend D., Schmitz K.P. [Poly-beta-hydroxybutyric acid (PHB) films and plates in defect covering of the osseus skull in a rabbit model]. *Laryngorhinootologie*, 2002, 81(5):351-356. [Article in German].

[53] Malm T, Bowald S, Karacagil S, Bylock A, Busch C. A new biodegradable patch for closure of atrial septal defect. An experimental study. *Scand. J. Thorac. Cardiovasc Surg*. 1992; 26(1): 9-14 (a).

[54] Malm T, Bowald S, Bylock A, Saldeen T, Busch C (1992). Regeneration of pericardial tissue on absorbable polymer patches implanted into the pericardial sac. An immunohistochemical, ultrastructural and biochemical study in the sheep. Scandinavian *Journal of Thoracic and Cardiovascular Surgery*, 26(1): 15–21 (b).

[55] Malm T, Bowald S, Bylock A and Busch C. Prevention of postoperative pericardial adhesions by closure of the pericardium with absorbable polymer patches. An experimental study. *The Journal of Thoracic and Cardiovascular Surgery*, 1992, 104: 600-607. (c)

[56] Malm T, Bowald S, Bylock A, Busch C, Saldeen T (1994). Enlargement of the right ventricular outflow tract and the pulmonary artery with a new biodegradable patch in transannular position. *European Surgical Research*, 26: 298–308.

[57] Duvernoy O, Malm T, Ramström J, Bowald S. A biodegradable patch used as a pericardial substitute after cardiac surgery: 6- and 24-month evaluation with CT. *Thorac. Cardiovasc. Surg*. 1995 Oct;43(5):271-274.

[58] Kawaguchi T., Tsugane A., Higashide K., Endoh H., Hasegawa T., Kanno H., Seki T., Juni K., Fukushima S, Nakano M. Control of drug release with a combination of prodrug and polymer matrix: antitumor activity and release profiles of 2',3'-Diacyl-5-fluoro-2'-deoxyuridine from poly(3-hydroxybutyrate) microspheres. *Journal of Pharmaceutical Sciences*, 1992, 87(6): 508-512.

[59] Baptist, J.N., (Assignor to W.R. Grace Et Co., New York), US Patent No. 3 225 766, 1965.

[60] Holmes P. Biologically produced (R)-3-hydroxy-alkanoate polymers and copolymers. In: Bassett DC (Ed.) *Developments in crystalline polymers*. London, Elsevier, 1988, Vol. 2: 1–65.

[61] Saad B., Ciardelli G., Matter S., Welti M., Uhlschmid G. K., Neuenschwander P., Suterl U.W. Characterization of the cell response of cultured macrophages and fibroblasts td particles of short-chain poly[(R)-3-hydroxybutyric acid]. *Journal of Biomedical Materials Research,* 1996, 30, 429-439.

[62] Fedorov M., Vikhoreva G., Kildeeva N., Maslikova A., Bonartseva G., Galbraikh L. [Modeling of surface modification process of surgical suture]. *Chimicheskie volokna* 2005, (6), 22-28. [Article in Russian].

[63] Rebrov A.V., Dubinskii V.A., Nekrasov Y.P., Bonartseva G.A., Shtamm M., Antipov E.M. [Structure phenomena at elastic deformation of highly oriented polyhydroxybutyrate]. *Vysokomol. Soedin*. (Russian) 2002, 44, 347–351 [Article in Russian].

[64] Kil'deeva N.R., Vikhoreva G.A., Gal'braikh L.S., Mironov A.V., Bonartseva G.A., Perminov P.A., Romashova A.N. [Preparation of biodegradable porous films for use as wound coverings] *Prikl. Biokhim. Mikrobiol.* 2006, 42(6): 716-720. [Article in Russian].

[65] Solheim E., Sudmann B., Bang G., Sudmann E. Biocompatibility and effect on osteogenesis of poly(ortho ester) compared to poly(DL-lactic acid). *J. Biomed. Mater.* Res. 2000, 49(2), 257-263.

[66] Bostman O., Pihlajamaki H. Clinical biocompatibility of biodegradable orthopaedic implants for internal fixation: a review. *Biomaterials,* 2000, 21(24), 2615-2621.

[67] Lickorish D., Chan J., Song J., Davies J.E. An in-vivo model to interrogate the transition from acute to chronic inflammation. *Eur. Cell. Mater*., 2004, 8, 12-19.

[68] Khouw I.M., van Wachem P.B., de Leij L.F., van Luyn M.J. Inhibition of the tissue reaction to a biodegradable biomaterial by monoclonal antibodies to IFN-gamma. *J. Biomed. Mater. Res*., 1998, 41, 202-210.

[69] Su S.H., Nguyen K.T., Satasiya P., Greilich P.E., Tang L., Eberhart R.C. Curcumin impregnation improves the mechanical properties and reduces the inflammatory response associated with poly(L-lactic acid) fiber. *J. Biomater. Sci. Polym*. Ed. 2005, 16(3), 353-370.

[70] Lobler M., Sass M., Schmitz K.P., Hopt U.T. Biomaterial implants induce the inflammation marker CRP at the site of implantation. *J. Biomed. Mater. Res*., 2003, 61, 165–167.

[71] Novikov L.N., Novikova L.N., Mosahebi A., Wiberg M., Terenghi G., Kellerth J.O. A novel biodegradable implant for neuronal rescue and regeneration after spinal cord injury. *Biomaterials,* 2002, 23, 3369–3376.

[72] Cao W., Wang A., Jing D., Gong Y., Zhao N., Zhang X. Novel biodegradable films and scaffolds of chitosan blended with poly(3-hydroxybutyrate). *J. Biomater. Sci. Polymer* Edn., 2005, 16(11), 1379–1394.

[73] Wang Y.-W., Yang F., Wu Q., Cheng Y.C., Yu P.H., Chen J., Chen G.-Q. Effect of composition of poly(3-hydroxybutyrate-co-3-hydroxyhexanoate) on growth of fibroblast and osteoblast. *Biomaterials*. 2005, 26(7), 755-761 (b).

[74] Ostwald J., Dommerich S., Nischan C., Kramp B. [In vitro culture of cells from respiratory mucosa on foils of collagen, poly-L-lactide (PLLA) and poly-3-hydroxy-butyrate (PHB)]. *Laryngorhinootologie*, 2003, 82(10), 693-699 [Article in German].

[75] Wollenweber M., Domaschke H., Hanke T., Boxberger S., Schmack G., Gliesche K., Scharnweber D., Worch H. Mimicked bioartificial matrix containing chondroitin sulphate on a textile scaffold of poly(3-hydroxybutyrate) alters the differentiation of adult human mesenchymal stem cells. *Tissue Eng*. 2006 Feb;12(2):345-359.

[76] Wang Y.-W, Wu Q., Chen G.Q. Attachment, proliferation and differentiation of osteoblasts on random biopolyester poly(3-hydroxybutyrate-co-3-hydroxyhexanoate) scaffolds. *Biomaterials*, 2004, 25(4), 669-675.

[77] Nebe B., Forster C., Pommerenke H., Fulda G., Behrend D., Bernewski U., Schmitz K.P., Rychly J. Structural alterations of adhesion mediating components in cells cultured on poly-beta-hydroxy butyric acid. *Biomaterials* 2001, 22(17): 2425-2434.

[78] Qu X.-H., Wu Q., Chen G.-Q. In vitro study on hemocompatibility and cytocompatibility of poly(3-hydroxybutyrate-co-3-hydroxyhexanoate) *J. Biomater. Sci. Polymer Edn*., 2006, 17(10), 1107–1121 (a).

[79] Pompe T., Keller K., Mothes G., Nitschke M., Teese M., Zimmermann R., Werner C. Surface modification of poly(hydroxybutyrate) films to control cell-matrix adhesion. *Biomaterials*. 2007, 28(1), 28-37.

[80] Deng Y., Lin X.S., Zheng Z., Deng J.G., Chen J.C., Ma H., Chen G.-Q. Poly(hydroxybutyrate-co-hydroxyhexanoate) promoted production of extracellular matrix of articular cartilage chondrocytes in vitro. *Biomaterials,* 2003, 24(23), 4273-4281.

[81] Zheng Z., Bei F.-F., Tian H.-L., Chen G.-Q. Effects of crystallization of polyhydroxyalkanoate blend on surface physicochemical properties and interactions with rabbit articular cartilage chondrocytes, *Biomaterials*, 2005, 26, 3537–3548.

[82] Qu X.-H., Wu Q., Liang J., Zou B., Chen G.-Q. Effect of 3-hydroxyhexanoate content in poly(3-hydroxybutyrate-co-3-hydroxyhexanoate) on in vitro growth and differentiation of smooth muscle cells. *Biomaterials*. 2006 May;27(15):2944-2950 (b).

[83] Shishatskaya E.I., Volova T.G. A comparative investigation of biodegradable polyhydroxyalkanoate films as matrices for in vitro cell cultures. *J. Mater. Sci-Mater.* M., 2004, 15, 915–923.

[84] Fischer D., Li Y., Ahlemeyer B., Kriglstein J., Kissel T. In vitro cytotoxicity testing of polycations: influence of polymer structure on cell viability and hemolysis. *Biomaterials* 2003, 24(7), 1121–1131.

[85] Nitschke M., Schmack G., Janke A., Simon F., Pleul D., Werner C. Low pressure plasma treatment of poly(3-hydroxybutyrate): toward tailored polymer surfaces for tissue engineering scaffolds. *J. Biomed. Mater. Res.*, 2002, 59(4), 632-638.

[86] Chanvel-Lesrat D.J., Pellen-Mussi P., Auroy P., Bonnaure-Mallet M. Evaluation of the in vitro biocompatibility of various elastomers. *Biomaterials* 1999, 20, 291–299.

[87] Boyan B.D., Hummert T.W., Dean D.D., Schwartz Z. Role of material surfaces in regulating bone and cartilage cell response. *Biomaterials* 1996, 17, 137–146.

[88] Bowers K.T., Keller J.C., Randolph B.A., Wick D.G., Michaels C.M. Optimization of surface micromorphology for enhanced osteoblasts responses in vitro. *Int. J. Oral. Max. Impl.*, 1992, 7, 302–310.

[89] Cochran D., Simpson J., Weber H., Buser D. Attachment and growth of periodontal cells on smooth and rough titanium *Int. J. Oral. Max. Impl.*, 1994, 9, 289–297.

[90] Sevastianov V.I., Perova N.V., Shishatskaya E.I., Kalacheva G.S., Volova T.G. Production of purified polyhydroxyalkanoates (PHAs) for applications in contact with blood. *J. Biomater. Sci. Polym. Ed.*, 2003, 14, 1029–1042.

[91] Seebach D., Brunner A., Burger H.M., Schneider J., Reusch R.N. Isolation and 1H-NMR spectroscopic identification of poly(3-hydroxybutanoate) from prokaryotic and eukaryotic organisms. Determination of the absolute configuration (R) of the monomeric unit 3-hydroxybutanoic acid from Escherichia coli and spinach. *Eur. J. Biochem.,* 1994, 224(2), 317-328.

[92] Reusch R.N. Poly-β-hydroxybutryate/calcium polyphosphate complexes in eukaryotic membranes. *Proc. Soc. Exp. Biol. Med.*, 1989, 191, 377–381.

[93] Reusch R.N. Biological complexes of poly-β-hydroxybutyrate. *FEMS Microbiol. Rev.*, 1992, 103, 119–130.

[94] Reusch R.N. Low molecular weight complexed poly(3-hydroxybutyrate): a dynamic and versatile molecule in vivo. *Can. J. Microbiol.*, 1995, 41(Suppl. 1), 50–54.

[95] Müller H.M., Seebach D. Polyhydroxyalkanoates: a fifth class of physiologically important organic biopolymers? *Angew Chemie*, 1994, 32, 477–502.

[96] Huang R., Reusch R.N. Poly(3-hydroxybutyrate) is associated with specific proteins in the cytoplasm and membranes of Escherichia coli. *J. Biol. Chem.* 1996, 271, 22196–22201.

[97] Reusch R.N., Bryant E.M., Henry D.N. Increased poly-(R)-3-hydroxybutyrate concentrations in streptozotocin (STZ) diabetic rats. *Acta Diabetol.*, 2003, 40(2), 91-94.

[98] Reusch R.N., Sparrow A.W., Gardiner J. Transport of poly-β-hydroxybutyrate in human plasma. *Biochim. Biophys. Acta*, 1992, 1123, 33–40.

[99] Reusch R.N., Huang R., Kosk-Kosicka D. Novel components and enzymatic activities of the human erythrocyte plasma membrane calcium pump. *FEBS Lett.*, 1997, 412(3), 592-596.

[100] Pavlov E., Zakharian E., Bladen C., Diao C. T. M., Grimbly C., Reusch R.N., French R. J. A large, voltage-dependent channel, isolated from mitochondria by water-free chloroform extraction. *Biophysical Journal*, 2005, 88, 2614–2625.

[101] Theodorou M.C., Panagiotidis C.A., Panagiotidis C.H., Pantazaki A.A., Kyriakidis D.A. Involvement of the AtoS-AtoC signal transduction system in poly-(R)-3-hydroxybutyrate biosynthesis in Escherichia coli. *Biochim. Biophys. Acta.*, 2006, 1760(6), 896-906.

[102] Wiggam M.I., O'Kane M.J., Harper R., Atkinson A.B., Hadden D.R., Trimble E.R., Bell P.M., Treatment of diabetic ketoacidosis using normalization of blood 3-hydroxy-butyrate concentration as the endpoint of emergency management, *Diabetes Care*, 1997, 20, 1347–1352.

[103] Larsen T., Nielsen N.I. Fluorometric determination of beta-hydroxybutyrate in milk and blood plasma. *J. Dairy Sci.*, 2005, 88(6), 2004-2009.

[104] Agrawal C.M., Athanasiou K.A. Technique to control pH in vicinity of biodegrading PLA-PGA implants. *J. Biomed. Mater. Res.*, 1997, 38(2), 105-114.

[105] Ignatius A.A., Claes L.E. In vitro biocompatibility of bioresorbable polymers: poly(l, dl-lactide) and poly(l-lactide-co-glycolide). 1996, 17:8, 831-839.

[106] Rihova B. Biocompatibility of biomaterials: hemocompatibility, immunocompatibility and biocompatibility of solid polymeric materials and soluble targetable polymeric carriers. *Adv Drug. Delivery Rev.*, 1996, 21, 157-176.

[107] Ceonzo K., Gaynor A., Shaffer L., Kojima K., Vacanti C.A., Stahl G.L. Polyglycolic acid-induced inflammation: role of hydrolysis and resulting complement activation. *Tissue Eng.* 2006, 12(2), 301-308.

[108] Chasin M. and Langer R. (eds), *Biodegradable Polymers as Drug Delivery Systems*, New York, Marcel Dekker, 1990.

[109] Johnson O.L., Tracy M.A. Peptide and protein drug delivery. In: Mathiowitz E, ed. *Encyclopedia of Controlled Drug Delivery*. Vol 2. Hoboken, NJ: John Wiley and Sons; 1999, 816-832.

[110] Jain R.A. The manufacturing techniques of various drug loaded biodegradable poly(lactide-co-glycolide) (PLGA) devices. *Biomaterials*. 2000, 21, 2475-2490.

[111] Gursel I., Hasirci V. Properties and drug release behaviour of poly(3-hydroxybutyric acid) and various poly(3-hydroxybutyrate-hydroxyvalerate) copolymer microcapsules. *J. Microencapsul.*, 1995, 12(2):185-193.

[112] Li J., Li X., Ni X., Wang X., Li H., Leong K.W. Self-assembled supramolecular hydrogels formed by biodegradable PEO–PHB–PEO triblock copolymers and a-cyclodextrin for controlled drug delivery. *Biomaterials* 2006, 27( ): 4132–4140.

[113] Akhtar S., Pouton C.W., Notarianni L.J. Crystallization behaviour and drug release from bacterial polyhydroxyalkanoates. *Polymer,* 1992, 33(1): 117-126.

[114] Akhtar S., Pouton C.W., Notarianni L.J. The influence of crystalline morphology and copolymer composition on drug release from solution cast and melting processed P(HB-HV) copolymer matrices. *J. Controlled Release*, 1991, 17: 225-234.

[115] Korsatko W., Wabnegg B., Tillian H.M., Braunegg G., Lafferty R.M. Poly-D-hydroxybutyric acid-a biologically degradable vehicle to regard release of a drug. *Pharm. Ind.* 1983, 45:1004-1007.

[116] Korsatko W., Wabnegg B., Tillian H.M., Egger G., Pfragner R., Walser V. Poly D(-)-3-hydroxybutyric acid (poly-HBA)-a biodegradable former for long-term medication dosage. 3. Studies on compatibility of poly-HBA implantation tablets in tissue culture and animals. *Pharm. Ind.*, 1984, 46:952-954.

[117] Kassab A.C., Piskin E., Bilgic S., Denkbas E.B., Xu K. Embolization with polyhydroxybutyrate (PHB) micromerspheres: in vivo studies, *J. Bioact. Compat. Polym.,* 1999, 14, 291–303.

[118] Kassab A.C., Xu K., Denkbas E.B., Dou Y., Zhao S., Piskin E. Rifampicin carrying polyhydroxybutyrate microspheres as a potential chemoembolization agent, *J. Biomater. Sci. Polym. Ed.*, 1997, 8: 947–961.

[119] Sendil D., Gursel I., Wise D.L., Hasirci V. Antibiotic release from biodegradable PHBV microparticles. *J. Control. Release* 1999, 59: 207–17.

[120] Gursel I., Yagmurlu F., Korkusuz F., Hasirci V. In vitro antibiotic release from poly(3-hydroxybutyrate-co-3-hydroxyvalerate) rods. *J. Microencapsul.*, 2002, 19:153–164.

[121] Turesin F., Gursel I., Hasirci V. Biodegradable polyhydroxyalkanoate implants for osteomyelitis therapy: in vitro antibiotic release. *J. Biomater. Sci. Polym. Ed.*, 2001, 12: 195–207.

[122] Turesin F., Gumusyazici Z., Kok F.M., Gursel I., Alaeddinoglu N.G., Hasirci V. Biosynthesis of polyhydroxybutyrate and its copolymers and their use in controlled drug release. *Turk. J. Med. Sci.*, 2000, 30: 535-541.

[123] Gursel I., Korkusuz F., Turesin F., Alaeddinoglu N.G., Hasirci V. In vivo application of biodegradable controlled antibiotic release systems for the treatment of implant-related osteomyelitis. *Biomaterials*, 2001, 22(1): 73-80.

[124] Korkusuz F., Korkusuz P., Eksioglu F., Gursel I., Hasirci V. In vivo response to biodegradable controlled antibiotic release systems. *J. Biomed. Mater. Res.*, 2001, 55(2): 217-228.

[125] Yagmurlu M.F., Korkusuz F., Gursel I., Korkusuz P., Ors U., Hasirci V. Sulbactam-cefoperazone polyhydroxybutyrate-co-hydroxyvalerate (PHBV) local antibiotic delivery system: In vivo effectiveness and biocompatibility in the treatment of implantrelated experimental osteomyelitis. *J. Biomed. Mater. Res.*, 1999, 46: 494–503.

[126] Khang G., Kim S.W., Cho J.C., Rhee J.M., Yoon S.C., Lee H.B. Preparation and characterization of poly(3-hydroxybutyrate-co-3-hydroxyvalerate) microspheres for the sustained release of 5-fluorouracil. *Biomed. Mater. Eng.*, 2001, 11: 89–103.

[127] Bonartsev A.P., Bonartseva G.A., Makhina T.K., Mashkina V.L., Luchinina E.S., Livshits V.A., Boskhomdzhiev A.P., Markin V.S., Iordanskii A.L. [New poly-(3-hydroxybutyrate)-based systems for controlled release of dipyridamole and indomethacin] *Prikl. Biokhim. Mikrobiol.* 2006, 42(6): 710-715.

[128] Salman M.A., Sahin A., Onur M.A., Oge K., Kassab A., Aypar U. Tramadol encapsulated into polyhydroxybutyrate microspheres: in vitro release and epidural analgesic effect in rats. *Acta Anaesthesiol. Scand.*, 2003, 47: 1006-1012.

[129] Bonartsev A.P., Livshits V.A., Makhina T.A., Myshkina V.L., Bonartseva G.A., Iordanskii A.L. Controlled release profiles of dipyridamole from biodegradable microspheres on the base of poly(3-hydroxybutyrate). *Express Polymer Letters*, 2007 Vol.1, No.12, p. 797–803 (b).

[130] Bonartsev A.P., Postnikov A.B., Myshkina V.L., Artemieva M.M., Medvedeva N.A. A new system of nitric oxide donor prolonged delivery on basis of controlled-release polymer, polyhydroxybutyrate. *American Journal of Hypertension*, 2005, 18(5A), p.A

[131] Bonartsev A.P., Postnikov A.B., Mahina T.K., Myshkina V.L., Voinova V.V., Boskhomdzhiev A.P., Livshits V.A., Bonartseva G.A., Iorganskii A.L. A new in vivo model of prolonged local nitric oxide action on arteries on basis of biocompatible polymer. *The Journal of Clinical Hypertension*, 2007, Suppl. A., V. 9, N. 5, p. A152 (c).

[132] Pouton C.W., Akhtar S. Biosynthetic polyhydroxyalkanoates and their potential in drug delivery. Adv. *Drug Deliver. Rev.*, 1996; 18:133–162.

[133] Kots A.Y., Grafov M.A., Khropov Y.V., Betin V.L., Belushkina N.N., Busygina O.G., Yazykova M.Y., Ovchinnikov I.V., Kulikov A.S., Makhova N.N., Medvedeva N.A., Bulargina T.V., Severina I.S. Vasorelaxant and antiplatelet activity of 4,7-dimethyl-1,2,5-oxadiazolo[3,4-d]pyridazine 1,5,6-trioxide: role of soluble guanylate cyclase, nitric oxide and thiols. *Br. J. Pharmacol.* 2000, 129(6): 1163-1177.

In: Renewable Resources: Obtaining, Processing... ISBN 978-1-6060692-534-8
Editors: R. Kozlowski, G. Zaikov and F. Pudel

Chapter 15

THE APPLICATION OF HEMP ESSENTIALS OILS IN FINISHING OF NONWOVENS

Wanda Konczewicz and Natalia Łukaszewska
Institute of Natural Fibres,
Wojska Polskiego 71B, 60-630 Poznan, Poland

ABSTRACT

Hemp essential oil finds a lot of applications in cosmetic, medicine, food industry. Moreover its repellent and microbiological properties allow for using it in for plant protection and also for developing new functional textile materials characterized with antibacterial activity also.

The study aimed at using microcapsules with hemp essential oil in finishing processes in nonwoven production. The trials were conducted on the essential oil from the Polish cultivar of hemp – SILESIA, extracted in 2005. The essential oil was micro encapsulated.

The microcapsules were applied by padding method. The experiments were conducted on chemobond and spunlace (hydroentanglement) 100% viscose and 100% polyester nonwovens.

The study comprised determination of physical-chemical parameters, the oil composition and the antimicrobial activity (MIC and MBC for selected strains of microorganisms) of the oil itself and the nonwoven with the microcapsules. It was observed that effective antibacterial activity for Gram+ bacteria was achieved for microcapsule application at 6 g/m^2 of the nonwoven.

INTRODUCTION

A considerable increase of hemp cultivation for fiber and seed can recently be observed. In case of hemp grown for textile (spinnable) fiber, hemp panicles are nothing more but a ballast which worsens stem retting conditions and hemp mechanization of post harvest treatments during retting process. One of the methods of hemp panicles utilization is using them for essential oil production. Hemp is characterized by specific smell which is a result of volatile mono- and sesquiterpenes present in hemp [2,7,9,10]. Terpenes are synthesized in hairy glands which can be found in high numbers on leaves surface.

Four cultivars of monoecious hemp developed at INF are currently registered in National Register of Original Cultivars: BIALOBRZESKIE (1968), BENIKO (1985) and SILESIA (2000) and TYGRA (2007). All these cultivars are characterized by cannabinoids (psychoactive substances) content in dry matter well below the threshold value (0.2 % of delta^9 THC).

Since 2000 the essential oil obtained from hemp panicle is investigated for its properties, application in market products as well as harvesting methods are studied.

Hemp essential oil finds growing application in industry. Due to its characteristic aroma it is used for production of soap, shampoo, creams, cosmetic oils, perfumes as well as food products such as beer, beverages, etc. It is also used in natural medicine (aromatherapy). The terpenes contained in hemp essential oil show also bacteriostatic properties comparable to thyme oil. This effect is the strongest towards Gram+ bacteria (*Staphylococcus* and *Streptococcus*) [1,5,6].

Hemp essential oil has also insect repelling properties which makes it useful as plant protection product. This action is associated to the content of terpenes mainly limonene and pinene terpenes.

The studies showed that essential oil shows a repellent action on cockchafers of large pine weevil *Hylobius abietis* (L.) and mosquito *Culex pipiens* [3,4].

Application of hemp essential oil on textiles allows obtaining new functional product with specific properties. Microencapsulating of hemp essential oil will allow for its application on the surface of the textiles and gradual emission of the essential oil.

Microcapsulation of essential oil flavors has led to many novelty applications, which make use of their fragrant and antibacterial properties. Textile applications using microcapsules would be a system that is easy to apply, does not affect the existing textile properties and has a shelf-life on a garment that allows normal fabric-care process to take place. However, microcapsules show low resistance to abrasion and mechanical washing, additionally conventional ironing and other heat input processes such as tumble-drying can cause a dramatic reduction of microcapsules [6,8].

Microcapsules can be used in production of nonwovens. Generally nonwovens are not washed; therefore there is no risk of microcapsule damage. Using the hemp essential oil in the finishing process of nonwovens allows for obtaining new functional products with antibacterial properties. Encapsulating the oil allows for distributing it on the nonwoven and guarantees its gradual release.

It is possible to use the micro capsules in such textile materials as: curtains/drapes, bed sheets (single use, based on linen nonwoven, used in beauty salons), handkerchiefs and clothing (in pharmaceutical, cosmetic and medical industries).

The aim of research was applied microcapsules with hemp essential oil in finishing processes in nonwoven production. The nonwovens with the applied oil were subjected to micro-biological evaluation, determining its anti-microbial activity.

MATERIALS AND METHOD

1. Hemp Essential Oil

1.1. Material

The material for the study was hemp essential oil obtained in 2005 by steam distillation of hemp panicles from Polish cultivar Silesia. Panicles were provided by the commercial plantation at INF Research Station in Petkowo. Panicles were harvested using machine for hemp harvesting especially designed by Institute of Natural Fibres in Poznan [5]. Essential oil was produced on an industrial scale by steam distillation of hemp panicles. Yield of hemp essential oil was about 2,6 l/ha.

1.2. GC Analysis

The analysis of the hemp essential oil composition was performed with gas chromatography according to the terpene standards (SIGMA-ALDRICH). For determination of terpene content the Autosystem XL gas chromatograph (Perkin Elmer) with SPB-5 column (SUPELKO) (0.32 mm x 30 cm, 0,25 μm film thickness) and FID detector. The analytical conditions were: injector temperature 220°C, oven temperature 60-240°C at 3°C/min, detector temperature 260°C, injections 1 μl, split ratio 1: and carrier gas – helium. Analysis was done in Istitute of Natural Fibres (Analytical Laboratory of Chromatography and Spectrophotometry).

1.3. Antimicrobial Activity

The studies covered testing the composition of hemp essential oil and its antimicrobial activity. Antimicrobial activity was investigated according to macro-dilution method in liquid media based on the method for antibiotic MIC (minimal inhibitory concentration) and MBC (minimal bactericidal concentration) evaluation, according to National Committee for Clinical Laboratory Standards (NCCLS): Methods for dilution antimicrobial susceptibility test for bacteria that grows aerobically - Approved standard 5th Ed. NCCLS Document, January 2000.

The tests covered determination of MIC and MBC values for selected strains of microorganisms: *Staphylococcus aureus*, *Escherichia coli*, *Pseudomonas aeruginosa*, *Candida albicans*, *Corynebacterium xerosis* and *Bacillus licheniformis*.

1.4. Micro-Encapsulation

Hemp essential oil was locked in microcapsules. Micro-encapsulation technique was applied using melamine for wall construction. The particle size of microcapsules was determined by scanning electron microscope (HITACHI S-3400N).

2. Application of Microcapsules on Textiles

2.1. Material

Microcapsules with hemp essential oil ware applied on three tipe of nonwovens: 100% viscose chemobond and spunlace (hydroentanglement), surface weight - 50 g/m^2 and 100% polyester spundbond nonwoven, surface weight - 25 g/m^2.

2.2. Method

The microcapsules were applied by padding method. The nonwoven was padded in water solution of the binding agent with addition of micro-capsules by the use of laboratory padding device (fulard). After padding the nonwoven was left for air-drying and then heated for 5 min. in a dryer at temperature of 135°C.

The following conditions of padding processes were kept:

- bath temperature: 25°C
- bath components:
- binders: Devabound C and E (Devan, Belgium)
- amount of microcapsules [AM] - 3; 6 and 10 [g/m^2]

Before padding the 100% polyester nonwovens were washed using three kinds of surfactants: Sulfolen (Rotta GmbH - Germany), Roksol ICE and PSW (PCC Rokita - Poland)- 2 g/dm^3.

2.3. Evaluation

Application of microcapsules on the nonwovens was tested by scanning electron microscope (HITACHI S-3400N) and microbiological screening tests (*Parallel Streak Method,* acc. AATCC Test Method 147-1998 in presence of synthetic sweat)in Institute of Natural Fibres (Metrological Textile Laboratory).

RESULTS AND DISCUSION

1. Hemp Essential Oil

The parameters of hemp essential oil were the following:

- appearance – liquid, clear
- colour – from light yellow
- aroma – typical for hemp
- light refraction index – $n_D^{20} = 1.4897$
- density – $d_{20} = 0.890 – 0.910$ [g/cm^3]
- miscibility with ethyl alcohol - represents a turbid mixture
- the content of Δ^9-tetrahydrocannabinol (Δ^9-THC) – less than 0.1%.

Analysis of chemical composition of hemp essential oil showed that trans-caryophyllene 35.58%, β-myrcene 18.45%, (+)-α-pinene 9.76%, terpinolene 7.40%, ocymene 6.38% represent the biggest share among terpenes contained in it – table 1. Analysis was done in Institute of Natural Fibres (Analytical Laboratory of Chromatography and Spectrophotometry).

Table 1. Terpenes content in hemp essential oil

No.		Terpenes	Content [%]
1		α-Pinene	9.76
2		Camphene	0.16
3	MONO-TERPENES	β-Pinene	4.03
4		β-Myrcene	18.45
5		α-Phellandrene	0.25
6		3-Carene	0.29
7		Limonene	2.07
8		1,8-Cineol	
9		Ocimene	6.38
10		Sabinene hydrate	0.06
11		Terpinolene	7.40
12		Linalool	0.10
13		Fenchol	0.06
14		Borneol	0.03
15		Bornyl acetate	0.04
16	SESQUI-TERPENES	trans-Caryophyllene	35.58
17		β-Humulene	0.03
18		α-Humulene	9.05
19		trans-β-Farnesene	3.99
20		Caryophyllene oxide	2.28

The results of antimicrobial activity of hemp essential oil are presented in Table 2. It was found that hemp essential oil shows very strong anti staphylococcus effect towards Gram+ bacteria (*Staphylococcus aureus*), weak antibacterial effect towards Gram- bacteria (*Pseudomonas aeruginosa* and *Escherichia coli*), and very weak antifungal effect towards *Candida albicans, Corynebacterium xerosis* and *Bacillus licheniformis*.

Table 2. The microbiological activity of hemp essential oil

Microorganismss	MIC mg/ml	MBC mg/ml
Staphylococcus aureus	0 87	7.10
Escherichia coli	113.75	227.50
Pseudomonas aeruginosa	227.50	227.50
Candida albicans	113.75	No fungicidal effect
Corynebacterium xerosis	112.50	450.00
Bacillus licheniformis	225.00	450.00

2. Application of Microcapsules on Textiles

Size of microcapsules containing hemp essential oil varied between 1-10 μm Figure 1. The photo of nonwovens with applied microcapsules is shown in Table 3 and 4. The results of antimicrobial properties are presented in Table 5.

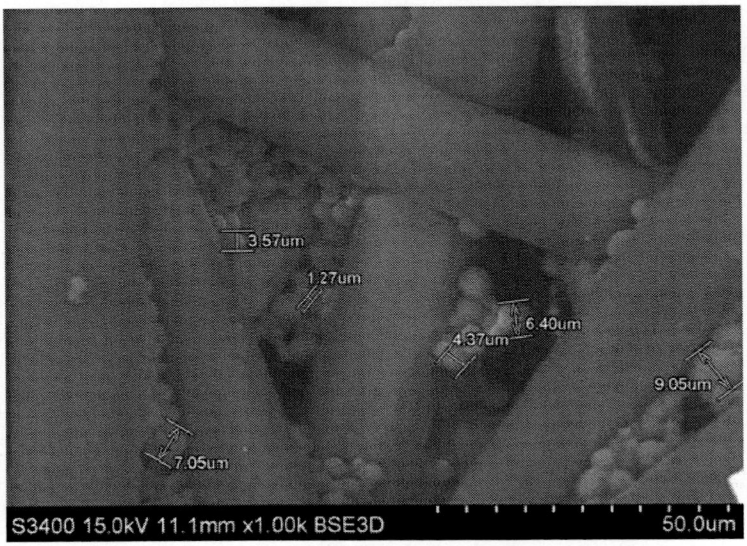

Figure 1. SEM photographs of microcapsules.

SEM photo showed that the best distribution of microcapsules on the nonwoven was achieved with the use of Roksol ICE, which thus was selected for further tests.

Table 3. SEM photographs of spundbond nonwoven 100% polyester with microcapsules – with the use of selected surfactants

Surfactant		
Sulfolen	Roksol ICE	Roksol PSW
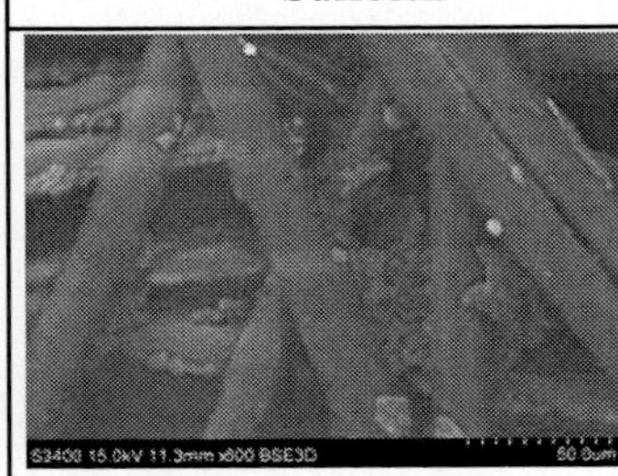	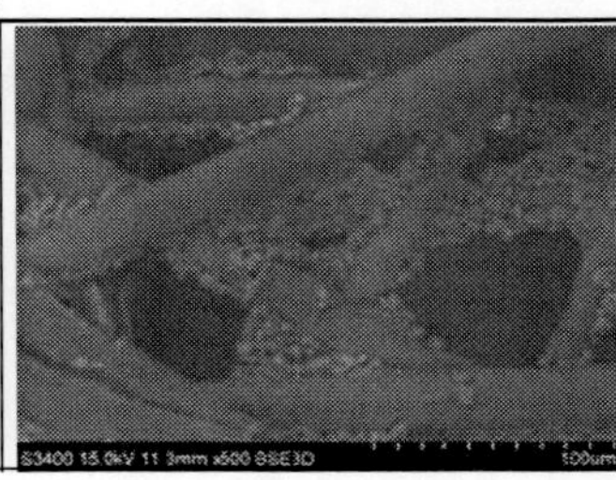	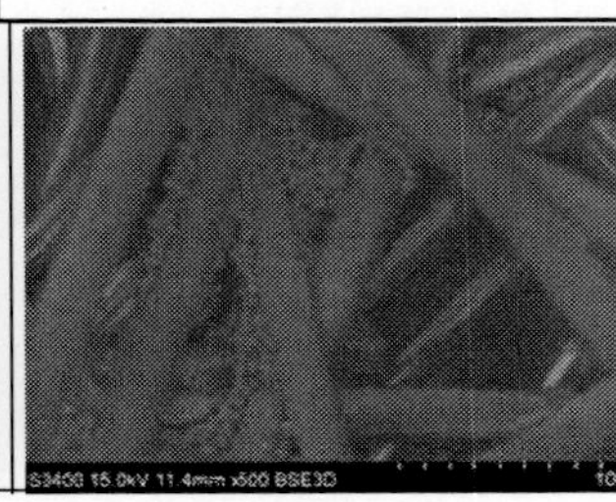

Analyzing microscopic photographs of nonwoven with microcapsules applied by padding (Table 4) method it is clear that microcapsules are distributed in conglomerates and as single capsules.

Table 4. SEM photographs of nonwovens with microcapsules

AM of microcapsules[g/m^2]		
3	6	10
	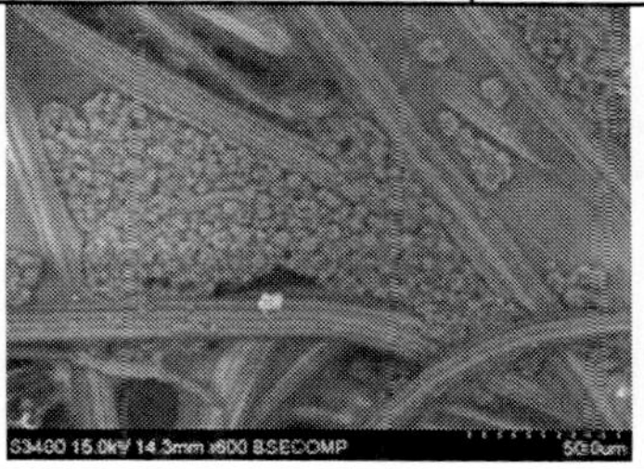	
spunlace nonwoven,100% viscose		

spunbond nonwoven100% polyester		

The presented results show that hemp essential oil applied to chemobond and spunlace nonwovens gives antibacterial effect only towards Gram positive bacteria.

The effective antibacterial activity for Gram+ bacteria was achieved for microcapsule application at 6 g/m^2 of the nonwoven. However, applying the oil on spunbond nonwoven in the amount of 3g/m^2 shows antibacterial effect towards *Micrococcus flavus* and *Staphylococcus aureus.* The increase of the oil dose results in growth inhibition in case of all the tested bacteria.

Microcapsules are joined together by bonding agent which fills voids between capsules. It is visible as a thin layer of bonding agent occurring in conglomerates. Photographs analyses showed also that microcapsules are visible on the surface of the nonwoven between threads as well as inside the nonwoven structure. However, the majority of capsules is present on the surface.

Obtained samples of nonwovens treated with essential oils applied by both methods are characterized by strong smell.

Table 5. Antimicrobial properties of nonwoven with applied of microcapsules

Microorganism growth under samples				
Bacterial	Control sample	AM [g/m^2]		
		3	6	10
chemobond nonwoven,100% viscose				
Bacillus licheniformis	+	+	-	-
Micrococcus flavus	+	+	-	-
Staphylococcus aureus	+	+	-	-
Escherichia coli	+	+	+	+
Pseudomonas aeruginosa	+	+	+	+
Klebsiella pneumoniae	+	+	+	+
spunlace nonwoven,100% viscose				
Bacillus licheniformis	+	+	+	+
Micrococcus flavus	+	+	-	--
Staphylococcus aureus	+	+	--	--
Escherichia coli	+	+	+	+
Pseudomonas aeruginosa	+	+	+	-
Klebsiella pneumoniae	+	+	+	+
spunbond nonwoven,100% polyester				
Bacillus licheniformis	+	+	+/-	+/-
Micrococcus flavus	+	+/-	+/-	+/-
Staphylococcus aureus	+	+/-	+/-	+/-
Escherichia coli	+	+	+/-	+/-
Pseudomonas aeruginosa	+	+	+/-	+/-
Klebsiella pneumoniae	+	+	+/-	+/-

- no bacterial growth under samples
-- growth inhibition zone up to 5 mm
+ bacterial growth (no bactericidal effect)
+/- weak growth inhibition

CONCLUSION

(1) Hemp essential oil shows good antibacterial properties against S*taphylococcus aureus* (Gram+).The best distribution of microcapsules on the spundbond nonwoven (100% polyester) was achieved with the use of detergent (Roksol ICE), which thus was selected for further tests.

(3) Hemp essential oil applied to chemobond and spunlace nonwovens (100% viscose) show very good antibacterial effect towards Gram positive bacteria. Effective antibacterial activity was achieved for microcapsules amount of 6 g/m^2 for Gram+ bacteria i.e.: *Staphylococcus aureus, Bacillus licheniformis* and *Micrococcus flavus.*

(4) Hemp essential oil on spunbond nonwoven (100% polyester) in the amount of 3g/m^2 shows weak antibacterial effect towards *Micrococcus flavus* and *Staphylococcus aureus.* The increase of the oil dose results in growth inhibition in case of all the tested bacteria.

(5) Applying microcapsules on nonwoven allows for obtaining material of antibacterial properties.

(6) Microcapsules can possibly be used in such textile materials as: curtains/drapes, bed sheets (single use, based on linen nonwoven, used in beauty salons), handkerchiefs, clothing (in pharmaceutical, cosmetic and medical industries).

The research of which results have been presented in this paper were conducted in FLEXIFUNBAR „Multifunctional Barriers for Flexible Structures (Textile, Leather, Paper)", Project, contract No. 505864 (NMP2-2004-505864).

REFERENCES

[1] Fournier G.; Paris M.R.; Fourniat M.C.; Quero A.M.. (1978). Activité bacteriostique d'huiles essentielles de Cannabis sativa L. *Annales pharmaceu-tiques françaises, 36* (11-12), 603-606.

[2] Hendriks H.; Malingré T.M.; Battermann S.; Bos R. (1975). Mono- and sesqui-terpene hydrocarbons of the essential oil of Cannabis sativa. *Phytochemistry,* 14, 814-815.

[3] Kaniewski R.; Konczewicz W.; Banach J.; Kuźmiński R. (2004). Cannabis Sativa as a Potential Source of Repellent. *The Proceedings of the 3rd Global Workshop (General Consultation) "Bast Fibrous Plants for Healthy Life"* Banja Luka, Bosnia and Herzegovina, October 24-28, 2004.

[4] Kaniewski R.; Konczewicz W.; Kuźmiński R. (2004). Utilizating Essential Oils Obtained from Hemp Panicles as a Raw Material for Production Repelling Insects. *The 10th International Conference for Renewable Resources and Plant Biotechnology,* Magdeburg, Germany, June 07-08, 2004.

[5] Kaniewski R. (2005). New self-propelled harvester for fibrous hemp. *Textiles for sustainable development*, Port Elizabeth, South Africa, October 23-27, 2005.

[6] Konczewicz W.; Zimniewska M.; Lukaszewska N. (2007). The application of hemp essential oils in finishing of linen fabrics. *4th Global Workshop (General Consultation) Of The FAO European Cooperative Research Network On Flax And Other Bast Plants: "Innovative technologies for comfort" to be held in University of Arad*, Romania, October 7-10, 2007. 163-168.

[7] Malingré T.; Herndriks H.; Battermann S.; Bos R.; Visser J. (1975). The essential oil of *Cannabis sativa*. *Planta medica, 28*, 56-61.

[8] Nelson G. (2002). Application of microencapsulation in textiles. *International Journal of Pharmaceutics, 242*, 55-62.

[9] Ross S.A.; ElSohly M. (1996). The volatile oil composition of fresh air-dried buds of *Cannabis sativa*. *J. Nat. Prod., 59*, 49-51.

[10] Turner C.E.; Elsohly M.A.; Boeren E.G. (1980). Constituents of *Cannabis sativa* L. XVII. A review of the natural constituents. *J. Nat. Prod., 43* (2), 169-243.

In: Renewable Resources: Obtaining, Processing...
Editors: R. Kozlowski, G. Zaikov and F. Pudel

ISBN 978-1-6060692-534-8

Chapter 16

CELLULOSIC NANOFIBRE COMPOSITES

M. J. A. van den Oever*, G. Perez Sanchez# and G. Yilmaz
Wageningen UR – AFSG,
6700 AA Wageningen, The Netherlands

ABSTRACT

Cellulosic nanofibres were made from sulphite bleached softwood using mechanical action. Fibre dimensions were evaluated with FE-SEM. The average diameter was found to be below 100 nm. Composites containing up to 50% of these cellulosic nano fibres and starch were prepared using a wet film casting method. Three routes for preparation were evaluated for their effect on mechanical performance of the resulting films. All resulting films were translucent and showed an increase in strength of a factor of 1.5 larger than for conventional flax-PP composites. The failure strain remained at the same level. Nanofibre inclusion did not result in a significant effect on the oxygen barrier properties of starch films up to at least 30 wt.% fibre in starch. Furthermore, the sensitivity of starch to water was reduced. Barrier properties against O_2 and H_2O were improved by addition of 4 wt.% of nanoclay. Nanoclay addition up to 4 wt.% had no significant effect on the films strength, stiffness and strain at failure.

INTRODUCTION

Recently, agrofibre and wood fibre reinforced composites are gaining ever increasing interest all over the world, becoming competitive with glass fibre reinforced plastics on a performance/cost basis. Annually, ca. 200 kton natural fibre reinforced composites find application in Europe [Nova, 2008]. Wood polymer composites (WPC's) find an outlet in decking and siding applications. Agrofibres mat reinforced thermoplastic (NMT) and thermoset composites are mainly used in automotive industry thus far. Agrofibre extrusion

* E-mail: martien.vandenoever@wur.nl; Tel.: +31 317 480105
# Current address: Endesa Energia S.A., Ribera del Loira n° 60, 28042 Madrid (Spain).

compounds for injection moulding applications are currently used in niche markets and are ready for larger scale market introduction [Carus, 2007].

Natural fibres are in fact composites themselves (also see Figure 1) [Bos, 2002; Fahlén, 2002]. The technical fibres are a composite of strong elementary plant cells (fibres), which on their turn are composed of nanofibres. These fibrils consist of crystalline cellulosic regions with estimated stiffness of around 130 GPa [Berglund, 2005] and high strength. The nanofibrils have a diameter down to below 50 nanometer scale range [Purz, 1998; Dufresne, 2000; Zimmermann, 2004] which makes them invisible when well dispersed in a transparent polymer [Iwamoto, 2005].

Obviously, cellulose nanofibres have the potential to form the basis of the next generation of natural fibre composites. However, they are well incorporated in plant and wood structures, witness the wide application of wood and plant fibres and witness the extensive procedures on gram scale used to produce cellulose nanofibres [Nakagaito, 2005; Chakraborty, 2005]. Therefore, the challenge is to efficiently extract these strong and stiff nanoscaled cellulosic fibres from plant tissues without damaging the fibre too much, while at the same time producing significant amounts with economical feasibility.

This presentation addresses the preparation and properties of cellulosic nanofibre composites based on softwood fibres in a thermoplastic starch matrix. Because cellulosic (nano)fibres tend to agglomerate after being formed, different routes of mixing the nanofibres and the polymer are discussed. Addition of nanoclay is discussed for its effect on barrier properties. Future prospects are included in the finalizing conclusions.

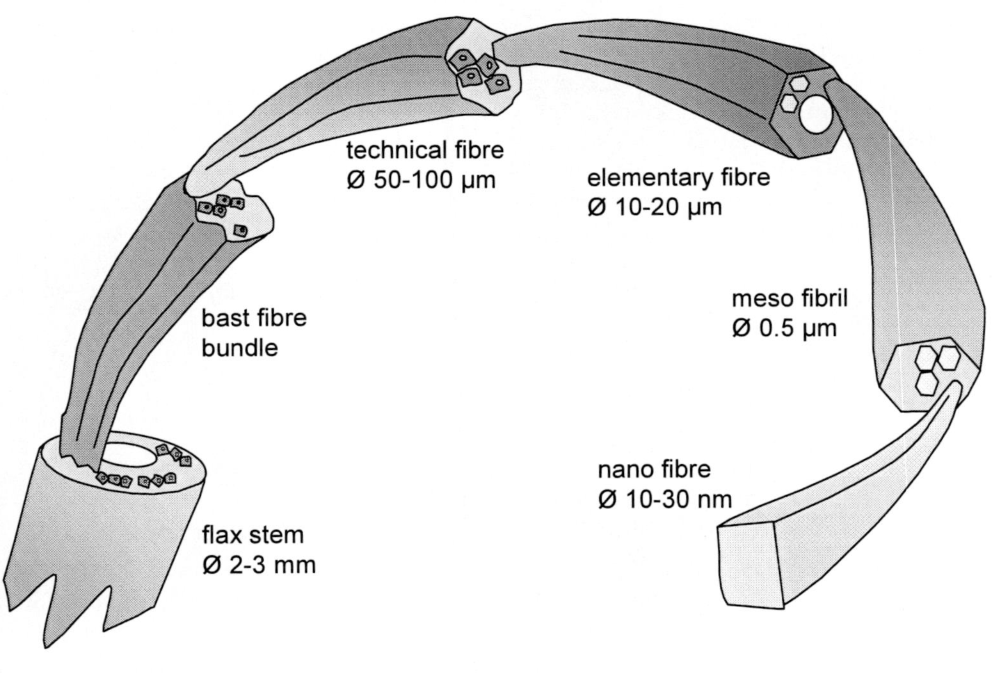

Figure 1. Schematic representation of flax bast fibres (after Bos et al., 2002).

EXPERIMENTAL

Material

Sulphite bleached softwood fibre (Ecocell) was supplied by Sappi Nijmegen, the Netherlands, in sheet form. Native potato starch (Avebé, The Netherlands) was gelatinised and plasticized with 20 wt.% glycerol (Chemproha Chemiepartner b.v., The Netherlands). Sodium Cloisite was supplied by Districhem bv, The Netherlands.

Nanofibre Production

Fibres were soaked in water at 1% dry matter content (DM) for 16 hours prior to disintegration in a Messmer Disintegrator MK.III C at 3,000 RPM for 10,000 rounds. This dispersion, containing 200 gram DM fibres, was refined in a Valley beater for 5 hours. The fibres were refined further in batches of 3 litre and 1.5% DM using an APV Gaulin homogenizer at 500 bar pressure for 5 passages on average.

Mixing Routes

Three routes for mixing the fibres and (gelatinized) polymer were applied: mixing before Valley beating of the fibres (route A); mixing after Valley beating, but before homogenization of the fibres (route B); after homogenization of the fibres (route C). For route A, mixing was performed in the Valley beater; for routes B and C an Ultra Turrax was used for 10 minutes at 20,500 RPM.

Composite Production

The obtained dispersions were vacuum extracted in order to remove air inclusions. The dispersions, 1.5% DM, were cast into polystyrene containers of 11*11*1 cm^3 and dried at 23°C and 50% relative humidity. The compositions that were prepared to study the effect of mixing route had a nanofibre content of 30 wt.%. Composites containing 10, 30 and 50 wt.% nanofibre in thermoplastic starch were made by route B. Plasticized starch with 20 wt.% glycerol was processed by applying the same refining procedure and cast to films as a reference. Additionally, films based on 30 wt.% nanofibre and 4 wt.% nanoclay in starch have been prepared by route B.

Test Methods

Scanning electron microscopy was used to evaluate the morphology of the prepared films. The analysis was performed at room temperature using a Jeol JSM-6300F (FE-SEM) scanning microscope. Samples evaluated with FE-SEM include fibres after Valley beating and after homogenization as well as a tensile fracture surface of a composite film produced by applying mixing route A.

The density of the films was calculated by dividing film weight by sample area and the average film thickness. The film thickness was determined using a TMI MI-21 with an accuracy of 1 μm. For evaluation of the mechanical properties, the composite films were cut into dogbone shaped samples with dimensions of the parallel part 35*15 mm^2. Samples were conditioned at 23°C and 50% RH for at least 1 week before the mechanical properties were determined. Uniaxial tensile tests were performed in 5-fold using a Zwick universal tester, type Z010 at a grip to grip separation of 75 mm and an extensometer gauge length of 20 mm. The crosshead speed was 1 mm/min for the modulus and 10 mm/min for the strength.

Transparency was determined using a Datacolor Elrepho 2000 with diffuse illumination via a photometer sphere (DIN 5033) and according to DIN 53147. Films of ca. 100 μm were used, 5 films were used to simulate an infinite stack of films. The difference between measured transparency values when using 4 or 5 films was insignificant. Water vapour permeability (WVP) was determined at 30, 50, 60 and 70 % relative humidity (RH). The WVP determination was made using a method based on the weight change of boxes containing dried silica gel that were covered with the composite films when stored at the indicated RH conditions. Oxygen transmission rate (OTR) was determined at 23°C using a MOCON OX-TRAN type 2/20 MH at RH values in the range 35–90%.

Results and Discussion

Film Morphology

The observations show that fibre diameter has been reduced significantly during Valley beating from ca. 10–20 μm for the starting softwood material to ca. 0.02–1 μm after beating (Figure 2). During homogenization, in particular the thick fibres have been refined further and average fibre diameter drops well below 100 nm (Figure 3). The theoretical yield of nanofibres is 100%, only small losses have been observed due to splashing.

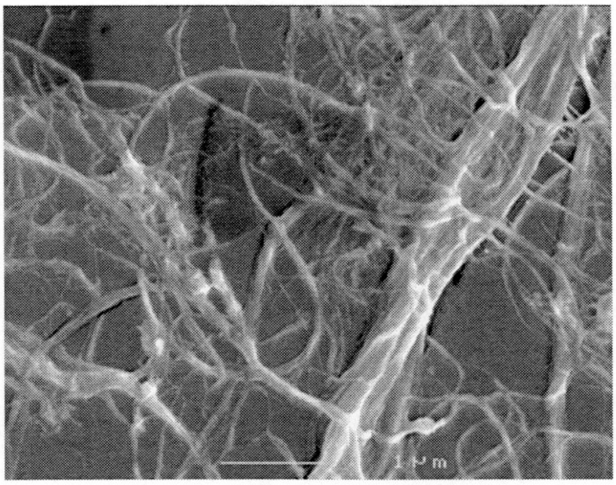

Figure 2. FE-SEM graph of softwood fibres, after Valley beating.

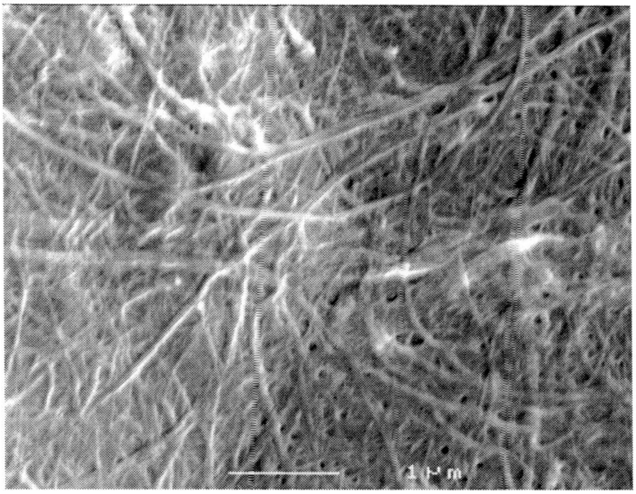

Figure 3. FE-SEM graph of softwood fibres, after Valley beating and homogenization.

The fracture surface of a cellulosic nanofibre-thermoplastic starch composite film indicates good fibre matrix adhesion (Figure 4). The structure looks very condensed, although some voids are visible which seem to be air inclusions. One other reason for the voids could be fibre pull out. The condensed structure is confirmed by the density of 1.27–1.41 g/cm^3 for the nanofibre-starch films when compared to 1.28–1.38 g/cm^3 for the starch films.

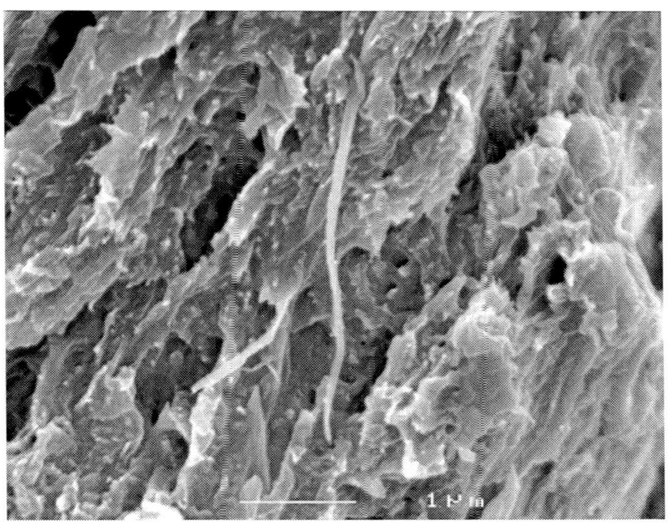

Figure 4. FE-SEM graph of fracture surface of a 30wt.% nanofibre/thermoplastic starch composite, produced by mixing route A.

Mechanical Properties

During the initial stages of the research, composites based on cellulosic nanofibres prepared from hemp bast fibre by route C exhibited very poor mechanical performance. This was thought to be due to fibre agglomeration, which hinders adequate impregnation with the polymer. The hypothesis was that if the polymer is able to impregnate the nanofibre as soon as it is formed (route A), less fibre agglomerates will occur in the composite than when ready nanofibres are mixed with a polymer (route C). And less agglomerates would mean less weak spots. However, the effect of mixing route on the strength performance of the composite films was small (Figure 5), in contradiction with this hypothesis. These data suggest that either ready made cellulosic nanofibres do not agglomerate or that fibre agglomerates have no significant negative effect on composites strength at the conditions applied.

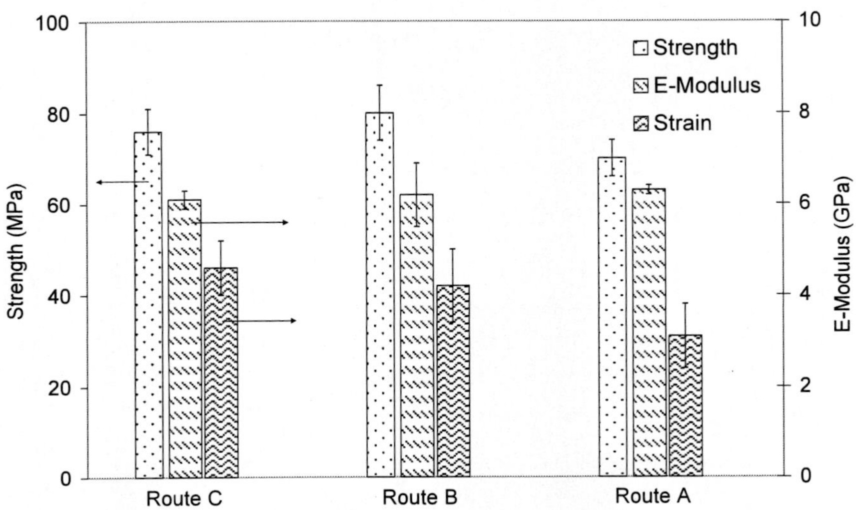

Figure 5. Effect of mixing route on tensile properties of 30 wt.% nanofibre-starch composites.

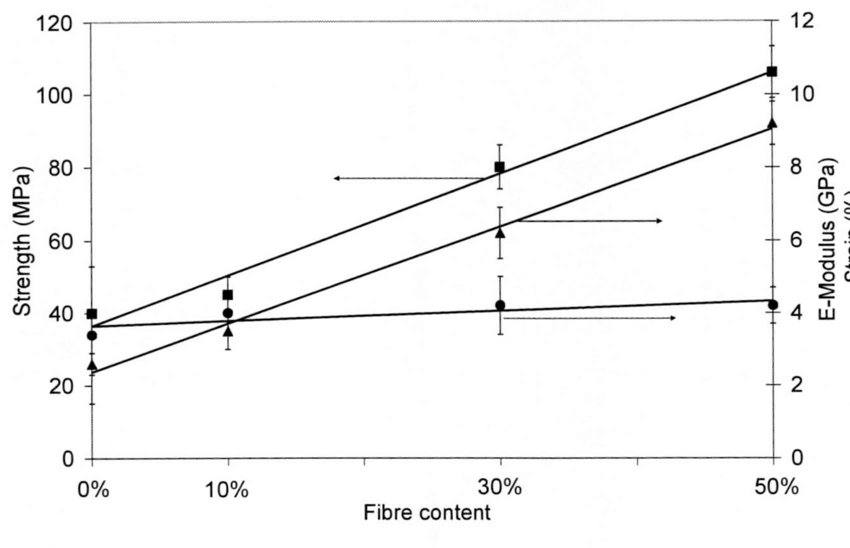

Figure 6. Tensile properties of nanofibre/starch composites produced by route B: strength (■), E-modulus (▲), Strain at failure (●).

Results of the tensile measurements are given in figure 6. Strength and stiffness of starch films increase by a factor of 2.7 and 3.5 respectively upon incorporation of 50% fibres. In particular the strength increase is higher compared to that of conventional flax-PP composites which show a tensile strength of 67 MPa only at 50 wt.% fibre content, although flax is supposed to be one of the strongest natural fibres available [Van den Oever, 2000; Carus, 2007]. It has been observed that the failure strain remained at the same level, whereas fibre reinforcement usually causes a decrease of the failure strain.

Initial soaking tests in water revealed that the nanofibres enhanced the structural integrity of starch films to a large extent.

Transparency

Although the films are based on substantial amounts of wood fibre, transparency is very high (Figure 7). Up to 30% nanofibre, 100 µm thick films can be looked through remarkably well.

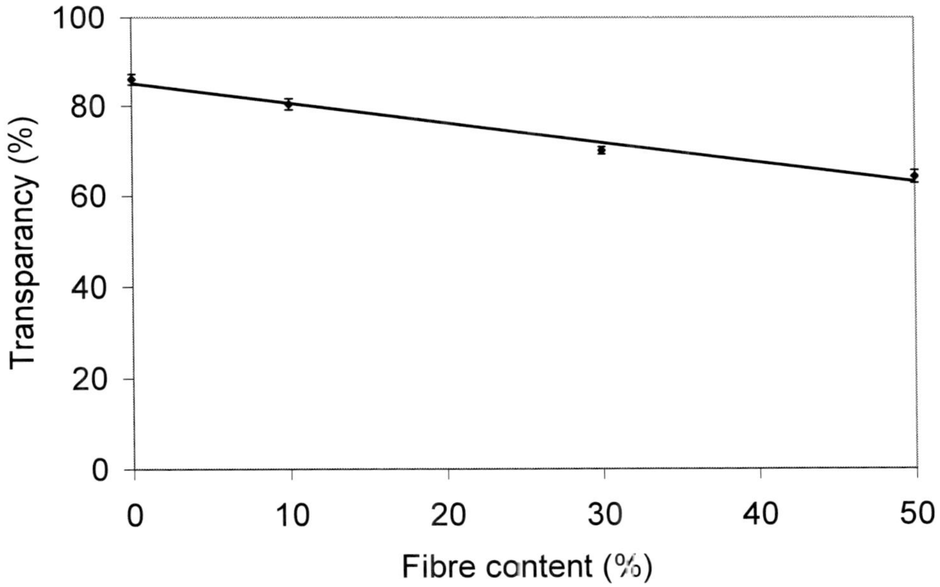

Figure 7. Transparency of nanofibre/starch composites produced by route B.

Permeabililty

Starch based films are known for their high oxygen barrier properties. This however is only the case in a limited RH range and minimal amount of plasticizers. The results show that the incorporation of cellulosic nanofibres has no significant influence the oxygen transmission rate (OTR) of starch (Figure 8). Regarding water vapour barrier properties, cellulosic nanofibres increase the barrier at increasing RH (Figure 9).

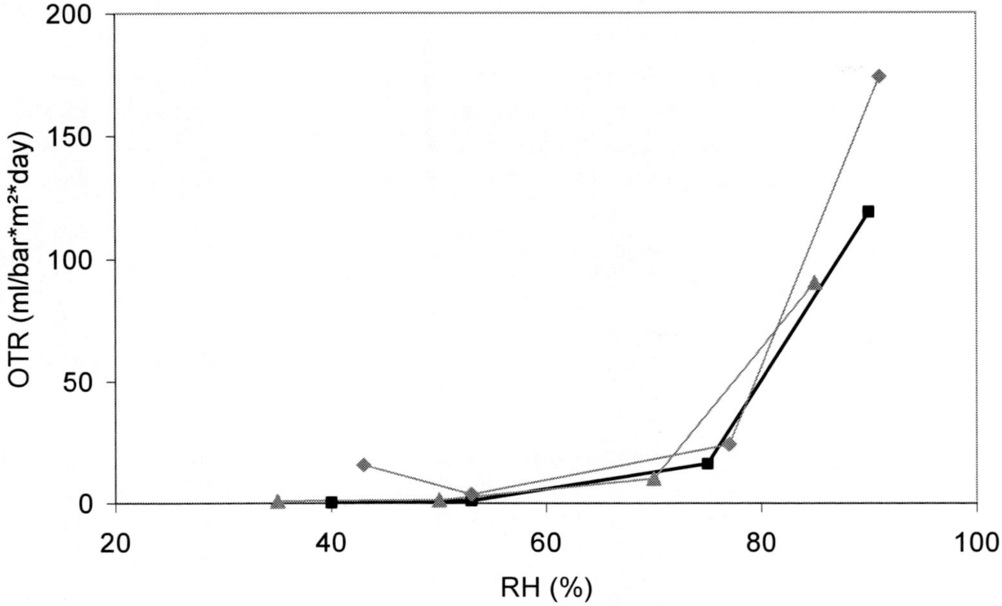

Figure 8. Oxygen transmission rate vs. relative humidity, normalized to values for 100 □m thick films: starch (■), 30% nanofibre-starch produced by route A (◆) and route B (▲).

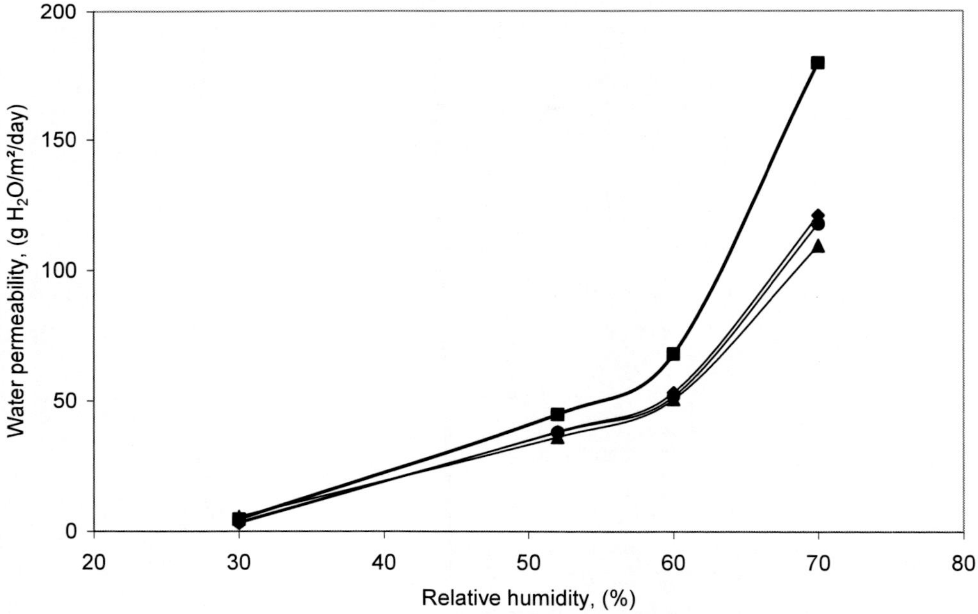

Figure 9. Water vapour permeability vs. relative humidity, normalized to values for 100 □m thick films: starch (■), 30% nanofibre-starch produced by route A (◆), route B (▲), route C (●).

Nanofibre/Nanoclay Composites

It appeared that addition of 4 wt.% of nanoclay further improved the barrier against O_2 and H_2O. However, addition of nanoclay had no effect on the films strength, stiffness and strain at failure.

CONCLUSION

The production of strong and stiff softwood based cellulosic nanofibre-starch films appears technically feasible. The strength improves a factor of 1.5 more than for conventional flax-PP composites. At the same time, the nanofibre-starch composites exhibit transparency up to 50% fibre content.

Up to at least 30 wt.% in starch, nanofibres do not deteriorate the excellent oxygen barrier properties of starch. Furthermore, the sensitivity of starch to water is reduced.

The results indicate that the problems such as water sensitivity and poor water vapour barrier properties which are associated with starch based materials can be addressed using a composites approach. This means that applicability of starch based materials not only limits to film applications, but also may be useful in coatings, paints, water resistant glues as well as 3D structures such as plates, packaging materials or even hybrid fibres.

ACKNOWLEDGEMENT

Financial support from the EU, within the framework of the integrated project Sustainpack, is acknowledged.

REFERENCES

Berglund L. Cellulose-Based Nanocomposites In *Natural Fibers, Biopolymers and Biocomposites*; Mohanty; Misra; Drzal; Eds, CRC Press: Boca Raton (FL), 2005; 807-832.

Bos, H.L.; Van den Oever, M.J.A.; Peters, O.C.J.J. *J Mat Sci* 2002, *37*, 1683-1692.

Carus, M.; Gahle, C.; Pendarovski, C.; Vogt, D.; Ortmann, S.; Grotenhermen, F.; Breuer, T.; Schmidt, C. *Studie zur Markt- und Konkurrenz-situation bei Naturfasern und Naturfaser-Werkstoffen (Deutschland und EU)*; Gülzower Fachgespräche, Band 26; Fachagentur Nachwachsende Rohstoffe e.V.: Gülzow, Germany, 2007, 139-144.

Chakraborty, A.; Sain, M.; Kortschot, M. *Holzforschung* 2005, *59*, 102-107.

Dufresne, A.; Dupeyre, D.; Vignon, M.R. 2000. *J Appl Pol Sci* 2000, *76*, 2080-2092.

Fahlén, J.; Salmén, L. *Plant Biology* 2002, *4*, 339-345.

Iwamoto, S.; Nakagaito, A.N.; Yano, H.; Nogi, M. *Appl Phys A* 2005, *81*, 1109-1112.

Nakagaito, A.N.; Yano, H. *Appl Phys A* 2005, *80*, 155-159.

Nova Institut *Reinforced Plastics* 2008, *52 (4)*, 18-25.

Purz, H.J.; Fink, H.P.; Graf, H. *Das Papier* 1998, *6*, 315-324.

Van den Oever, M.J.A.; Bos, H.L.; Van Kemenade, M.J.J.M. *Appl Comp Mat* 2000, *7*, 387-402.

Zimmermann, T.; Pohler, E.; Geiger, T. *Advanced Engineering Materials* 2004, *6*, 754-761.

In: Renewable Resources: Obtaining, Processing... ISBN 978-1-6060692-534-8
Editors: R. Kozlowski, G. Zaikov and F. Pudel

Chapter 17

FLUIDIZED BED EXTRACTION OF SPICE PLANTS USING SUPERHEATED STEAM

Carmen Behns, Wolfgang Behns, Hartmut Haida, Sabine Schlüsselburg and Lothar Mörl
Otto-von-Guericke-University Magdeburg, Institute of Process Equipment and Environmental Technology, 39016 Magdeburg, Germany

ABSTRACT

New results of experimental investigations on the fluidized bed extraction of plant materials using superheated steam are presented. This process enables e. g. the production of ethereal oils from leaves or seeds of appropriate plants. In particular the influence of the steam mass flow on the extraction is pictured. Thyme leaves serve as example for spice plants. The investigations are performed by means of a fluidized bed pilot plant ND 200 with steam circulation. The extraction rate is estimated using an FID-signal which is representing the total hydrocarbon concentration in the vapour phase.

INTRODUCTION

The process of the fluidized bed extraction of spice plants using superheated steam has been presented in former NAROSSA-conferences [1]. It can be used for the production of essential oils from suitable pretreated plant materials.The experimental investigations of this process have been continued. Some new results are presented in the following.

EXPERIMENTAL INVESTIGATIONS AND RESULTS

The experimental investigations are performed by means of a fluidized bed pilot plant DN 200 with steam circulation – Figure 1. The extraction is a batch process. The gas distributor bottom can be turned so that the whole bed mass can be removed immediately. A portion of the vapor phase is condensed. The condensate is a mixture of water and oil. An oil phase can be decanted. The watery phase also contains essential oils.

Figure 1. Fluidized bed extraction pilot plant DN 200 with steam circulation.

Figure 2. Sight glass below the condenser with the condensate and a swimming oil phase.

Figure 2 shows the condensate and a swimming oil phase in the sight glass below the condenser. The course of extraction can be observed measuring the total hydrocarbon concentration in the vapor phase by means of a flame ionization detector (FID, company Eismann and Stöbe) – Figure 3. Dried thyme leaves separated from the stems have been used as plant material for the extraction experiments. This material was also pretreated by pelletizing in order to achieve a better fluidization behaviour and so that the extractability is not restricted.

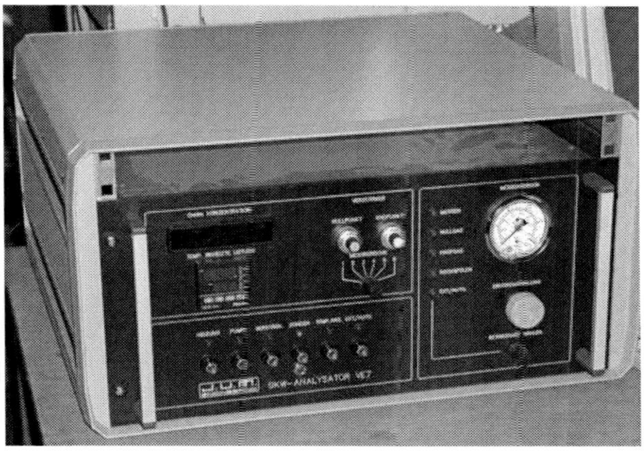

Figure 3. Flame ionisation detector (FID) for the measurement of the total hydrocarbon concentration in the vapour phase.

A laboratory flat die pelleting press (company Amandus Kahl, type 14-175) was brought into action for this pretreatment - Figure 4. The diameter of the cylindrical press channels in the die-plate was 4 mm and the length 25 mm.

Figure 4. Laboratory flat die pelleting press.

Figure 5 shows the produced pellets from raw material with different water content. The best pelletizing could be observed in case of material with the water content 18 %. This pellets had the highest hardness and the highest oil content after the pelletizing process - Figure 6. So they were used as input material for the experiments.

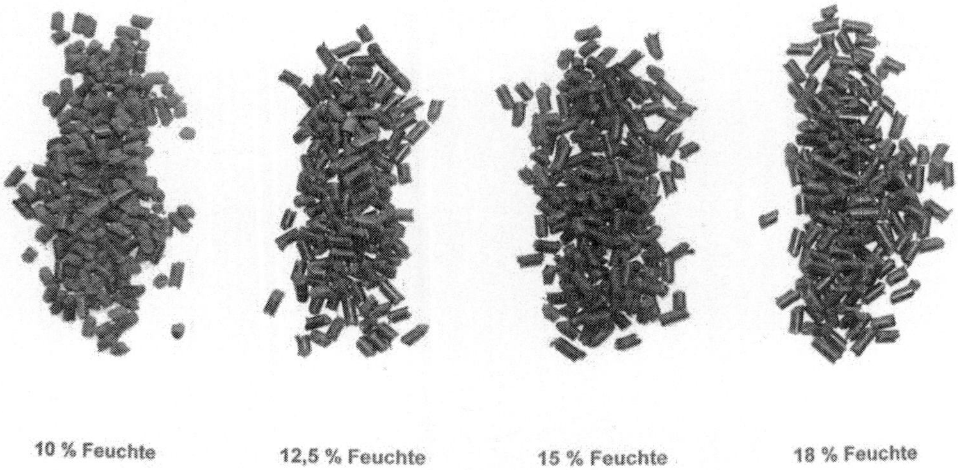

Figure 5. Pellets produced from thyme leaves with different water content.

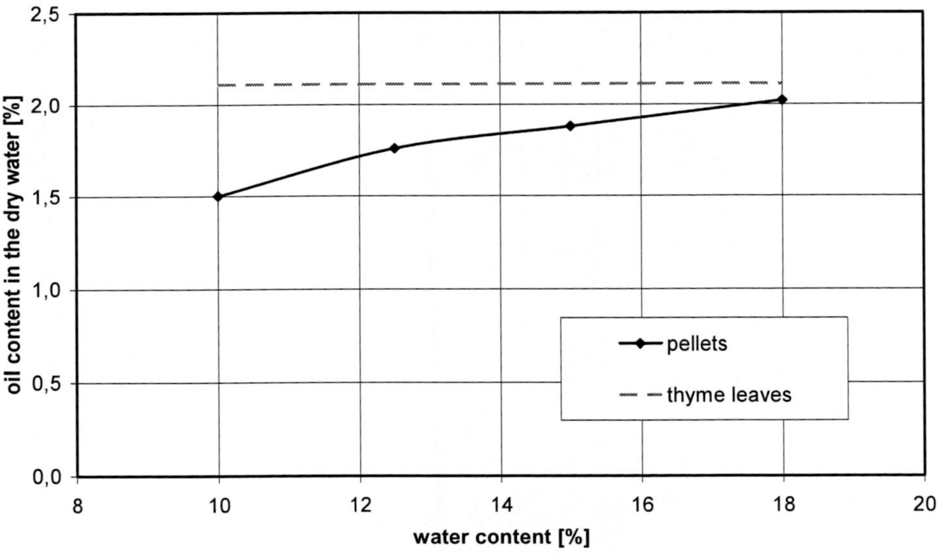

Figure 6. Oil content of pellets from thyme leaves with different water content.

Fluidized bed masses of 400 g thyme leaves or 1000 g pellets have been experimentally extracted. The experimental investigations have focused on the comparison of the extractability of dried thyme leaves and pellets and on the dependency of the extraction effectivity from the residence time in the fluidized bed and the steam flow rate.

Figure 7 represents the FID-value versus time. After a steep increase the signal is slowly decreasing. The signal tends to zero in less than 10 minutes.

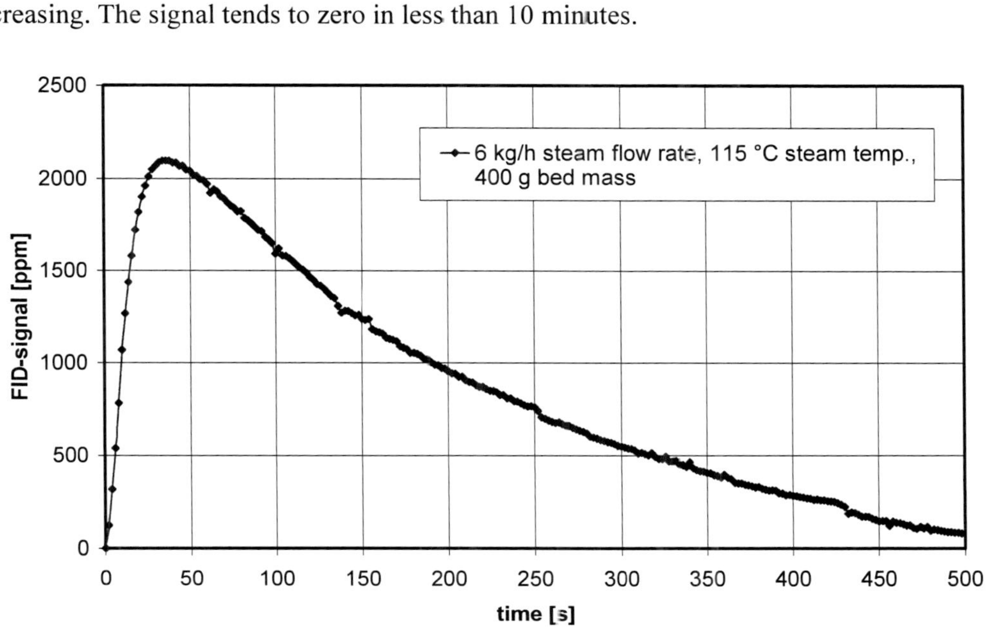

Figure 7. Typical plot of the FID-signal during the extraction of thyme leaves.

This is corresponding with the reduction of the oil content and the increase of the extraction efficiency. The interrelationship for thyme leaves is shown in Figure 8 resulting from some experiments with the same parameters.

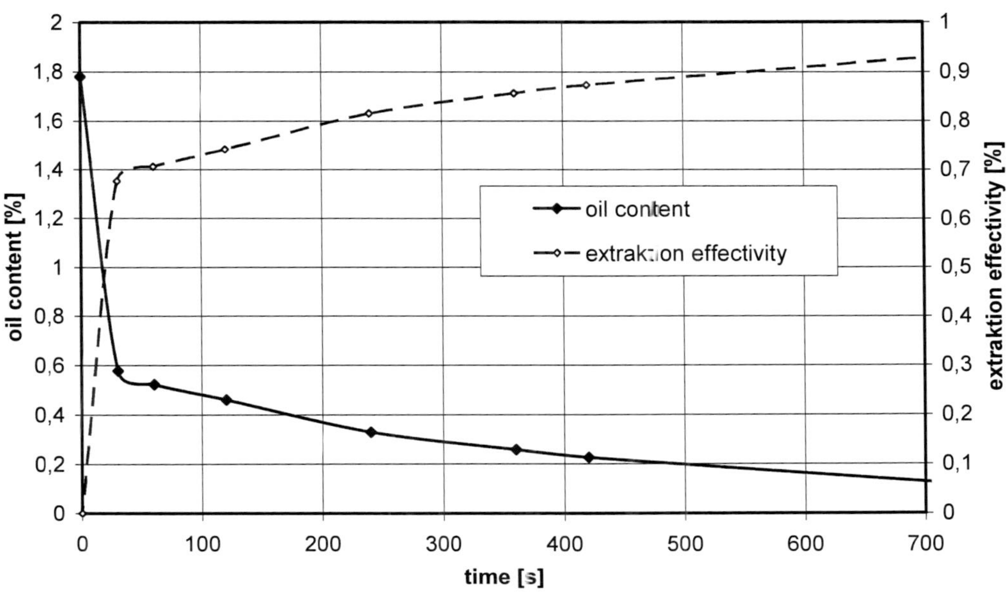

Figure 8. Typical plot of the oil content and the extraction effectivity of thyme leaves (see parameters in Figure 7).

The experimental investigations lead to the understanding that the steam flow rate in the circulation affects the extraction effectivity strongly. Figure 9 represents the FID-value versus time in case of two steam flow rates for the extraction of thyme leaves. The signal peak is higher but the signal decrease is slower in case of the smaller flow rate. The same oil content reduction can be achieved in a distinct shorter period with the higher steam flow rate.

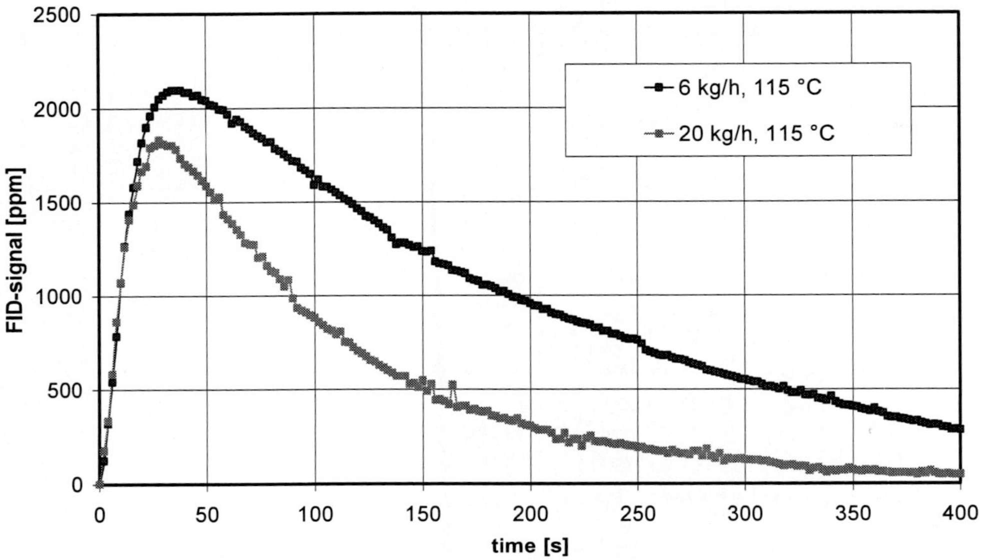

Figure 9. Comparison of the FID-signals during the extraction of thyme leaves for two steam flow rates.

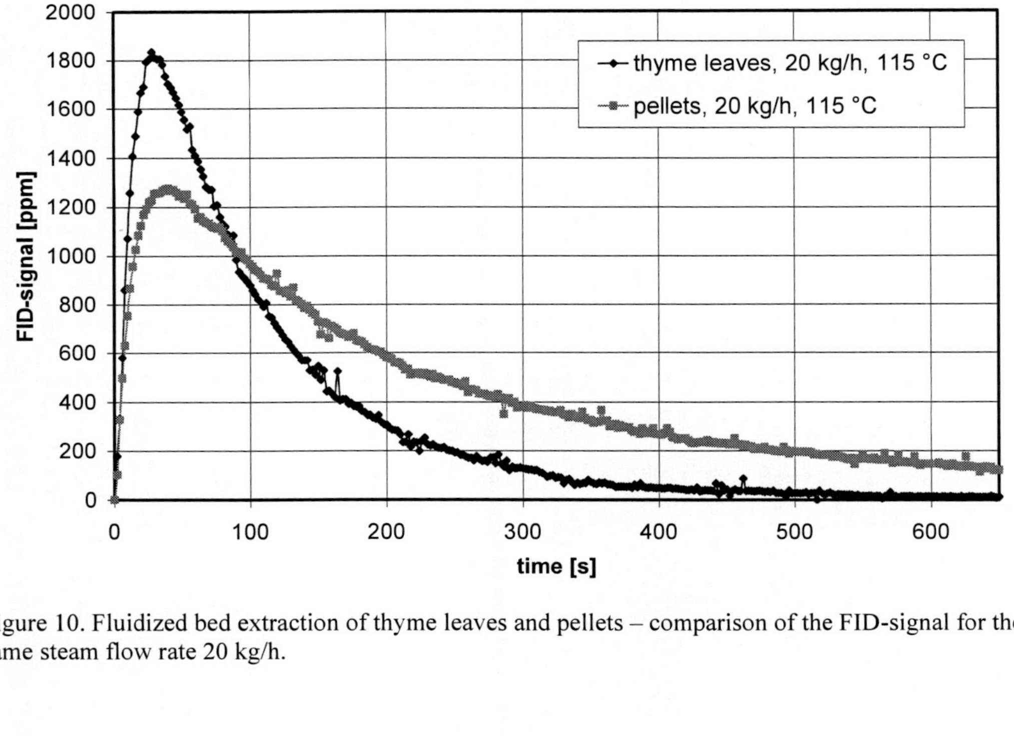

Figure 10. Fluidized bed extraction of thyme leaves and pellets – comparison of the FID-signal for the same steam flow rate 20 kg/h.

The extraction of thyme leaves and pelletized thyme leaves was compared using the high steam flow rate (20 kg/h). The measured FID-value versus time is shown in Figure 10. In case of the pellets the signal peak is clear smaller and also the decrease is slower. So the time period is clear higher to achieve the same extraction effectivity. This is arising out of the higher mass transfer resistances in the pellets. The mass transfer area is reduced by the pelletizing. But it must be observed that the input mass of the pilot plant in case of pellets is more than twice compared to the thyme leaves.

Conclusion

Essential oils of good quality can be produced from suitable plant materials by fluidized bed extraction using superheated steam. The process period is much shorter in comparison to conventional steam distillation. A high extraction efficiency is achieved. The residual matter is dried and free of solvents.

The pretreatment of the plant material is very important for the material handling and the extraction efficiency. The pelleting can be recommended as a pretreatment process. The material handling and the fluidization behaviour are improved. The bed mass can be increased. The disadvantage is the reduction of the extraction efficiency in case that the plant leaves contain the essential oils. The oil content is also reduced by this pretreatment.

The steam temperature influences the extraction efficiency only in a small scale. The temperature 115 °C emanates from the investigations as optimal. The steam flow rate has considerable effects on the extraction efficiency. The optimal steam flow rate is derived from economic aspects.

Reference

[1] Mörl, L. et al.In *Renewable resources and plant biotechnology*; Kozlowski, R., Zaikov, G. E., Pudel, F. Eds.; Nova Science Publishers - New York, NY : - 2006, pp 69-76.

In: Renewable Resources: Obtaining, Processing... ISBN 978-1-6060692-534-8
Editors: R. Kozlowski, G. Zaikov and F. Pudel

Chapter 18

RED AND PURPLE NATURAL DYESTUFSS AS AN EXCELENT UV BARRIER

Katarzyna Schmidt-Przewoźna* and Jakub Kowaliński
Institute of Natural Fibres, 60-630 Poznan, Poland

ABSTRACT

This chapter is a study of red colours and shades obtained from natural source an their application on linen and silk fabrics . Although most natural dyes come from plant sources, some of historically important dyes were derived from shellfish and insects.

The presentation concerns research carried out as part of the research project concerned with the scientific basis of the cultivation of the natural dyeing plants and creation of the collection of data on existing natural dyes and production process. The results of our study will present large variation in hue and the influence of the type of mordants on colour. Mordants and dyes will have different effects depending on the fibres used (protein or cellulosic). The method of preparation of the fibre will also influence the final colour. The examples used in illustrating this paper will show the range variations to be achieved while studying the extracts from the: Madder Rubia tinctorium L., Safflower Carthamus tinctoria L., Madder Rubia tinctoria L., Indian madder Rubia cardifolia L., Sappanwood (red wood) Calsapinia sappan and insects dyes: Kerria lacca Kerria laccae, Cochineal Coccus cacti in the combination with a broad spectrum of alternative mordants.

1. ETYMOLOGICAL MEANING OF RED AND HISTORICAL BACKGROUND

Red colour plays a great emotional and esthetic role, which genesis dates back to the ancient times. Its positive value is heritage of ancient crimson, symbolizes imperial and priestly power. Red is flame, blood, passion and fertility, power, royalty, majesty and

* Ph: 48 61 8455828, Fax: 48 61 841 78 30, kasia@inf.poznan.pl

sacrifice. The term red comprised also: red itself, crimson and purple. Each age has a different perception of red colour.

Red as a colour has a bright past - „colorantus" in Latin and „colorado" in Spanish, „krasnyj" in Russian all mean both „colourful" and „red". In folk culture everything red is considered beautiful. Besides it is colour of magic, challenge and provocation.

In the ancient Rome crimson robes were reserved for the Emperor and the most important officials. It is also associated with Greek-Roman gods of sun (Ares, Dionizos, Hefajstos). In the ancient times it was one of the three ritual colours along with black and white. It was used for painting the body and dyeing clothing – it served as a symbol of vital power. To honour Mars, faces of Roman leaders were painted red and their bodies were covered with red togas to celebrate their triumph. Red was believed to protect against demons and danger; this attribute has persevered until present days in traditional folk customs.

Crimson Symbolizes Justice and Royal Power

Red dyes are extracted from either animal or plant raw material – mostly insects and various part f plants – flower, leaves, roots and bark.

One of the main sources of crimson were sea snails from Mediterranean: *Murex brandaris, Murex trunkulus, Purpurea haemostoma.* The first records on dyeing with Tyros crimson date back to 1600 r. B.C. The Fenitians first discovered the dyeing properties of secretion of the mussels. On Crete in Knossos Palace, slates were discovered with inscriptions from 15 century B.C. with the words - crimson *(porpureja)* –Tyrian purple and plant red *(phoinikos).*

Tyrian purple, the colour of emperors, bishops and magnates caused almost complete extinction of the mussels in the ancient times.

Crimson used to symbolize royalty, magnates and noblemen. In Europe red colour was extracted from *Porophyropora polonica* L., which was very famous in the middle ages and Reinessance both in Europe and Asia. The insect used to live on the roots of *Scleranthus perrennis* which grew in northern Europe and Poland. This insect was of great economic importance for Poland. Magnates and noblemen wore robes dyed with *Porophyropora polonica* L., which gave very bright and fast red colour. The poorer classes had to satisfy themselves with red from madder, although it was not equally bright.

Since the discovery of America, the importance of *Porophyropora polonica* L gradually became smaller, since it was lees efficient source of red dye than Cochineal *Cocus cacti* brought from The New World. The insects occurred on some cacti species in Mexico. The dyestuff of animal origin, known in the ancient times and the Middle ages in the Mediterranean area was *Kermes vermilio* (Planchon 1864). The insect lived on oak trees *Quercus coccifera* and *Quercus ilex.* It was often mistaken for *Porophyropora polonica* L.

The dye that revolutionized dyeing is cochenile obtained from dried females of the insect *Coccus cacti.* These insects live in Mexico, South america, India and Algier. The active ingredient of the dye constitutes carmine acid. The same dye is present both in *Kermes vermilio* and *Porophyropora polonica* L but in lower concentrations.

The most widely known plant dye is madder *Rubia tinctorum* L., commonly referred to as Turkish red. It is one of the oldest dyestuffs – it was used in ancient Egypt, India and Persia. It was grown in India 3000 years B.C.. Madder allows to achieve a wide range of colours from warm pinks, through reds, carmine and purples. Madder root contains alizarine, that is the red dyestuff. One of the red dyes often used in Poland came from trunks of the Indian tree *Acacia catechu.*

„Orseil", extracted from lichens, was dyestuff brought from Sardinia and Corsica in barrels. Another plant source of red dye was a South american red tree called "fernabuko" *Corealpinia barsilensts.*

Blackberry *Vaccinum myrtillu* was one of Polish plants giving fast fading effects, safflower *Karthamus tinctorius*, saffron *Crocus sativus* and red autumn stems of nettle.

2. Chemical Classification of Red Natural Dyes

Almost all natural red dyes have a basic quinone structure. They are mainly antraguinones – alizarin, purpurin, munjistin, laccaic and carmine acid, or napthoguinones – alkannin, and benzoquinonic dye – catrhamin.(11)

4. Materials and Methods

4.1. Fabrics

100% linen after boiling
100% silk after boiling
Method of preparation of the fabric for dyeing:
Linen and silk fabric are washed in soda soap.

4.2. Natural Dyestuffs

Natural dyes are dyes obtained from plants, animals, vegetable. They are being looked as as an eco-solution and eco friendly dye. Therefore, natural dyes were selected for the study.

Natural sources of red colour used in our treatments are:

- Madder Rubia tinctorum L.,
- Cuth Accacia catechu,
- Logwood Haematoxylum campechianum L.,
- Sappanwood (red wood) Calsapinia sappan
- insects dyes: Kerria lacca Kerria laccae, Cochineal Coccus cacti.

4.3. Extraction of Dye

The plants are crused to small pices and soaked in hot water overnight, boiled one hour and filtrated.

Mordants:

In our methods we used:

- Sodium carbonate anhydrous
- Copper sulphate
- Citric acid
- Iron - ferrous sulphate
- Alum - potassium aluminium sulphate

5. Development of Colour on Linen and Silk Fabric

Equipment:
Laboratory dyeing machine: EASYKROME UGOLINI

Fabrics:

Fabric A - 43002 thin linen
Fabric B - 30187 thick linen
Silk A – silk knitwear 100%
Silk B –silk shantung 100%

Table 1. Cochineal dyeing results- no mordant, Copper sulphate

No.	Fabrics	Dyestuff	Mordant	Colour	Remarks
1.	FabricA linen	Cochineal	No mordant	Misty rose	Unevenly dyeing
2.	Silk B	Cochineal	No mordant	Rosy brown	Evenly dyed, bright colour
3.	FabricA linen	Cochineal	Copper sulphate	Green gray	Evenly dyed
4.	Silk B	Cochineal	Copper sulphate	Dim gray	Evenly dyed

Table 2. Cochineal dyeing results – washing soda, citric acid

No.	Fabrics	Dyestuff	Mordant	Colour	Remarks
1.	FabricA linen	Cochineal	washing soda	Medium orchid	Evenly dyed, good colour
2.	Silk B	Cochineal	washing soda	Dark violet	Evenly dyed, good colour
3.	FabricA linen	Cochineal	Citric acid	Dark salmon	Evenly dyed, good colour
4.	Silk B	Cochineal	Citric acid	maroon	Evenly dyed, good colour

Table 3. Sappanwood *Calsapinia sappan* dyeing results – no mordant

No.	Fabrics	Dyestuff	Mordant	Colour	Remarks
1.	FabricA linen	Sappanwood	No mordant	Burly wood	Evenly dyed
2.	FabricB linen	Sappanwood	No mordant	Burly wood	Evenly dyed
3.	FabricB linen premordant	Sappanwood	No mordant	Deep burly wood	Evenly dyed
4.	Silk A	Sappanwood	No mordant	Burly wood	Evenly dyed

Table 4. Sappanwood *Calsapinia sappan* dyeing results – copper sulphate

No.	Fabrics	Dyestuff	Mordant	Colour	Remarks
1.	FabricA linen	Sappanwood	Copper sulphate	Burly wood	Evenly dyed
2.	FabricB linen	Sappanwood	Copper sulphate	Burly wood	Unevenlu dyeing
3.	FabricB linen premordant	Sappanwood	Copper sulphate	Burly wood	Evenly dyed
4.	Silk A	Sappanwood	Copper sulphate	Dark goldenrot	Evenly dyed

Table 5. Sappanwood *Calsapinia sappan* dyeing results – ferrous sulphate

No.	Fabrics	Dyestuff	Mordant	Colour	Remarks
1.	FabricA linen	Sappanwood	Ferrous sulphate	Burly wood	Evenly dyed
2.	FabricB linen	Sappanwood	Ferrous sulphate	Burly wood	Evenly dyed
3.	FabricB linen premordant	Sappanwood	Ferrous sulphate	Dark burly wood	Evenly dyed
4.	Silk A	Sappanwood	Ferrous sulphate	Dark khaki	Evenly dyed

Table 6. Sappanwood *Calsapinia sappan* dyeing results – potassium aluminium sulphate

No.	Fabrics	Dyestuff	Mordant	Colour	Remarks
1.	FabricA linen	Sappanwood	Alum	bisque	Evenly dyed
2.	FabricB linen	Sappanwood	Alum	bisque	Evenly dyed
3.	FabricB linen premordant	Sappanwood	Alum	bisque	Evenly dyed
4.	Silk A	Sappanwood	Alum	Light cream	Evenly dyed

Table 7. Sappanwood *Calsapinia sappan* dyeing results – washing soda

No.	Fabrics	Dyestuff	Mordant	Colour	Remarks
1.	FabricA linen	Sappanwood	washing soda	Papaya whip	Evenly dyed
2.	FabricB linen	Sappanwood	washing soda	Papaya whip	Evenly dyed
3.	FabricB linen premordant	Sappanwood	washing soda	Papaya whip	Evenly dyed
4.	Silk A	Sappanwood	washing soda	Cream	Evenly dyed

Table 8. Sappanwood *Calsapinia sappan* dyeing results – citric acid

No.	Fabrics	Dyestuff	Mordant	Colour	Remarks
1.	FabricA linen	Sappanwood	citric acid	Light peach puff	Evenly dyed
2.	FabricB linen	Sappanwood	citric acid	Light peach puff	Evenly dyed
3.	FabricB linen premordant	Sappanwood	citric acid	Light peach puff	Evenly dyed
4.	Silk A	Sappanwood	citric acid	sienna	Evenly dyed

Table 9. Kerria lacca dyeing results – no mordant

No.	Fabrics	Dyestuff	Mordant	Colour	Remarks
1.	FabricA linen	Kerria lacca	No mordant	Light heather	Unevenly dyeing, patchy dyeing
2.	FabricB linen	Kerria lacca	No mordant	heather	Evenly dyed
3.	FabricB linen premordant	Kerria lacca	No mordant	Dark heather	Good colour, Unevenly dyeing
4.	Silk A	Kerria lacca	No mordant	Deep maroon	Unevenly dyeing, patchy dyeing

Table 10. Laka dyeing results – copper sulphate

No.	Fabrics	Dyestuff	Mordant	Colour	Remarks
1.	FabricA linen	Kerria lacca	Copper sulphate	Dark violet	Unevenly dyeing, patchy dyeing
2.	FabricB linen	Kerria lacca	Copper sulphate	Dark violet	Unevenly dyeing, patchy dyeing
3.	FabricB linen premordant	Kerria lacca	Copper sulphate	purple	Unevenly dyeing, patchy dyeing
4.	Silk A	Kerria lacca	Copper sulphate	Brown violet	Unevenly dyeing, patchy dyeing

Table 11. Laka dyeing results – ferrous sulphate

No.	Fabrics	Dyestuff	Mordant	Colour	Remarks
1.	FabricA linen	Kerria lacca	Ferrous sulphate	Brown silver	Unevenly dyeing
2.	FabricB linen	Kerria lacca	Ferrous sulphate	Brown silver	Unevenly dyeing,
3.	FabricB linen premordant	Kerria lacca	Ferrous sulphate	gray	Unevenly dyeing
4.	Silk A	Kerria lacca	Ferrous sulphate	Brown gray	Unevenly dyeing

Table 12. Laka dyeing results – potassium aluminium sulphate

No.	Fabrics	Dyestuff	Mordant	Colour	Remarks
1.	FabricA linen	Kerria lacca	Alum	Light heather	Unevenly dyeing, patchy dyeing
2.	FabricB linen	Kerria lacca	Alum	Light heather	Unevenly dyeing, patchy dyeing
3.	FabricB linen premordant	Kerria lacca	Alum	Dark heather	Unevenly dyeing, patchy dyeing
4.	Silk A	Kerria lacca	Alum	heather red	Unevenly dyeing, patchy dyeing

Table 13. Laka dyeing results – washing soda

No.	Fabrics	Dyestuff	Mordant	Colour	Remarks
1.	FabricA linen	Kerria lacca	Washing soda	Dark violet	Unevenly dyeing, patchy dyeing
2.	FabricB linen	Kerria lacca	Washing soda	Dark violet	Unevenly dyeing, patchy dyeing
3.	FabricB linen premordant	Kerria lacca	Washing soda	Deep dark violet	Unevenly dyeing
4.	Silk A	Kerria lacca	Washing soda	Claret-coloured	Unevenly dyeing

Table 14. Laka dyeing results – citric acid

No.	Fabrics	Dyestuff	Mordant	Colour	Remarks
1.	FabricA linen	Kerria lacca	Citric Acid	coral	Unevenly dyeing, patchy dyeing
2.	FabricB linen	Kerria lacca	Citric Acid	Dark pink coral	Unevenly dyeing, patchy dyeing
3.	FabricB linen premordant	Kerria lacca	Citric Acid	coral	Unevenly dyeing, patchy dyeing
4.	Silk A	Kerria lacca	Citric Acid	chestnut	Unevenly dyeing

Table 15. Laka dyeing results – no mordant

No.	Fabrics	Dyestuff	Mordant	Colour	Remarks
1.	FabricA linen	Kerria lacca	No mordant	Light violet	Unevenly dyeing, patchy dyeing
2.	FabricB linen	Kerria lacca	No mordant	Light violet	Unevenly dyeing, patchy dyeing
3.	FabricB linen premordant	Kerria lacca	No mordant	Light violet	Unevenly dyeing, patchy dyeing
4.	Silk A	Kerria lacca	No mordant	Deep red	Very good colour, Unevenly dyeing, patchy dyeing

Table 16. India madder dyeing results – no mordant

No.	Fabrics	Dyestuff	Mordant	Colour	Remarks
1.	FabricA linen	India madder	No mordant	Pink bisque	Evenly dyed
2.	FabricB linen	India madder	No mordant	Pink bisque	Evenly dyed
3.	FabricB linen Premordant	India madder	No mordant	Pink bisque	Evenly dyed
4.	Dzianina jedwab 100%	India madder	No mordant	Light copper-coloured	Evenly dyed

Table 17. India madder dyeing results – copper sulphate

No.	Fabrics	Dyestuff	Mordant	Colour	Remarks
1.	FabricA linen	India madder	Copper sulphate	Peach	Evenly dyed
2.	FabricB linen	India madder	Copper sulphate	Ligot salmon	Evenly dyed
3.	FabricB linen premordant	India madder	Copper sulphate	Peach	Evenly dyed
4.	Silk A	India madder	Copper sulphate	Dark goldenrot	Evenly dyed

Table 18. India madder dyeing results – ferrous sulphate

No.	Fabrics	Dyestuff	Mordant	Colour	Remarks
1.	FabricA linen	India madder	Ferrous sulphate	Navajo white	Evenly dyed
2.	FabricB linen	India madder	Ferrous sulphate	Navajo white	Evenly dyed
3.	FabricB linen premordant	India madder	Ferrous sulphate	Dark navajo white	Evenly dyed
4.	Silk A	India madder	Ferrous sulphate	Light brown	Evenly dyed

Table 19. India madder dyeing results – potassium aluminium sulphate

No.	Fabrics	Dyestuff	Mordant	Colour	Remarks
1.	FabricA linen	India madder	Alum	Light pink	Evenly dyed
2.	FabricB linen	India madder	Alum	Light salmon	Evenly dyed
3.	FabricB linen premordant	India madder	Alum	Light Peach	Evenly dyed
4.	Silk A	India madder	Alum	Light goldenrot	Evenly dyed

Table 20. India madder dyeing results – washing soda

No.	Fabrics	Dyestuff	Mordant	Colour	Remarks
1.	FabricA linen	India madder	washing soda	Peach puff	Evenly dyed
2.	FabricB linen	India madder	washing soda	Peach puff	Evenly dyed
3.	FabricB linen premordant	India madder	washing soda	Peach puff	Evenly dyed
4.	Silk A	India madder	washing soda	Deep peach puff	Evenly dyed

Table 21. India madder dyeing results – citric acid

No.	Fabrics	Dyestuff	Mordant	Colour	Remarks
1.	FabricA linen	India madder	Citric acid	Cream	Evenly dyed
2.	FabricB linen	India madder	Citric acid	Cream	Evenly dyed
3.	FabricB linen premordant	India madder	Citric acid	Cream	Evenly dyed
4.	Silk A	India madder	Citric acid	Light brown	Evenly dyed

CONCLUSIONS

- In our experimental work we have obtained beautiful shades of red colour: light pink, pink, indian red, light coral, red, crimson, orchid, coral, hot pink, purple, magneta, violet.
- Different mordants give differences in colours.
- The changes of hue and shades in application of the same dyes were a result of using different mordants
- In treatments with mordants as: alum, copper sulphate, citric acid, those colours were obtained: lilac, dark lilac, violet, pink, rose, dark rose, peach.
- Without mordants it is not possible to obtain good colour with use of Cochineal.
- The same natural dyestuffs can give different colour on different fabric (silk and linen)
- Application of mordants on linen guarantied better colour reasullt specialy on linnenB samples

UPF Protection on Linen and Silk

The Laboratory of Physiological Influence of Textiles on Human Body has done research to compare the result of UPF protection on linen and silk samples dyed with natural dyes.

The transmission, absorption, and reflection of UV radiation are in turn dependent on the fibre, fabric construction (thickness and porosity) and finishing. Many dyes used in finishing process, absorb UVR. Darker colors of the same fabric type (black, navy, dark red) will usually absorb UVR more than light pastel shades and consequently will have a higher UPF rating. In this study were analyzed all shades of red and purple applied on silk and linen.

Determination of the UVR transmission of a dry textile was done in accordance to Australian/New Zealand Standard and British Standard for sun protection clothing with the use of Cary 50 Solascreen apparatus.

Table. UPF classification system (according to the Australian Standard)

UPF RANGE	UVR protection category	UPF Ratings
15 - 24	very good but insufficient protection	15, 20
25 - 39	very good protection	25, 30, 35
40 – 50, 50+	excellent protection	40, 45, 50, 50+

The study wads concluded according to: PN– EN ISO 13758-1:2002

Fabric A - 43002 thin linen
Fabric B - 30187 thick linen
Silk A – silk knitwear 100%
Silk B –silk shantung 100%

The result of the study are presented in the table 22.

Table 22. The results of UPF on linen and silk samples dyed by natural dyestuffs

Samples	Natural Dyestuff	Mordant	UVA	UVB	UPF
Fabric A- linen		No mordant	13,828	19,205	5
Fabric B- linen		No mordant	3,966	2,878	20
Silk A		No mordant	21,911	27,249	0
Fabric A - linen	Laca	No mordant	7,948	7,821	10
Fabric A - linen	Laca	Copper Sulphate	3,270	3,334	20
Fabric A - linen	Laca	Alumn	8,115	7,821	10
Fabric A - linen	Laca	Washing soda	2,951	2,771	30
Fabric A - linen	Laca	Citric acid	6,043	6,297	10
Fabric B - linen	Laca	Copper Sulphate	1,326	1,299	45
Fabric B - linen	Laca	Washing soda	1,823	1,714	35
Fabric B - linen	Laca	Citric acid	1,966	2,133	35

Samples	Natural Dyestuff	Mordant	UVA	UVB	UPF
Silk A	Laca	No mordant	15,825	14,704	5
Silk A	Laca	Copper Sulphate	13,961	13,096	5
Silk A	Laca	Alumn	12,001	11,602	5
Silk A	Laca	Washing soda	13,346	11,422	5
Silk A	Laca	Citric acid	14,499	13,370	5
Fabric A - linen	India Madder	No mordant	5,536	5,433	15
Fabric A - linen	India Madder	Washing soda	5,889	5,460	10
Fabric B - linen	India Madder	No mordant	0,922	0,992	50+
Fabric B - linen	India Madder	Washing soda	1,058	1,089	50+
Jedwab A- linen	India Madder	No mordant	11,564	11,392	5
Jedwab A- linen	India Madder	Washing soda	15,704	16,267	5
Fabric A - linen	Madder	Premordant, alum	2,221	2,275	25
Fabric A - linen	Cochineal	Alum, citric acid	11,021	9,519	5
Silk B	Cochineal	Alum + citric acid	7,097	8,230	10
Fabric A - linen	Cochineal	Alum+ ferrouse sulphate	8,375	8,293	5
Silk B	Cochineal	Alum+ferrouse sulphate	3,645	3,885	20
Fabric A - linen	Cochineal	Alum	8,171	8,177	10
Silk B	Cochineal	Alum	4,824	5,125	15
Fabric A - linen	Cochineal	Citric acid	10,165	10,163	5
Silk B	Cochineal	Citric acid	1,857	1,781	35

CONCLUSION

- The value of UPF linen and silk fabrics depends on: product structure, density of thread, thickness, kind of used dyestuffs, colour and kind of fabrics
- The result of the comparison of UPF on linen and silk fabrics:

Excellent UVR protection we obtained on samles:

- India madder (linen Fabric B) – no mordant: +50
- India madder (linen Fabric B) –washing soda:+50
- Laka (linen Fabric B) – copper sulphate: 45
- Laka (linen Fabric B) – washing soda: 35

Very good protection we obtained on samples:

- Cochineal (silk B) - citric acid: 35
- Cochineal (silk B) - alum+ ferrous sulphate: 35

- Laka (linen Fabric A)- copper sulphate 20
- Laka (linen Fabric A) – washing soda 30
- Madder (linen Fabric A) – premordant 25

3. Comparison of the dyeing results with natural dyes shows good protection against UFV radiation on many linen samples of dense structure - fabric B and one example Fabric A –linen dyed by Madder *Rubia tinctorium* L., with special pretreatments.
4. The worse result were obtained on silk A. The white silk have UPV – 0. Dyeing treatments changed the result to 5. Better result were obtained on silk B (shantung) , the dyeing process changed UPF 5-35.
5. In comparison of the dyeing results by Cochineal on linen A and silk B by different mordants, the better result were obtained on Silk B.

REFERENCES

[1] Cantrell Mark – "*Sun – screens with style* " Textile Month February 2000.

[2] Hilfiker R., Kaufmann W., Reinert G.,Schmidt E.– „*High sun protection factors treated with UV-absorbers*" - 77th World Conference, Finland, 1996.

[3] Pailthorpe M.T. – "*Sun Protection and Apparel Textiles*" – The 3rd Asian Textile Conference, 1995.

[4] Rupp J., Bohringer A., Yonenaga A., Hilden J. – „*Textiles for protection against harmful ultraviolet radiation*" – *International Textile Bulletin* nr 6 November 2001.

[5] Sun protective clothing – Evaluation and classification – Australian / New Zealand Standard.

[6] Zimniewska M: "Linen and Hemp Fabrics as a Natural Way of Sun Protection" 2nd Global Workshop of The FAO European Cooperative Research Network on Flax and Other Bast Plants, "*Bast Plants in the New Millennium*", 3-6.06. Borovets, Bułgaria. (2001)

[7] Schmidt-Przewoźna K *Natural Dyes used in Polish Workshops in the 17-18thcentury*, DHA 22 Archetype Publications, London, U.K, 2004.

[8] Schmidt-Przewoźna K, Zimniewska M (2005) "*The effect of natural dyes used for linen fabrics on uv -blocking*"Arch –tex Conference , Krakow.

ENERGETIC USE OF BIOMASS

In: Renewable Resources: Obtaining, Processing… ISBN 978-1-6060692-534-8
Editors: R. Kozlowski, G. Zaikov and F. Pudel

Chapter 19

STUDY ON THE UTILIZATION OF JATROPHA CURCAS L. FOR POWER GENERATION PURPOSES AT THE ISABELA ISLAND (GALAPAGOS - ECUADOR)

***Christoph Wieland*[*a], *Dieter Sonnenholzner*[ab] and *Andreas Schuster*[a]**

[a]Technische Universität München, Lehrstuhl für Energiesysteme (Institute of Energy Systems) / TUM Boltzmannstr.15, 85747 Garching b. München

[b]Ergal Project, Ministry of Electricity and Renewable Energy of Ecuador, Eloy Alfaro N29-50 y 9 de Octubre, Edificio Correos del Ecuador, Quito, Ecuador

ABSTRACT

Nowadays the Ecuadorian Government is working on a series of projects to guarantee the environmental security of the Galapagos Islands. It is intended to change the power generation towards hybrid systems operated by renewable energies. In the frame of these projects, a low penetration hybrid system based on photovoltaic (400 kWp) and intern combustion engines (350 - 450 kWel) for the Isabela Island is investigated. It is intended to use vegetal oil as fuel derived from the Jatropha curcas seeds. This application could bring benefits for socioeconomically depressed rural areas at the Ecuadorian mainland by introducing them into the value chain of Jatropha oil production. More than 200’000 ha of marginal land suitable for Jatropha cultivation have been detected near the Ecuadorian coastal regions. The performed study covers social, environmental and technological aspects, ranging from the propagation of the Jatropha curcas tree in Ecuador, to the small scale production of the Jatropha oil in rural areas and the power generation with intern combustion engines at the Isabela Island. Properties of the Ecuadorian Jatropha curcas seeds, the technologies and its difficulties have been analyzed in order to quantify the Ecuadorian case.

* Corresponding Author (wieland@es.mw.tum.de)

ACRONYMS

DED	Deutscher Entwicklungsdienst (German Development Service) – Ecuador.
DIN	Deutsches Institut für Normung (German Institute for Standardization)
ELECGALAPAGOS	Electricity generation company of Galapagos
ERGAL	Energías Renovables para las Islas Galápagos (Renewable Energy for the Galapagos Islands). Ministry of Electricity and Renewable Energy Ecuador.
Genset	Generation Set.
IICA	Inter-American Institute for Cooperation on Agriculture (Ecuador).
INIAP	Instituto Nacional Autónomo de Investigaciones Agropecuarias (National Autonomous Institute of Agricultural Investigation, Ecuador)
MEER	Ministry of Electricity and Renewable Energy of Ecuador
TUM	Technische Universität München.

INTRODUCTION

The Archipelago of Galapagos has 14 islands and 9 Islets situated approx. 965 km away from the Ecuadorian coast in South America. Since 1978 it has been considered by the UNESCO as a World Heritage Site, due to its unique biodiversity. Every year the islands demand around 43.5 Million liters of fossil fuels, of which 7.8 Millions (18%) are designated for power generation. That's because electricity is mostly being generated with diesel in all the inhabited Islands: San Cristobal (also wind 2.4 MW), Isabela, Floreana (also solar 25kWp) and Santa Cruz. In 2001 and 2002 two ecological disasters occurred when two tank ships carrying fuel from the mainland to the islands ran aground and spilled around 700'000 liters of fossil fuels causing irreparable damages to the Islands. As a consequence of that, in 2003 the Ecuadorian Ministry of Environment prohibited the environmental licenses for the construction of new fossil fuel power plants in the Galapagos Islands.

Facing this new challenge the Government of Ecuador started the project "Zero Fossil Fuels for Galapagos", which will gradually replace the use of fossil fuels with renewable energies in the Galapagos Islands within a period of 10 years (see Figure 1). For this purposes Galapagos has an average solar potential of 5.5 kWh/m^2/day and wind velocities of around 6.8 m/s.

Still the necessity for base load power generation, demands a predictable and stable energy source. Thereby in the absence of other common stable sources in the islands, the generation through combustion with biomass liquid fuels still seems to be the only reasonable way to generate the base load (avoiding also high costs on expensive power storage technologies). Willing to accept that the substitution of fossil fuels for power generation through vegetable oil would not only reduce the threat of a new catastrophe in the Galapagos Islands (lower contamination risk up to 95% biodegradable in 21 days [3]), but would also imply a reduction of 20'400 tons of CO_2 per year. Among other renewable energy projects for the Islands, a hybrid system with photovoltaic generation and vegetable oil combustion is planned for the Isabela Island.

In this work a proposal for Isabela has been designed, using the vegetable oil and environmental advantages of a promising energy plant: the Jatropha curcas L.

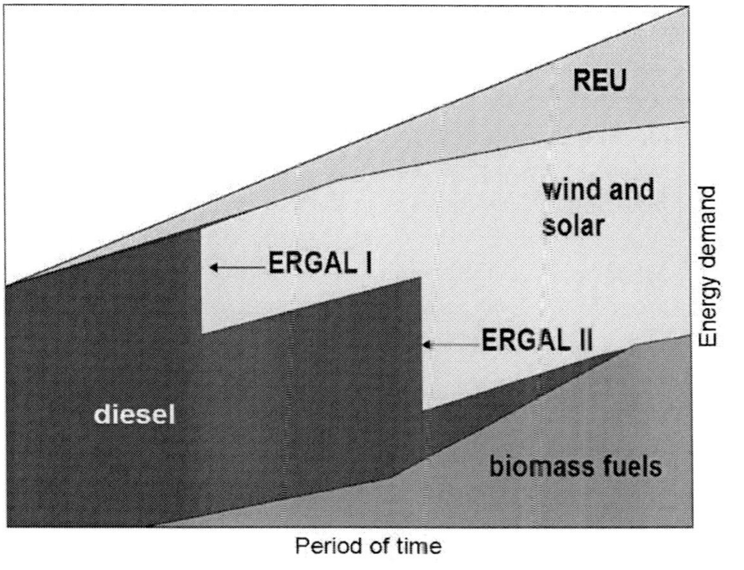

Figure 1. Substitution of fossil fuels through clean energies in two phases (ERGAL I and ERGAL II) as planned by ERGAL. Period of 10 years starting 2005. REU: Rational Energy Use [1].

Galapagos Actual DevelopmentInhabited Islands: S. Cristobal, S. Cruz, Isabela and Floreana		
Population	[Habitants]	19'600
Pop. Growth	[%/year]	2
Total fossil fuel demand	[Mio. l/year]	43.5
Fuel demand growth	[%/year]	12.9
Diesel for power gen.	[Mio. l/year]	7.8
Net power gen.	[MWh/year]	26'200
Diesel gen.	[%]	96.3
Solar (Floreana)	[%]	0.1
Wind (S. Cristobal since Sep 07)	[%]	3.7
Pow. gen. Growth (since 2001)	[%/year]	9.7
Elec. coverage	[%]	99
Diesel cost at Galapagos	[USD/liter]	0.86
Diesel Gen. Performance at Gal.	[kWh/l]	3.22
Total government subsidy (Diesel and Electricity)	[USD/kWh]	0.82

Figure 2. Actual development of the Galapagos Islands (Balance for 2007). More than 60% of the fuel demand is for the maritime transport. Diesel and electricity have a high subsidy at the island. The power demand growth for 2007 was only 3.2% [2] [4] [8].

2. Jatropha Curcas L. in Ecuador

The Jatropha curcas L. specie found in Ecuador is believed to be a native plant. It exists mostly in coastal regions of Ecuador as endemic specie. There are four Provinces where a considerable presence of Jatropha has been known by the locals for decades: Manabí, Los Ríos, Esmeraldas and Loja (also existent in Galapagos but as introduced specie). The areas were Jatropha curcas was sighted, have all the same properties: Low precipitations (200 – 500 mm/m^2 year), high evapotranspiration (1300 – 1500 mm/m^2 year), average temperature of 24 -26 °C and solar global radiations of 4.0 – 5.0 kWh/(m^{2}d).

The botanical knowhow about Jatropha curcas from local farmers in some coastal regions of Ecuador is good. Most of them grew having the plant in their farms, but mostly without knowing more purposes for the plant other than the use as a "live fence" to delimit terrain for cattle or property, known as "Cerca Viva" (also rarely used for the production of soap or some alternative medical treatments for stomach or skin wounds). Around 40 km of these Jatropha hedges and near to 200'000 ha of less fertile and degraded soil have been detected at the south of Manabí [3]. Therefore the Ecuadorian government is planning to make use of this capacity, to supply the first pilot projects to generate power with Jatropha oil in the Galápagos Islands. The following observations were made after visiting plantations and interviewing experts (INIAP) [5] and farmers in the south of Manabí. Seeds and oil from this region were used for this study.

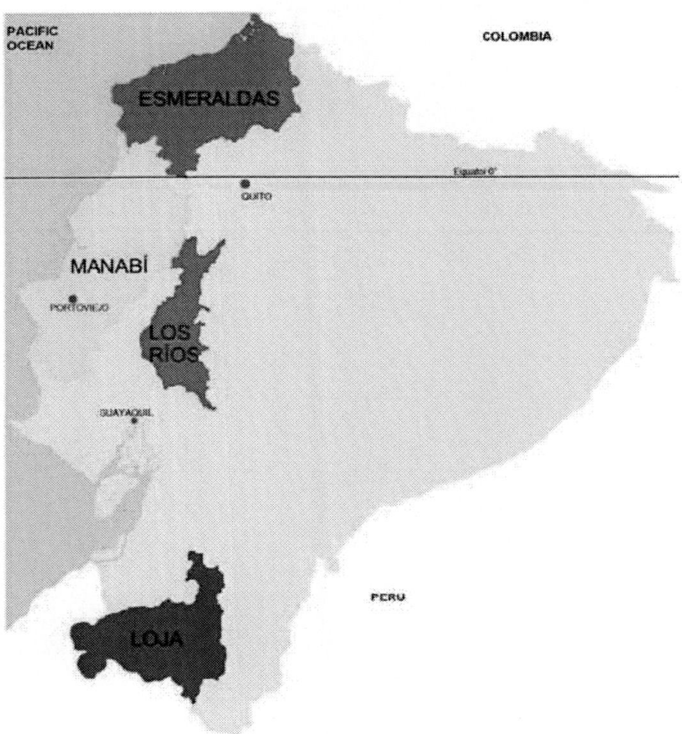

Figure 3. Map of Ecuador showing the four provinces where considerable presence of Jatropha curcas has been detected.

2.1. Manabí's Variety: Details and Observations

Species and Origin

Local famers and researchers in the field [5], assume that the Jatropha curcas is native in Ecuador. No evidence has been found yet, that applies for the theory that it was brought from Central America. Also the seeds found in Ecuador are believed to be more toxic than the ones in Central America. Even the oldest local farmers remember the tree since they were kids (also another variety of Jatropha is known by the locals as "Jatropha Pinpinfolia" (probably "Jatropha gossypifolia L."). No mutations are expected to occur among the Jatropha curcas in Ecuador because the propagation has always been done with cuttings. Still the soils in Ecuador are very different and a mutation could have been possible.

The botanical description of the plant is very similar to others made from Jatropha curcas in Africa [7] [6] (introduced from Central America into Africa and Asia in the 19th century).

Soil

In Manabí it appears that Jatropha curcas develops well in soils with at least 50 cm of depth [3] and with good drainage properties. Soils in Manabí are poor in Nitrogen, medium in Phosphor, and high on Potassium.

Roots

The root system of Jatropha curcas has 3 to 4 peripheral roots and one taproot (only with seedlings, not with cuttings).

Propagation

With cuttings the blossoming occurs 4 month after plantation and the first production of fruits is after 5 – 6 Months (cuttings are approx. 50 cm long). With seedlings fruits appear after 12 or 18 Months.

A few rare cases have been detected at the INIAP where the seedlings gave fruits after 6 months. Normally there were 3 seeds in every fruit. Depending if planted in bags (to be later transferred to the soil) or directly into the soil, seedlings had in each case approx. 90% and 80% of germination success. Seedlings planted directly in the soil grew faster than those in bags.

Breeding

The INIAP considers that for the harvesting the optimum is around 2 - 3 m height with a stem of 15 cm diameter. Therefore trees in an early state are trimmed at the stem to obtain more co-florescence, paraclade and to force so a maximum height. That way more fruits can be produced, the production per hectare increases (fruits only grow at the terminals) and the tree is easier to harvest. Also the technique called "Pinching" is applied (removing terminal buds or tips to stimulate branching and fruit production). The INIAP is trying with spacing from 3 m x 2 m with cuttings (fruits production after 5 months); 3x3 and 1x0.5 (rows x plants) with seedling. Jatropha curcas plants have been growing in live fences for decades with separations of 0.5 m or less without a problem. Depending on soil conditions, type of cultivation (cuttings or seedlings) and purposes of cultivation (oil or Erosion control), the recommended spacings[1] are [8]: 3 x 3 (1111 plants/ha); 2.5 x 2.5 (1600 plants/ha); 2 x 2 (2500 plants/ha).

Figure 4. Local farmer building a "Cerca Viva" from Jatropha with cuttings, near to the township of Colón.

Harvest and Productivity

Some experiences in Ecuador imply that the Jatropha tree reaches its maximum production per year in areas with precipitations of 500 mm/year and stays constant until 1500 mm/year where the production declines again. In the driest regions the production could be of around 0.5 ton seeds/year/ha. Best case is 5 – 6 tons seeds/year/ha. The yield of Jatropha trees is known as progressive. For example in the average, a production of 510 kg/year of seeds and 153 liters of oil/year per hectare in the first year is expected. In the 5th year, a 100% of yield per hectare of around 3400 kg seeds/year and 1020 liters of oil/year is expected (with 2500 plants/ha at 2x2 spacing) [3]. In live fences the production is higher due to fertilizers used for other crops (up to 5 kg seeds/tree/year). The first test at the INIAP in Manabí show a production of 90 g seeds/tree (144 kg seeds/ha) in the first 9 months under the following conditions: Soil type: Vertic Ustropepts; Average Temperature: 24.6 °C; Relative humidity: 82%; Precipitations: 550 mm/year; Altitude: 44m above sea level; spacing 2.5 x 2.5; no trimming; no irrigation.

Harvesting in Ecuador is done normally once a year at the end of the rainy season (April/May). If the rainy season was prolonged or the humidity in the air is high, there could also be a 2nd production in the same year (harvest in November/December before start of the rainy season). The seeds in Manabí are harvested when the fruit is dried out and still hanging from the tree. This is due to the fact that when the fruit dries, it can easily be destroyed with the hands to obtain the seeds inside. The time of harvesting for the fruit has an influence on the oil content and quality. Testimonies of the oldest farmers in the region imply that the plant could be productive for approx. 50 to 60 years.

[1] Physical separation between cultivated plants; a x b a: distance between rows b: distance between plants in meters.

Wood

The wood of the Jatropha curcas could be described similar to other local trees like the Erythrina crista-galli ("Ceibo"). The wood is light and has a high humidity to stand long dry seasons. It burns bad and is not used commercial.

Plagues

Although no potential plagues have been detected yet in Ecuador for Jatropha, during the selection process, some seeds were found empty (no Endosperm) and had a small hole (0.5 - 1 mm) at the Caruncle, probably from a small insect. In a few cases an insect was still in the seed. In approx. 15 kg. of dry seeds almost 100 g were damaged. There is also the suspicion about the Californian Trips ("Frankliniella occidentalis") that could be attacking the leaves of the Jatropha in Ecuador [5].

The Seeds

The locals in Manabí say that digesting 4 to 6 seeds work as a purgative (no cases of mayor toxication are known in Ecuador yet). Being this an endospermic seed, usually created by nature to store and provide food, the Jatropha curcas seeds in Ecuador have no edible purposes for any species known yet. All the components of the tree and fruits are toxic. The oil content of the seeds, depending on the ripeness of the fruit, the precipitations and soil, is around 34%, which is compared to other international references at the average under mentioned environmental conditions.

1st year	2nd year	3th year	4th year	5th year
15%	35%	65%	80%	100%

Figure 5. Average Jatropha seeds yield growth [3].

Figure 6. Jatropha curcas plant at the INIAP in Manabí. Cultivation: Cuttings, 6 months old, terminal buds pinched twice, monthly irrigation with furrow. Shorter but more vitality (more blossoming, branches and fruits).

Observations on 20 randomly picked dry seeds from Manabí were made (collected from "live fences" from different producers in Manabí). Figure 5 shows the seeds from the Raphe side (flat side) and the back side. Also the caruncle is visible. Figure 6 shows the longitudinal cut of the seed, the seed coat and the endosperm.

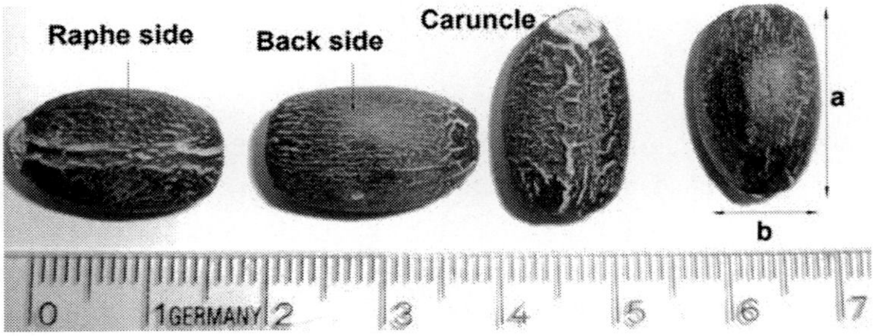

Figure 7. Analyzed Jatropha seeds. Ruler shows cm.

Fgure 8. Analyzed Jatropha curcas dry seeds from Ecuador.

Morphological analysis of the Ecuadorian seeds			
Color	Dark brown to black		
Form	Almost an ellipsoid, more flat on the raphe side and curved in the back.		
Long (a) (see Figure 5)	[mm]	15.84 - 19.90	17.89
Wide (b)	[mm]	9.84 - 11.88	11.01
Thick (c) (see Figure 6)	[mm]	7.80 - 9.90	9.13
Seed coat thickness	[mm]	0.4 - 0.6	0.45
Weight	[mg]	468.9 - 993.7	769.56
Weight without seed coat	[mg]	284.9- 664.17	492.98
Weight ratio (coat/endosperm)		38/67	
Volume	[mm^3]	5320 - 9640	7575
Volume without seed coat	[mm^3]	4175 - 7529	5938
Volume ratio (coat/endosperm)		172/625	
Density endosperm	[kg/m^3]	83.75	
Density seed coat	[kg/m^3]	171.16	

Figure 9. Morphological analysis of the Jatropha curcas seeds. Averages made from observations on 20 seeds from Manabí Ecuador.

The cotyledon and the radicle are also visible. The Volume of the seeds was approximated with the form of an Ellipsoid. The seed coat has an average volume of 21.6% and its weight represents around 36.1% of the seed.

Elemental composition of the Ecuadorian seeds		
C	[ma-%]	56.76
H	[ma-%]	8.59
N	[ma-%]	3.2
S	[ma-%]	0.32
O*	[ma-%]	26.16
Ash (P, Mg, Ca, etc. *Also O)*	[ma-%]	4.97
Gross calorific value	[kJ/kg]	26'684

Figure 10. CHNS composition of the seeds. Ash content after DIN CEN/TS 14775. * Oxygen of ash not included.

Figure 11. Left: An exceptional Jatropha curcas tree at the INIAP cultivated with seedlings and pinching, giving fruits already after 6 months. Right: Dried out fruit hanging from a Jatropha curcas tree.

2.2. Jatropha Curcas Oil Ecuador

The properties of the crude Jatropha oil worldwide vary depending on different environmental, technological and genetic aspects (soil, precipitations, breeding, harvesting etc). There is no quality management standard for the production yet, to obtain and oil with the same properties in every case. Thus, Jatropha curcas oil in Ecuador might not be the same as in India, Cape Verde or even among different producers in Ecuador. For example when harvesting, the content and quality depends on the ripeness of the fruit (if harvested too early the endosperm in the seeds would not be finished, affecting so the oil content in the seeds). Also the extraction process would have an influence on the properties (phosphorus, calcium and magnesium are expected to be mostly in the seed coat). There is also a difference between chemical extraction and using the cold press method) [5] [3]. All these parameters make it even harder to maintain a standard in decentralized oil productions.

Oil Components and Properties

As a triglyceride vegetable oil, the Jatropha oil is composed principally from a mix of fatty acids, esters (glycerin) and free fatty acids.

Properties	Jatropha curcas oil Ecuador		Standard DIN 51605 min-max
Density	[Kg/m^3]	910	900 - 930
Viscosity (kinematic)	[mm^2/s]	34.7	0 - 36
Total Impurities	[mg/kg]	47	0 -24
Phosphorus content	[mg/kg]	31.4	0 - 12
Sulfur content	[mg/kg]	-	0 -10
Water content	[mg/kg]	557	0 - 750
Ca and Mg content	[mg/kg]	43.9	0 - 20
Lower Heat Value	[kJ/kg]	36'396	36'000
Upper Heat Value	[kJ/kg]	39'529	-

Figure 12. Components and properties of the oil pressed from Jatropha curcas seeds from Manabí - Ecuador. (*) is the min. recommendable for the lower heat value for vegetal oil in adapted diesel engines (not a standard) [3].

Fatty acids content Jatropha curcas oil Ecuador			
Saturated (21%)			
Lauric	C12:0	[%]	< 0.1
Mystiric	C14:0	[%]	1.0
Palmitic	C16:0	[%]	14.1
Stearic	C18:0	[%]	6.3
Arachidic	C20:0	[%]	0.3
Mono- Unsaturated (38%)			
Palmitoleic	C16:1	[%]	0.1
Oleic	C18:1	[%]	37.8
Poly – Unsaturated (40%)			
Linoleic	C18:2	[%]	39.7
Linolenic	C18:3	[%]	0.2

Figure 13. Fatty acids composition of the Ecuadorian Jatropha curcas oil [3].

These values were obtained through experiments performed in Ecuador by the DED and ERGAL [3] and others at the Technische Universität München (Institute of Energy Systems). The values show some similarities with other experimental results in Ecuador and Germany as well. Still some properties, e.g. the phosphorus and the impurities content vary dramatically, sometimes below the standard limit and sometimes above (sometimes even same oil in different labs). These values depend much on the characteristics of the cultivation the harvesting and the extraction or cold pressing of oil. Therefore these values are difficult to keep within the standards, and the refining of the oil is needed in any case to guarantee the accomplishment of the standards.

The fatty acids with the highest ratio are the linoleic, oleic, palmitic and stearic. These acids will have the most influence on the properties of the oil. E.g. the toxicity of the oil will be higher because of the high palmitic and the oleic acid content, the flash point is raised by the stearic and the oleic and the Iodine value by the linoleic. Almost all the values are within international references for Jatropha curcas oil. Only the Oleic and the Linoleic acids are 8 points above and below the average. With this high number of unsaturated fatty acids (78%), the iodine raises with the autoxidation tendency of the oil. This results in a lower cetane number which diminishes the combustion quality but also lowers the cloud point (palm oil 50% and Castor oil 98% of unsaturated fatty acids).

3. Proposal for Ecuador's Sustainable Development around Jatropha

Although Jatropha curcas already exists in the Galapagos Islands (introduced specie), the environmental risk of developing energy crops at the islands is too high because of the very complex and fragile ecosystem. That is why the whole production of Jatropha oil needs to take place at Ecuador's mainland. For this matter a region at the south of the province of Manabí has been chosen. Around 200'000 ha of degraded land (useful for Jatropha) and more than 40 km of existing Jatropha live fences were the main reasons to choose for this region.

3.1. Social Sustainable Development

In a land where the basic basket of goods is around USD$ 495 and the average income per family is around USD$ 370 per month, every source of economical development is of great help. The situation is even worse in Manabí and its rural areas. Manabí is a province with an area of around 18'960.6 km^2 and a population of 1.3 million (rural 612'834). With 54% of poverty index, the province is one of the poorest in Ecuador (rural 82 %). If planned well introducing part of this rural agricultural population (80'000 families) into the value chain of Jatropha oil production could mean an opportunity to improve their economics. The rural development proposal is separated in three phases:

1st Phase: For the first biomass project with Jatropha in Galapagos (Floreana Island), the demand would be around 19.800 liters of Jatropha oil per year. This amount could be partially covered with the existing 40 km of Jatropha live fences. These 40 km could produce around 24'000 liters of oil (34% oil content, 1 5 m trees distance, and 3 kg seeds tree/year). Of course it is impossible to use the full potential. But if 40 % are used then the other 60% of the demand could be covered with Diesel for this first phase (dual engines as described in 4). A Jatropha live fence surrounding a hectare could produce around 800 kg/year for which an income of USD$ 112 could be achieved (14 USD for 100 kg). The oil can be produce with chemical extraction (industrial method) or with a cold pressing (also decentralized). The cost of a small cold press for approx. 1.20 liters oil/hour (6 kg cake/hour) from around 7kg of seeds is around USD$ 12'000 -15'000 (not all the oil content can be obtained with this method). The filtering and refining of the oil could be done in a more centralized production unit. The harvesting of the seeds is done manually and therefore the production is planned only for small units (even as a family business where the members of a family will do the harvesting).

2nd Phase: Once the hybrid systems with Jatropha at Galapagos are fully developed (only Isabela and Floreana considered. St. Cruz is also planned but not included here yet), the demand for vegetable oil would reach 300'000 to 900'000 liters of Jatropha oil/year (depending on the configuration of the hybrid systems). This demand could be covered with around 800 to 2500 ha, that could also be cultivated by the same farmers that started with the project (covering the whole fuel demand for Galapagos would need around 47 mill. liter of vegetable oil [2]). The production of Jatropha curcas does not demand much agricultural infrastructure, what makes it easier for all the farmers to access the production. Many small farmers at the south of Manabí own some hectares of degraded soils that are not suitable for any other crops but Jatropha.

3rd Phase: In the third phase, it is intended to incentivize the farmer organizations through government loans, to improve their production and aim the production for other international markets like the Biodiesel (of course without affecting the supply security of the islands).

Figure 14 shows the time periods for every phase. Jatropha trees need at least 5 years to produce the maximum. Therefore the phase 3 needs to be planned at least 5 years after commencing with phase 2. The production from the live fences in Phase 1 will support the Phase 2 until reaching enough production to guarantee the supply security of Galapagos.

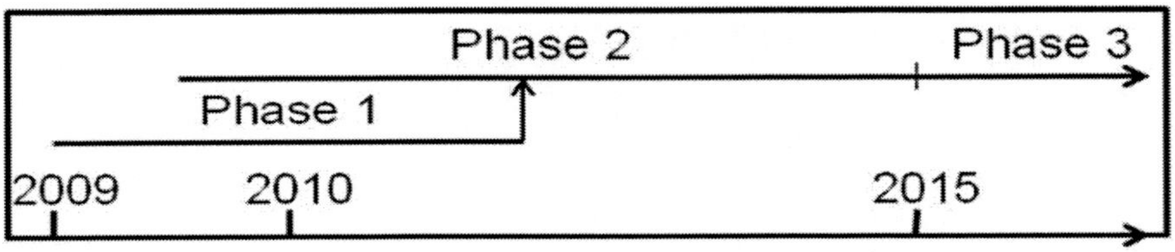

Figure 14. Planned time periods for the 3 phases of the Jatropha oil production.

Conditions for the Sustainable Development

- The phases will be accompanied by economical incentives and loans from the government to the farmers. Educational seminars will be offered to inform the farmers about the opportunities and how to improve the quality and production of the oil.
- Although a highly decentralized production is desired, this is often not the case due to economical and technological reasons. For the engines a certain fuel quality standard must be achieved. A highly decentralized system diminishes the quality of the oil. This is the main argument for a more centralized production.
- After the 2nd phase the proposal is to promote the cultivation of Jatropha for the reactivation of degraded soils. This will help degraded land to gain value again. For these purposes intensive cultivation should not be applied (also no irrigation).
- The production cost of a refined liter of Jatropha oil should be around USD$ 1.1. This value has been estimated with the actual prices and it tends to be lower when the production rises. Still the cost for the state of an imported liter of diesel is around USD$ 0.6 from which USD$ 0.3 are subsidized. This would mean that the subsidy

(normally flowing to oil companies) could now flow into the production of Jatropha oil until it reaches more competitive price levels.

- At least in the 1st phase the production should be done in small scale production units (SPU familiarly business). For this case the farmers will be organized in groups to build one small oil extraction unit (OEU), having an oil press near their farms that they will share. The oil extraction unit should be close to the seeds production unit, to return the cake to the farmer which will be used as a fertilizer (the harvesting at the SPU should be done by the family members). And so the OEU will be organized again to form a quality management and refinery unit (QRU will be the responsible unit for the quality of oil and the seeds production. Also for the administration of all the other units).

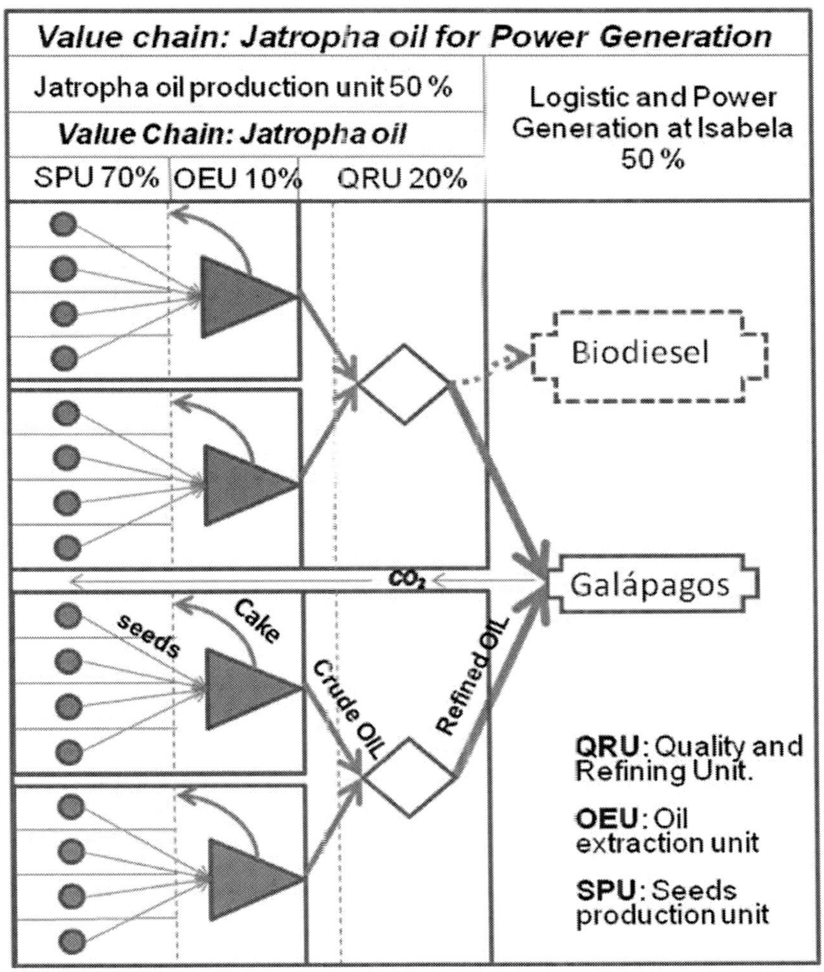

Figure 15. Value chain for the power generation at the Isabela Island and for the production of refined Jatropha oil at the mainland. The chart organizes the production units for every step. The SPE farmers (small scale seeds production units) will participate in the ownership of the oil extraction units (OEU) and the quality management and refining unit (QRU).

3.2. Resources and Environmental Sustainability

Among other problems that the world is facing like hunger, global warming and poverty, Ecuador is also one of the most deforested countries in South America and Manabí the most in Ecuador. Therefore when dealing with energy plants some conditions must be fulfilled to avoid that “the new solution” becomes “the new problem”.

1st Controversy: Alimentary security

Cause: Displacement of basic alimentary crops through more lucrative crops. The high price o n fossil fuel makes the use of alimentary crops to be more lucrative as an energy crop.

Condition 1: Taking food out of the picture (especially basic alimentary goods).

2nd Controversy: Deforestation

Cause: The search for new more fertile soils, needed for some crops to produce higher yields puts pressure on other natural forests and causes deforestation.

Condition 2: Suitability for less fertile soils.

3rd Controversy: Resources mishandling

Cause: Bad soil management and intensive cultivation leads to the degradation of the soils and could even cause desertification. Also the excessive use of water is leading to a shortage

Condition 3: Sustainable use of resources.

4th Controversy: Pollution

Cause: Not every biomass energy crops is necessarily a clean energy source. Sometimes more CO_2 or NO_x is emitted when producing or using biogenous fuels compared to the utilization of common fossil fuels.

Condition 4: Positive energy and emissions balance.

5th Controversy: Negative social impact

Cause: The unequal distribution of resources leads to an unequal access to the economic opportunities. When this situation is mixed with the energy crops it could lead to latifundism or to poor working conditions of the hand labor.

Condition 5: Accessibility for the resources, needed technology and production in small scale.

6th Controversy: Incentives vs. regulations

Cause: In a free market economy, the high oil prices will impede the accomplishment of all the conditions mentioned above.

Condition 6: A government intervention is needed to regulate who, where, when and how to produce the biofuels.

Jatropha curcas, like in every other new unexplored technology, has still many open questions (e.g. effect of Jatropha as monoculture). Still it could be said that Jatropha could fulfill all the conditions mentioned above, at least when compared to other conventional energy crops. Jatropha is suitable to be cultivated in degraded soils without irrigation, fertilizers, pesticides and less precipitation. The cultivation is easy (no need for expensive infrastructure) and it is economically accessible for the majority of the population. Jatropha has no alimentary purposes whatsoever because of its toxicity (not even the press cake, which could then be used as a fertilizer for soil. Still the option of reducing the toxicity is still open). Thus, Jatropha qualifies as a much better energy crop than the local competitors (palm oil), which would demand more resources and cause some of the controversies mentioned above.

If the conditions mentioned above are fulfilled, (including fair trade), Jatropha is currently not economically competitive compared to other oil plants like the palm oil tree which would have a higher yield and better price (relative because the needed land for palm oil tree is more expensive). Still Jatropha is one of the most productive oil plants and the potentials as a biomass fuel (biogas production with the cake, biodiesel from the oil, oil as a fuel or the seeds as a solid biomass fuel) have not been fully investigated and developed yet. But all these expectations will be clarified only with the investigations and results achieved during the next years.

Energy Equivalency

1 l Diesel 2 ~ 1.08 l Jatropha Oil
1 kg Diesel 2 ~ 1.18 kg Jatropha Oil

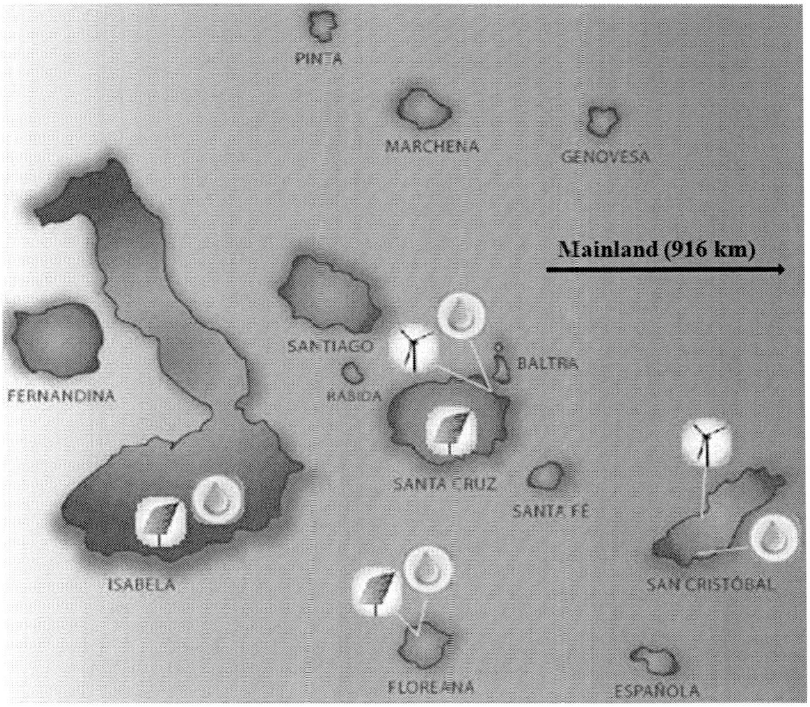

Figure 16. The Archipelago of Galápagos showing the Isabela Island and the projects under development by ERGAL [2].

4. Power Generation

As planned by ERGAL also all the fossil fuels for the Isabela Island must be replaced with alternative energy sources. The first step at Isabela is the substitution of power generation. For this Island a hybrid system with photovoltaic and Jatropha oil power generators is proposed.

4.1. The Isabela Island: Actual Situation

Demographic Situation

Situated under the Equator, the Isabela Island is the biggest Island in the Archipelago of Galapagos (5’368 km^2). With around 1800 habitants, it represents only 9% of the Galapagos population. In the 1990’s when Isabela began to use its touristic potential a dramatic increase was registered, but was stopped in 2001 by government regulations. Since then it has stabilized at a 2% population growth per year.

Weather

From January to May the average temperatures are between 24 and 27 °C. The rest of the year has average temperatures of min. 20 °C to max. 24 °C. (very rare min. 17 °C). The sun rises around 6:00 and sets at 18:00 giving approx. 12 hours (+/- 30 min) of light per day. With a 9 hour solar day, the total average global solar radiation per year is around 2028 kWh/m^2/year. The maximum average radiation for the hot seasons (March) is about 1000 W/m^2 at 12:00 pm and the maximum average in the cold seasons (July) is about 600 W/m^2 at 11:00 am.

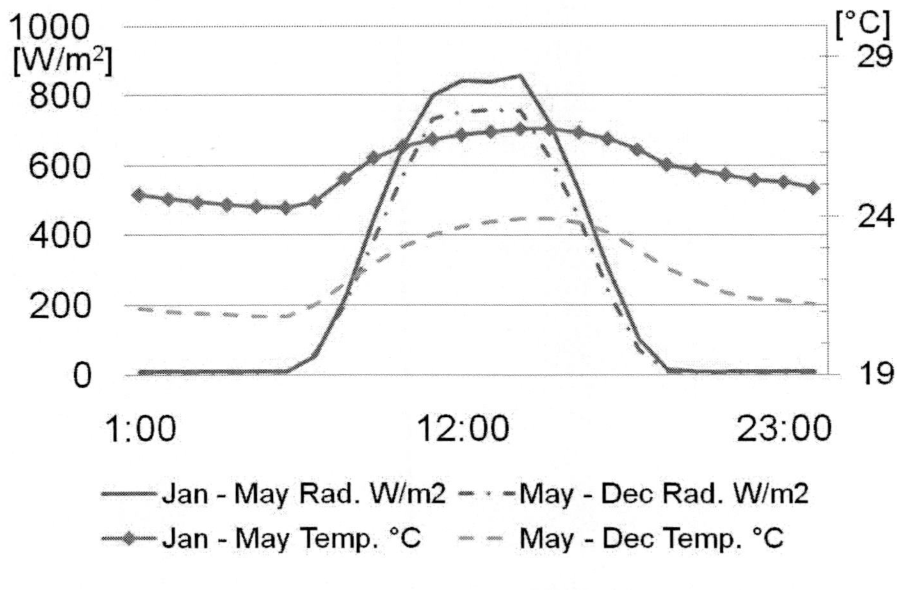

Figure 17. Average daily solar global radiation and temperature for the hot season (Jan – May) and the cold season (May – Dec).

Power Generation

Nowadays the power generation at the Isabela Island is done with Diesel Gensets. The Gensets generate around 1’889 MWh/year with a peak generation of 401 kW. For this around 640’948 liters of diesel per year are needed.

Demand and Growth

The power demand is higher in the hot season, mainly because the air conditioners and the refrigerators demand more power. The power demand growth has been calculated without the inclusion of a new water desalination plant or other new mayor consumers.

Year		2010	2015	2020	2025
Pupulation	[habitants]	1959	2209	2491	2808
Power demand +/- 10%	[MWh/year]	2084	2627	3313	4176

Figure 18. Predictions for population and power demand growth.

4.2. Proposal for the Hybrid System

At Isabela only marginal wind potential has been found [4]. Also the environmental impact of a wind turbine at Galapagos is of greater consequences due to birds and the changing panorama. Therefore a low penetration system with photovoltaic generation and adapted diesel engines (to run with Jatropha oil) is proposed. The Gensets should run on Jatropha oil as well with Diesel (Duality in case of a shortage of Jatropha oil). If the actual power generation at Isabela is replaced with renewable energies, an emission reduction of 1’669 tCO_2/year could be achieved.

Jatropha Oil Generators

Not much experiences with Jatropha oil has been achieved yet. Still the properties of the Jatropha oil (see Chart 5 and Chart 6) are similar to the ones from other oils like soya oil, from which vast information about tests in engines is available. This and the standards for vegetable oil according to DIN 51 605 give us a starting point to expect the following results:

- Generators in the needed dimensions for the Isabela island (300-500 kW), are only available as common rail or high pressure injection systems. These are very sensitive to the use of alternative fuels.
- To guarantee the power generation security, the Gensets should be able to work with Jatropha oil and Diesel as well (Dual Gensets).
- To adapt the diesel generators for Jatropha oil, the same components as for other vegetable oils (e.g. fuel feeding system, density is also the main difference) must be rebuilt.
- Due to the different fatty acids composition, Jatropha will have a different ignition timing, which must be regulated by software to reach an optimal combustion.

- The high Phosphor, Calcium and Magnesium content will cause some coking and abrasion (higher at part load). Still these elements can be reduced with the refining of the oil.
- Most of the coking and problems occur when functioning at part load or at the cold starting of the Gensets. At full load also some problems with the injectors are presented. It is advised to avoid idling or part load functioning.
- The Gensets will have an electrical efficiency of around 39% (at max. load).
- Free fatty acids would dramatically increase the abrasion in the combustion chamber. FFA's are hard to separate from the oil in decentralized refining. Therefore it should be kept under control with proper harvesting and oil production methods.
- The emissions will depend mostly on the combustion properties, but still lower SO2 and high NOx (higher at part load, similar to other oils) are expected.
- A higher autoxidation with vegetable oils is expected. Therefore a more frequent fuel filter change has to be done. Also an oil change has to be made more often.
- The expected live time of the engines will be reduced to approx 10 years.

Solar Generation

The solar power generation was chosen because of the great solar potential at the Galapagos Islands and the easy implementation and modular expansion. The proposed panels have a performance of 180 Wp/panel with an electrical efficiency of 13.7 %.

Hybrid Configuration

The proposed hybrid system is a low penetration system. That is why no costly batteries are needed and the Jatropha engines should provide the base load. The hybrid system was configured with the following guidelines (*in this case the installation of a possible water desalination plant for Isabela has not been considered*):

- Low cost of the system.
- Optimal use of the solar potential.
- Dimensioning for 10 years (2010 -2020).
- Low fuel consumption.
- Avoid the Gensets to work at part load.
- The Gensets should regulate the peaks.
- The Gensets should be able to cover the whole demand without solar generation.
- If possible only one Genset should work at the time.

Starting 2009, the power generation must be designed for a period of 10 years (approx. Gensets live time). After this period the hybrid system could be extended with more solar and more biofuel power generation. The cost of the system will be defined by the participation of the PV components (higher cost/kW). Therefore the system must be dimensioned for the needed demand during the peaks and the solar power generation must be used when available. To avoid that the Gensets work at very low part load (causes most of the damage in the Gensets), two Gensets will be alternated depending on the base load demand.

For the Isabela Island some configurations (here just two) have been simulated, with the estimated future demand and the solar radiations for two selected days with the highest and the lowest demand (IPSEpro Simtech technologies):

Configuration A

The configuration A is based in the initial proposal made for Isabela. The idea is to have a higher solar generation capacity and to avoid an upgrade of the solar power generation system in the next 20 years (solar generation capacity will be over the needed dimension for the first years). This would imply a lower usage of the Gensets and less Jatropha oil to be transported from the mainland.

Configuration B

The 2nd proposal is based on a higher usage of the Gensets with more fuel efficiency and a lower solar power generation. The idea is to avoid a low part load utilization of the Gensets and to reduce the costs produced by the solar generation (also expensive energy storage devices). The proposed combined functioning of the Gensets will allow the Gensets to avoid cold starts and work with high part loads. To maintain an optimal part load functioning of the Gensets, a modular upgrade of 75 kWp (total 225 kWp) for the solar power generation is needed in 2015 (if the two Gensets work constantly together the system could work at low part loads without an upgrade until 2022). With this configuration the minimum part load of the Gensets has an average of 40% at start and then it increases with the years.

	ConFigure A	ConFigure B
Solar [kWp]	400	150
Genset 1 [kWel]	350	360
Genset 2 [kWel]	400	446
Cost	-	++
Solar share	++	-
Fuel Consumption	++	+
System Efficiency	-	+
Part-load functioning	+	++
Capacity usage	+	++

Figure 19. Two calculated configurations for the power generation hybrid at Isabela (kWp at STC and kWel at PRP).

To maintain high part load ratios the Gensets will operate either as “Master” or “Slave” Genset depending on the demand of the year and season (January to March: highest demand and July to September: lowest demand). The Master will be running constantly and activate the Slave when needed. The master should not be turned off, and share the power demand with the slave to maintain part load ratios over 50% (worst case).

For the configuration B in the year 2010 around 500’000 of liters of Jatropha oil per year will be needed (max. 1800 l/day); for 2015 600’000 l/year (max. 2300 l/day); and for 2020 800’000 l/year (max. 3000 l/day). Figure 12 to 0 shows from 2010 until 2020, the simulation of the hybrid system with the configuration B for the two extreme cases of the year: Day with the lowest demand, and day with the highest demand, with the correspondent solar radiations.

Figure 21 shows the maximum power generation case for 2010, with only 150 kWp of solar power and with the Genset 1 as the master. In Figure 14 an upgrade of 75 kWp to solar power generation has been added. The hybrid has now more solar share (225 kWp) and the

Genset 2 is working as the master (max. case). Figure 16 and 0 show both cases for 2020 where the Gensets are used at its maximum and the 225 kWp solar as well.

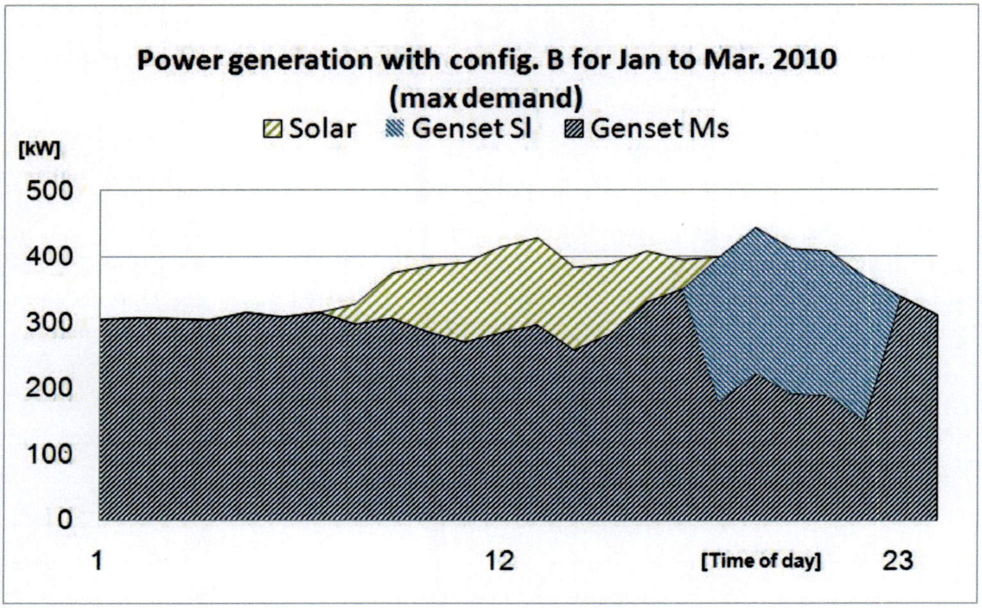

Figure 20. Power generation with the proposed configuration B for a day in the first quarter of the year 2010 (maximum case).

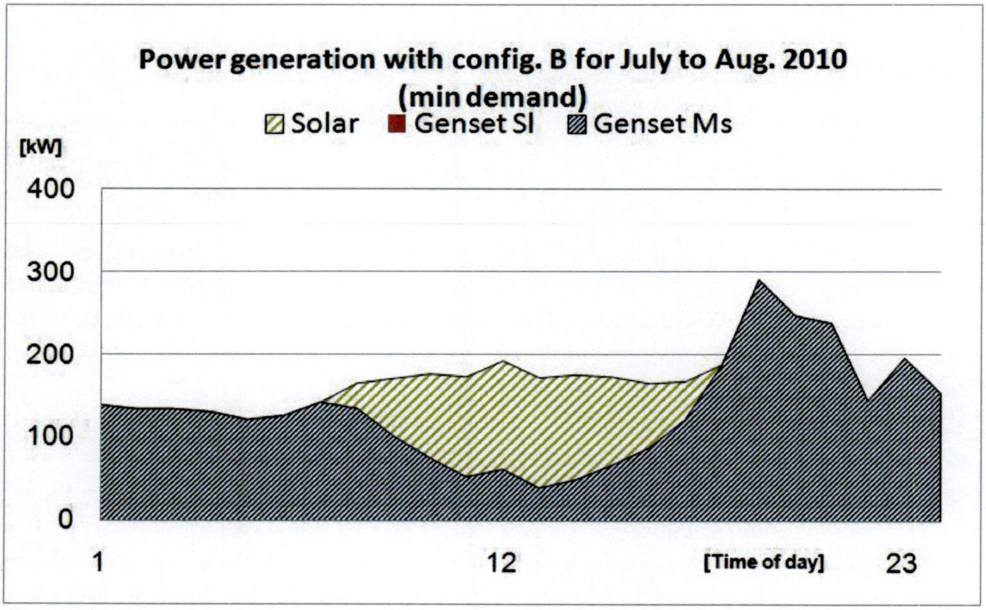

Figure 21. Power generation with the proposed configuration B for a day in the third quarter of the year 2010 (minimum case).

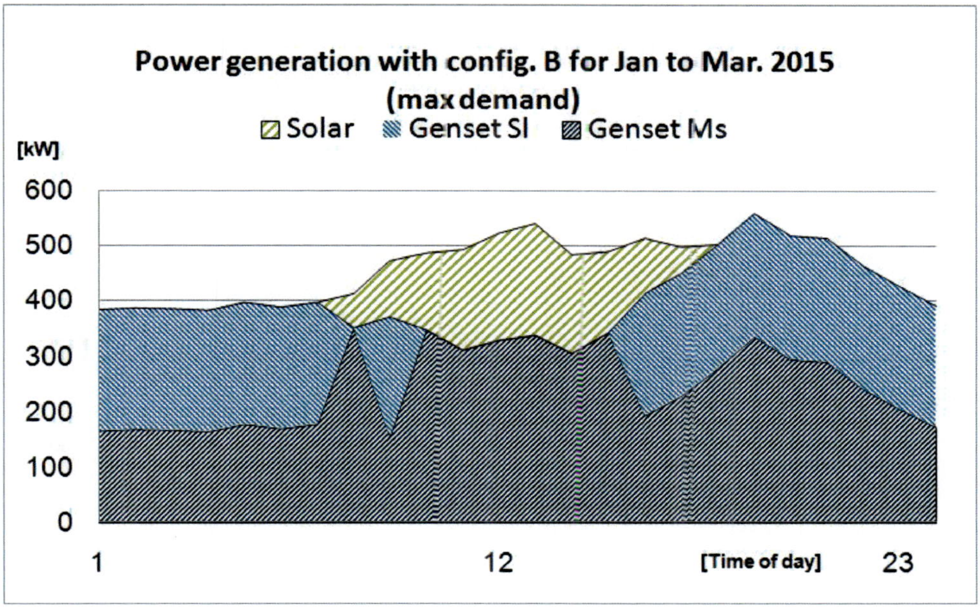

Figure 22. Power generation with the proposed configuration B for a day in the first quarter of the year 2015 (max case).

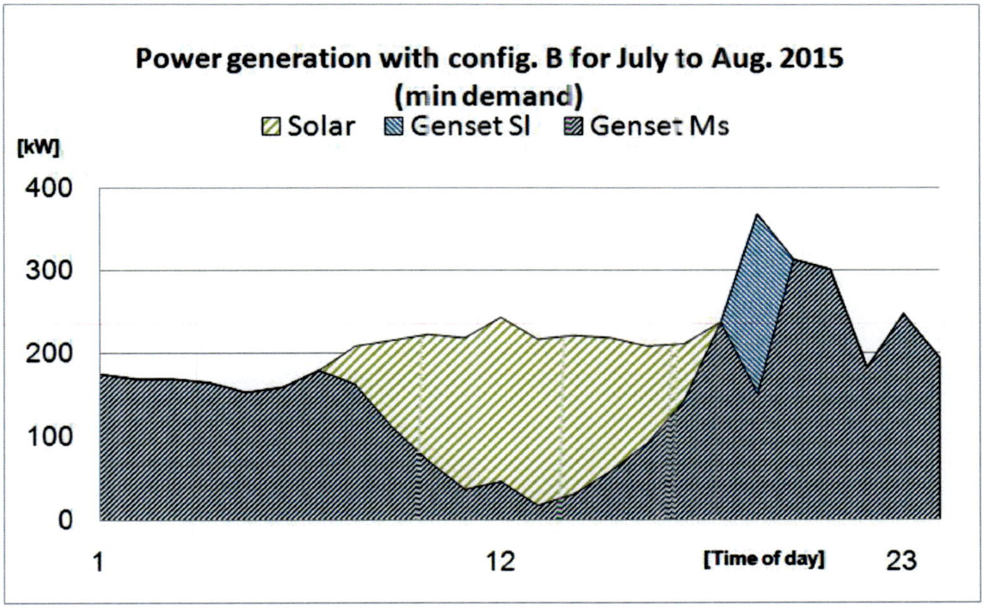

Figure 23. Power generation with the proposed configuration B for a day in the third quarter of the year 2015 (min case).

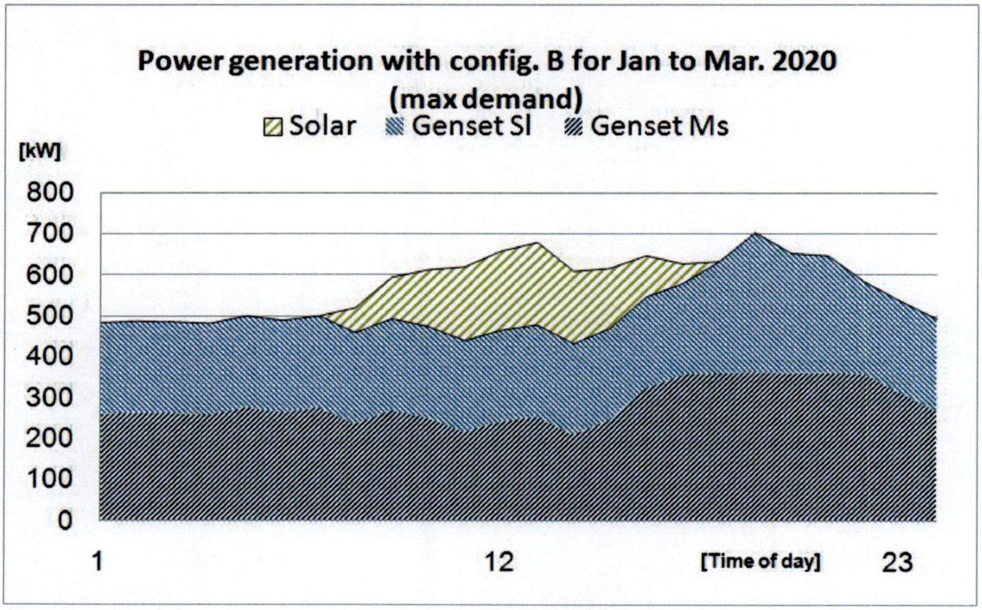

Figure 24. Power generation with the proposed configuration B for a day in the first quarter of the year 2020 (max case).

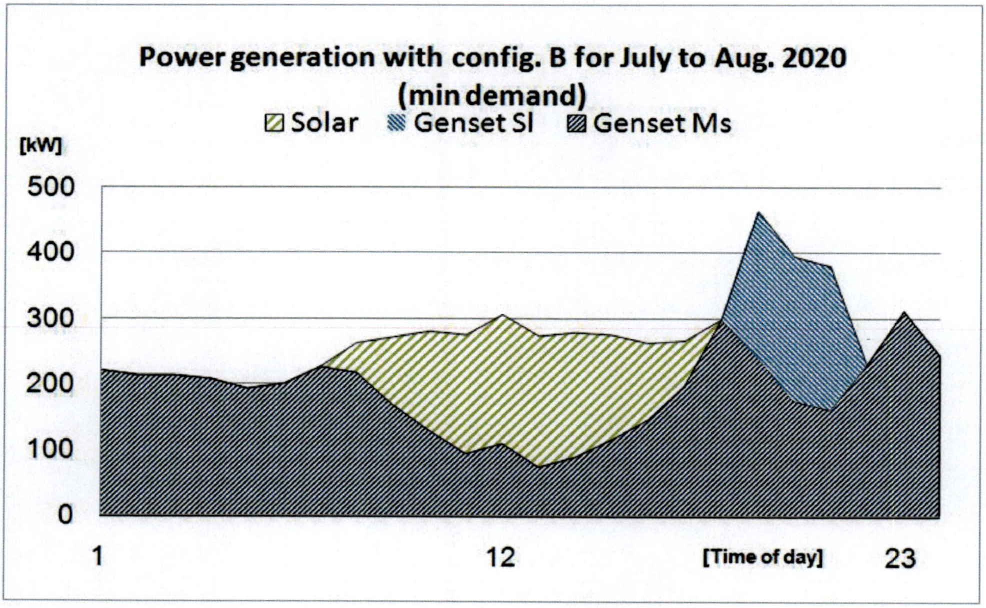

Figure 25. Power generation with the proposed configuration B for a day in the first quarter of the year 2020 (min case).

Conclusion

Due to the high prices of fossil fuels, the incentive for the cultivation of Jatropha is coming faster than the regulations needed for it. With the lack of an environmental regulation, the advantages of Jatropha curcas compared to other energy plants become of less importance. The knowhow and technology for cultivation and production of the oil is available also for decentralized productions. There are still some improvement potentials through the breeding of the plant and the extraction of oil. Especially, quality management guidelines for the production are needed to guarantee a standard oil quality everywhere. Still the more decentralized the production, the more difficult it becomes for the oil to maintain the needed standard. The cultivation of the Jatropha curcas tree and the production of the oil can be done easily. These two production units in the Jatropha oil value chain have a low capital cost and are accessible to the majority of the farmers, offering so an important income for their economies. The technology for power generation is also known from other vegetable oils and is expected to be similar with Jatropha. Still the experience with Jatropha is needed to validate these expectations. The oil consumption, the average part load of the Gensets and the cost of the hybrid system for Isabela, will depend on the final configuration and dimensioning of the hybrid system. Although the Jatropha oil needs to be transported from the mainland to the island (resulting in a low energy balance), the Jatropha oil seems to be the best option to generate the base load at Isabela and so to avoid the environmental risk in case of a spilling.

Acknowledgement

The authors of this paper would like to thank all the persons and institutions that made this work possible: ERGAL (Ing. Carlos Jácome and Ing. Arturo Villavicencio), INIAP (Ing. Hediberto Mendoza), IICA (Dr. Peter Schwiebert), DED (Ing. Matthäus Hoffman), MEER (Ing. Patricia Recalde) and the personal from the Chair of Energy Systems at the Technische Universität München (Dipl. Ing. Mathilde Saule).

References

[1] Villavicencio, A.. Electrificación en Base a Renovables en Patrimonios de la Humanidad: El caso de las Islas Galápagos, Presentación Taller de América Latina y el Caribe, Fuentes Renovables de Energía y Mitigación de Pobreza, ERGAL, April 2007.

[2] Villavicencio, A.; Jácome, C. ERGAL: Data compilation and interviews concerning the ERGAL projects, April-July 2008.

[3] Servicio Alemán de Cooperación Social-Técnica (DED). Estudio de factibilidad para el suministro de energía eléctrica para el proyecto de agua potable en la en la isla floreana a partir de biocombustible; Proyecto Ecu/02/G41 ERGAL, Quito-Ecuador 2007.

[4] Hosier, R.; Vallee, C. GEF, Ecuador: Renewable Energy for Electricity Generation—Renewable Electrification of the Galapagos Islands, *Project Report*, ECU/02/G31, September 2001.

[5] Mendoza, J. H. (Director of the Horticultural Program, INIAP). Personal interview at the INIAP; Ecuador, Manabí (National Autonomous Institute of Agricultural Investigation, Ecuador), 23.1.2008.

[6] Münch, E.; Kiefer, J. (1986). Die Purgiernuss (Jatropha curcas L.); Universität Hohenheim, Institut für Pflanzenproduktion in den Tropen und Subtropen; Stuttgart - Hohenheim, Diplomarbeit.

[7] Heller, J.. Untersuchungen über genotypischen Eigenschaften und Vermehrungs- und Anbauverfahren bei der Purgiernuss (Jatropha curcas L.); Verlag Dr. Kovac, Hamburg 1992.

[8] Avila, J.G. de (1949). Breves Instrucões para a Cultura da Purgumeira; Reparticão Técnica dos Servicos Agricolas; Florestais e Pecários; Impresa Naciona de Cabo Verde, Praia.

[9] Petrocomercial (2008). Reporte anual de operaciones, 2007. - URL: www.petrocomercial.com (Abril 2008).

This paper is based on the Diploma Thesis (Master Thesis): “A Review on the Potentials of the Jatropha curcas L. for Power Generation and Sustainable Development of Poor Communities. Case Study: Ecuador and the Isabela Island Galapagos”. Under development at the Technical University of Munich (Chair of Energy Systems), May 2008.

In: Renewable Resources: Obtaining, Processing... ISBN 978-1-6060692-534-8
Editors: R. Kozlowski, G. Zaikov and F. Pudel

Chapter 20

ENERGY POTENTIAL OF CROP RESIDUES IN NIGERIA

***S. O. Jekayinfa*[*1, 2] *and V. Scholz*[2]**

[1]Department of Agricultural Engineering, Ladoke Akintola University of Technology, P.M.B. 4000, Ogbomoso, Oyo State, Nigeria
[2]Leibniz-Institute for Agricultural Engineering (ATB), Max-Eyth-Allee100, D - 14469 Potsdam, Germany

ABSTRACT

Nigeria is one of the countries in the world with the lowest per capita energy consumption far below the world average as compared with those of developed economies such as North America, Western Europe and Japan where consumption far outstrips the world average. This is not unconnected with high dependence on fuel wood for domestic energy needs, and high incidence of wastage and underutilization of agricultural wastes as a source of energy. There is therefore the urgent need to source for other sustainable means of meeting the energy needs of the teeming Nigerian population. An assessment of energy potential of selected residues from maize, cassava, millet, plantain, groundnuts, sorghum, oil palm, palm kernel and cowpeas in Nigeria was therefore conducted. Data used for analysis was sourced directly and from relevant published work. It is estimated that nearly 58 million tonnes of these residues were potentially available in the year 2004 with energy potential of about 20.8 million tonnes oil equivalents. The residue availability for 2010 is projected to be about 80 million tonnes. These residues, when converted to energetically usable forms, can substitute or complement the fossil energy sources in Nigeria by more than 80%.

Keyword: Renewable energy, agricultural residues, energy potential, Nigeria.

* Corresponding Author: jekaysol@yahoo.com

1. INTRODUCTION

Nigeria is the most populous country in Africa with a 2006 population estimate of about 140 million people (NPC, 2007). Nigeria is one of the countries in the world with the lowest per capita energy consumption far below the world average as compared with those of developed economies such as North America, Western Europe and Japan where consumption far outstrips the world average (Maduka, 2005). This is not unconnected with high dependence on fuel wood for domestic energy needs, and high incidence of wastage and underutilization of agricultural wastes as a source of energy. Nigeria has a land area of 98.3 million hectares. At present about 34 million hectares or 35% are under cultivation. There is no doubt that considering the vast area of uncultivated land coupled with the natural fertility of its soil, Nigeria has great agricultural potentials. In Nigeria, agriculture has remained the largest sector of the economy. It generates employment for about 70% of Nigeria's population and contributes about 40% to the Gross Domestic Product (GDP) with crops accounting for 80%, livestock 13%, forestry 3% and fishery 4%. It plays significant roles in the nation's economic development. Nigeria is rich in both fossil fuels such as crude oil, natural gas, coal, and renewable energy resources like solar, wind, biomass, biogas, etc.

The national energy supply in Nigeria is at present almost dependent on fossil fuels and firewood which are depleting fast. Ikuponisi (2004) reported the proportion of each energy source in the total energy reserves in Nigeria. The report indicated the estimates of crude oil, natural gas, coal and lignite, tar sands and large scale hydropower. Figure 1 shows Nigeria's energy consumption pattern by sources. Only about 40% of the population is connected to the national electricity grid. There is therefore the urgent need to source for other sustainable means of meeting the energy needs of the teeming Nigerian population.

Literature is replete with many attempts to estimate global and national production and use of residues, but with large variations. These include residues availability in Philippines (Elauria et al., 2005), Karnataka (Ramachandra, 2004), and selected Asian countries (Bhattacharya et al., 2005). Other related reports include Jekayinfa and Omisakin (2005), El. Saeidy et al. (2003), Jekayinfa and Scholz (2007).

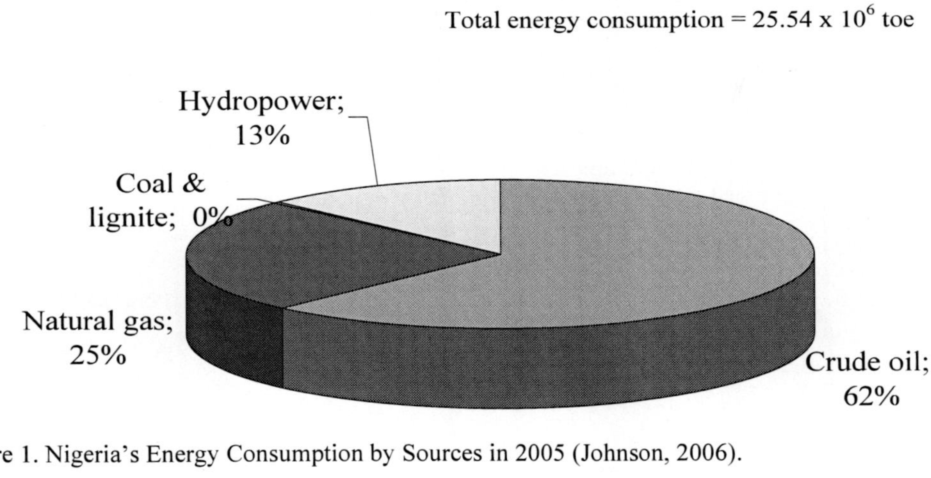

Figure 1. Nigeria's Energy Consumption by Sources in 2005 (Johnson, 2006).

Assessment of available agricultural residues is helpful in revealing its status and helps in taking conservation measures and ensures a sustained supply to meet the energy demand. The present study investigates the energy potential of energetically available crop residues in Nigeria.

2. Materials and Method

The amount of agricultural residues produced differs from crop to crop and is subject to seasons, soil type, and other farming conditions. The agricultural residue produced at a particular time is directly related to the corresponding crop production, and the ratio between the main crop produced and its residue. Therefore, for known quantities of crop production, it is possible to estimate the amounts of agricultural residues produced using the residue to product ratio (RPR). The amount of residue used for purposes other than energy was estimated using facts obtained from the literature and other secondary sources. The estimated values were assumed to be a certain percentage of the total annual generation from each of the nine selected major crops in Nigeria.

As a result of increasing production levels of most crops due to improved agricultural farming techniques, the availability of agricultural residues has also been changing with time. The time variation of the total crop produced can be expressed in various models (linear, semi-log, log-linear and polynomial) to develop functional relationships between crop production and total availability of agricultural residue for different selected crops. The general equation fitted for different crops is of the following nature:

$$Y_c = c_0 + c_1x + c_2x^2 \tag{1}$$

$$Y_R = Y_c \times RPR \tag{2}$$

where Y_c represents the amount of crop production in thousand tonnes/year, x the time in years (beginning with 2000), Y_R the total availability of agricultural residue for the same crop, c_0, c_1 *and* c_2 the regression coefficients. The RPR and heating values for the residues investigated in this study have been reported by Koopmans and Koppejan (1997), Jekayinfa and Omisakin (2005), Suwannakhanthi (2004) and Bhattacharya et al. (2005).

3. Results and Discussion

In this study, 17 residues from 9 major crops in Nigeria were investigated. The crops are maize, cassava, millet, plantain, groundnuts, sorghum, oil palm, palm kernel and cowpeas. Table 1 presents the outputs of the major crops in Nigeria between 2000 and 2004 as reported by FAO (2007).

Table 1. Yields of major crops in Nigeria (10^6 t)

Major crop	Year				
	2000	2001	2002	2003	2004
Maize	4.10	4.60	4.69	4.78	4.78
Cassava	32.01	32.07	32.75	32.91	38.18
Millet	6.15	5.53	6.10	6.28	6.28
Plantain	1.97	2.00	2.06	2.10	2.42
Groundnuts	2.90	2.68	2.70	2.80	2.94
Sorghum	7.71	7.08	7.70	8.03	8,03
Oil palm	2.47	2.47	2.50	2.52	4.78
Palm kernel	8.22	8.50	8.50	8.63	8.70
Cowpeas	2.15	2.17	2.17	2.23	2.32

Data for crop production available from FAO statistics (see http://www.fao.org).

Table 2 presents the estimated quantity of agricultural residues generated in Nigeria using 2004 production data. The total potential residue available in Nigeria in 2004 is about 70 million tonnes. An attempt was also made to estimate the energetically available residues from 2004 production data and quantify these available residues as detailed in Table 3. Because of other non-energy uses to which residues are put, 5 out of the 17 selected residues in this study are not completely available energetically.

Table 2. Estimated amounts of agricultural residues generated in Nigeria in 2004

Major crop	Crop yield (10^6 t)	Type of residue	RPR*	Residue amount (10^6 t)
Maize	4.78	Cob	0.273	1.30
		Stalk	2.000	9.56
		Husk	0.200	0.96
Cassava	38.18	Stalks	0.062	2.37
		Peelings	0.250	9.54
Millet	6.28	Stalks	1.750	11.00
Plantain	2.42	Peels	0.40	8.44
		Trunks/leaves	0.50	1.05
Groundnuts	2.94	Husks/shells	0.477	1.40
		Straw	2.300	6.76
Sorghum	8.03	Straw	1.750	14.05
Oil palm	4.78	Shell	0.065	0.31
		Fibre	0.140	0.67
		Empty bunches	0.230	1.10
Palm kernel	8.70	Shells	0.45	3.92
		Cake	0.25	2.18
Cowpea	2.32	Shells	1.75	4.05
Total				69.56

Data for crop production available from FAO statistics (see http://www.fao.org).
* Residue to Product Ratio.

The energetically available share of the residue generated from each crop is estimated from published data and personal communication with experts. Estimates have indicated that the 2004 annual energy potential of the crop residues (about 58 million tonnes) is equivalent to energy of approximately 878 PJ or 20.81 million toe of fuel-oil, which equals about 82% of Nigeria's annual (2005) total energy consumption (Johnson, 2006).

Technically, the conversion of these crop wastes and residues for energy is possible. However all the problems relating to several steps in the harvesting and conversion sequence need to be solved for sustainable and economical outputs on a long term basis. For Nigeria's current energy situations, crop residues can be used in their raw form or in form of chips, chunks, briquettes, etc for cooking, water-heating and other combustion systems. They can also be used in any other forms (e.g. biogas and alcohol) for combustion and gasification systems.

Although some of these technologies are not familiar and popular in Nigeria, however their full adoption and utilization will solve some of the identified energy problems plaguing every sector of the Nigerian economy. With the ever increasing rate of production of the major crops and same RPR, it would not be difficult for Nigeria to meet its soaring energy demand should energy from crop residues be earnestly harnessed as indicated in this work.

The future projections of the available quantities of residues from the 9 selected crops in the year 2010, the equation developed for each crop from regression analysis and the RPR values for each residue are presented in Table 4. From the projection, it is estimated that about 81 million tonnes of various residues could be energetically available in 2010. This gives an increase of 16% compared to year 2004 figure.

Table 3. Energetically available residues and their energy equivalents in Nigeria, 2004

Residue	Heating value (MJ/kg)	Total amount (10^6 t)	Energetically available amount		
			Share (%)	Weight (10^6 t)	Energy (10^6 toe)
Maize cob	16.63	1.30	100	1.30	0.51
Maize stalk	15.51	9.56	100	9.56	3.51
Maize husk	15.56	0.96	100	0.96	0.35
Cassava stalk	13.38	2.37	20	0.47	0.15
Cassava peelings	10.61	9.54	30	2.86	0.72
Millet stalks	15.51	11.00	30	3.30	1.21
Plantains peelings	12.56	8.44	100	8.44	2.51
Plantains trunks/leaves	15.48	1.05	100	1.05	0.39
Groundnuts husks/shells	15.56	1.40	100	1.40	0.52
Groundnuts straw	14.40	6.76	100	6.76	2.30
Sorghum straw	14.40	14.05	100	14.05	4.79
Oil palm shells	21.10	0.31	100	0.31	0.16
Oil palm fibre	19.94	0.67	100	0.67	0.32
Oil palm empty bunches	19.41	1.10	100	1.10	0.51
Palm kernel shells	21.10	3.92	50	1.96	0.99
Palm kernel cake	10.61	2.18	0	0	0
Cowpeas shells	19.44	4.05	100	4.05	1.87
Total					20.81

All mass units referring to fresh matter.

Table 4. Projection of residue availability in Nigeria for 2010

Residue	RPR*	Regression equation coefficients**				Share	Available amount
		c_0	c_1	c_2	R^2	(%)	(10^6 t)
Maize cob	0.273	153	4132	0	0.74	100	1.59
Maize stalk	2.00	153	4132	0	0.74	100	11.62
Maize husk	0.20	153	4132	0	0.74	100	1.16
Cassava stalk	0.062	707	-2924	34578	0.90	20	1.09
Cassava peelings	0.25	707	-2924	34578	0.90	30	6.60
Millet stalks	1.75	54	-216	6109	0.43	30	5.42
Plantains peelings	0.35	101	1808	0	0.77	100	1.39
Plantains trunks/leaves	0.40	101	1808	0	0.77	100	1.17
Groundnuts husks/shells	0.477	57	-323	3147	0.93	100	3.09
Groundnuts straw	2.30	57	-323	3147	0.93	100	14.92
Sorghum straw	1.75	69	-254	7717	0.53	100	23.15
Oil palm shells	0.065	14	2454	0	0.90	100	0.17
Oil palm fibre	0.14	14	2454	0	0.90	100	0.37
Oil palm empty bunches	0.23	14	2454	0	0.90	100	0.60
Palm kernel shells	0.45	1278	2618	0	0.47	50	3.76
Palm kernel cake	0.25	1278	2618	0	0.47	0	0
Cowpeas shells	1.75	39	2091	0	0.84	100	4.41
Total							80.51

* Residue to Product Ratio

** Regression equation of the form: $Y_c = c_0 + c_1 x + c_2 x^2$

To be able to fully utilize the energy potential of these resources, current and future efforts of the energy sector of the country must be focussed on their development and efficient use. The energy policy makers in the country must put in place, efficient machinery for harnessing these biomass materials by providing sustainable human, financial and technical resources. The government should establish a comprehensive approach, a master plan and a sustainable institutional and policy framework to develop and disseminate biomass energy technologies.

The place of well co-ordinated and adequately funded research, development and demonstration (R, D andD) in renewable energy technology cannot be over-emphasised. This can be achieved by encouraging private sector participation and fostering international collaboration for research and development of biomass utilization. Also, it is worth noting that the use of alternative energy technologies such as can be derived from crop residue conversion takes place in an environment where factors such as technology, political will, economics, and personal motivation are all required for its adoption.

Conclusions

1. There is availability of large amounts of non-plantation biomass resources in Nigeria for modern energy applications. These resources are residues from maize, cassava, millet, plantain, groundnuts, sorghum, oil palm, palm kernel and cowpeas
2. It is estimated that more than 70 million tonnes of agricultural residues were potentially produced in the year 2004 out of which only 58 million tonnes are energetically available. The residue availability projection for 2010 is about 80 million tonnes.
3. The estimated energy value of the energetically available residue using 2004 production data is 878 PJ or 20.81 million toe of fuel-oil, which equals about 82% of the present Nigeria's annual energy consumption.

Acknowledgements

The first author is grateful to Alexander von Humboldt Foundation, Germany for her financial support through the award of Georg Forster Fellowship in the course of this work.

References

Bhattacharya, S.C., Abdul Salam, P , Runqing, H., Somashekar, H.I., Racelis, D.A., Rathnasiri, P.G. and Yingyuad, R. 2005. An assessment of the potential for non – plantation biomass resources in selected asian countries for 2010. *Biomass and Bioenergy* 29(3): 153-166.

Elauria, J.C., Castro, M.L.Y., Elauria, M.M., Bhattacharya, S.C. and Abdul Salam, P. 2005. Assessment of sustainable energy potential of non-plantation biomass resources resources in the philippines. *Biomass and Bioenergy* 29( ):191-198.

El-Saeidy, E., Scholz, V. and Hahn, J. 2003. Energetic use of crop residues considering especially cotton stalks. *Proceedings of The International Conference on New Methods, Means and Technologies for Application of Agricultural Products*. Institute of Agriculture Engineering LUA Raudondvaris, 18-19 September, 2003. p.27-32.

FAO 2007. Agricultural Database. http://www.fao.org.

Ikuponisi, F.S. 2004. Status of renewable energy in nigeria : background brief for energetic solutions. Presented at an *International Conference on Making Renewable Energy a Reality*. November 21-27, 2003. Abuja/ Portharcourt/ Calabar, Nigeria.

Jekayinfa, S.O. and Omisakin, O.S. 2005. The energy potentials of some agricultural wastes as local fuel materials in nigeria. *Agricultural Engineering International: The CIGR- E-Journal*. Manuscript EE 05 003 . cigr-ejournal.tamu.edu.

Johnson, E. 2006. Nigeria energy data, statistics and analysis - oil, gas, electricity, coal. http://www.eia.doe.gov/emeu/cabs/Nigeria/pdf.pdf.

Koopmans, A. and Koppejan, J. 1997. Agricultural and forest residues generation, utilization and availability. Paper presented at the *Regional Consultation on Modern Applications of Biomass Energy*, 6-10 January 1997, Kuala Lumpur, Malaysia.

Maduka, J.O. 2005. The Energy and gender nexus in africa. A Paper Presented at the *ADB Finesse Africa Program Regional Consultative Workshop*. Tunis, Tunisia 23-25 February 2005.

NPC (National Population Commission of Nigeria), 2007. Results of 2006 population census in Nigeria. http://www.punchng.com/.

Ramachandra, T.V., Kamakshi, G. and Shruthi, B.V. 2004. Bioresources status in Karnataka. *Renewable and Sustainable Reviews* 8(2):1-47.

Suwannakhanti, N. 2004. Overview for Thailand's renewable energy focusing on biomass energy.http://www.ptm.org.my/icra/ICRA%20papers/BIOMASS/MsWord/Thailand_Biomass.de.

In: Renewable Resources: Obtaining, Processing...
Editors: R. Kozlowski, G. Zaikov and F. Pudel
ISBN 978-1-6060692-534-8

Chapter 21

INTEGRATED PROCESS SYSTEMS TO RECOVER ENERGY FROM BIOMASS IN FUEL CELLS (PROBIO)

Sascha Thomas, Eyck Schotte, André Herrmann, M. Sc. Ling He and Vicky Mosch

Business Unit Process and Plant Engineering,
Fraunhofer Institute for Factory Operation and Automation,
Magdeburg, Germany

ABSTRACT

One of the main future challenges on the world energy market will be guaranteeing a stable supply of energy and establishing independence from fossil fuels. One way to increase the proportion of renewable energy sources in the overall energy pool is to increase the utilization of lignocellulosic biomass, e.g. wood. In the frame of collaboration between the Max Planck Society and the Fraunhofer-Gesellschaft (ProBio), the combination of biomass (wood) gasification and the conversion of fuel gas into electrical energy in fuel cells are examined. The research project will analyze a system consisting of a biomass gasification unit employing fluidized bed technology, purification stages and solid oxide and polymer electrolyte membrane fuel cells. New concepts for economically cleaning the fuel gas produced shall be developed and evaluated experimentally. Various combinations and couplings of the mass and energy flows shall be explored theoretically. The formation of undesired components affecting fuel cell operation and leading to decreased fuel cell efficiency can be prevented by using additives in the gasification process. The fuel gas will be purified by optimally combined innovative high temperature cleaning processes. The parameters characterizing each individual process shall be determined experimentally and applied in the process simulation of the integrated system. The most promising combination in terms of cost effectiveness shall be selected for a prototype producing 1-2 kW of electrical power.

In this paper, simulation and experimental results regarding wood gasification and high temperature fuel gas cleaning are discussed in detail.

1. INTRODUCTION

One of humanity's greatest challenges is averting a global climate catastrophe with untold consequences for the development of habitants let alone the future of the human race. The latest study issued by the United Nation's Intergovernmental Panel for Climate Change (IPCC) substantiated that humans and thus not nature are responsible for the earth's warming. While climate change may be unstoppable, its impact can however certainly be lessened, i.e. can and must be brought under control. Efforts in the field of alternative energies will again intensify tremendously in the coming year to achieve this goal.

One potential use of renewable energy is the gasification of biomass, which, when combined with fuel cell technology to recover energy from the generated fuel gas, holds great potential to generate electrical energy sustainably and highly efficiently. Given their high electrical efficiencies and ideal conditions for cogeneration, low noise levels and low emissions, fuel cells will likely be accorded greater importance for the distributed supply of power in the future. Moreover, the use of renewable materials as the source material for fuel gas generation will increase independence from nations exporting fossil fuels and thus enhance the security of energy supply as well as ensure the energy generated emits little CO_2 and is environmentally compatible.

The Max Planck Institute for Dynamics of Complex Technical Systems, the Fraunhofer Institute for Factory Operation and Automation IFF and the Fraunhofer Institute for Ceramic Technologies and Systems IKTS in Germany are collaborating in the project ProBio to develop and engineer a system consisting of a gasification unit to generate fuel gas, cleaning stages to process it and high and low temperature fuel cells that utilize it. Both experimental studies to test the process steps integrated in the planned overall system and theoretical studies of the optimal interconnection of the individual processes are planned. Incorporating the experimental findings, the closely collaborating project partners intend to simulate individual system components and the coupling of their material and energy within the overall system in models. Figure 1 illustrates the development cycle of the overall system.

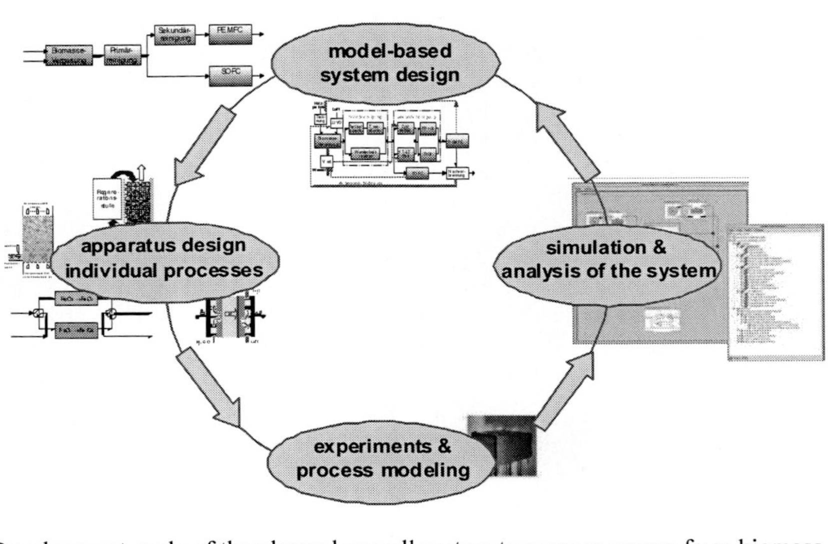

Figure 1. Development cycle of the planned overall system to recover energy from biomass in fuel cells.

Since its Process and Plant Engineering Business Unit specializes in the thermochemical conversion of organic materials, the Fraunhofer IFF is primarily working on the generation of fuel gas from biomass by means of fluidized bed gasification as well as on tests of primary gas cleaning and conditioning. New reactor concepts will be applied, e.g. a moving bed reactor for combined removal of dust and tar and other cleaning stages for the removal of sulfur and halogen compounds.

In the following chapter, simulation and experimental results of wood gasification in a lab-scale bubbling fluidized bed system, tar and dust removal in a moving bed reactor and sulphur and halogenide removal in a fixed bed reactor will be described in greater detail.

2. Results of Experiments and Simulation

2.1. Wood Gasification

2.1.1. Experimental Set-Up

The wood gasification experiments have been made in a lab-scale gasifier system. It consists of a solid feeding system, a bubbling fluidized bed, a gas analysis system and a post-combustion unit. The system flowchart is shown in Figure 2.

The solid feeding system is a twin-screw volumetric feeder with a hopper pressurized with N_2 to avoid the return of the hot gas from the gasifier. The screw type is chosen based on the physical properties of biomass, i.e. particle size, shape, and bulk density. Cooling water channel is installed in the stainless-steel screw pipe to prevent the biomass pyrolysis inside the pipe. Before experiments, the feeder is calibrated for the selected biomass to determine the output as a function of screw rpm to ensure the accurate mass flow rate.

The bubbling fluidized bed has three sections: gas preheating zone, bed and freeboard. Before entering the bed, the gas will be heated to 500°C in the preheating zone. A porous plate is used as the gas distributor and pressure drop of the plate is determined beforehand. The bed has 5cm inner diameter and 28cm height, with the biomass feeding point at 7cm above the distributor. The freeboard is widened to 8cm inner diameter and with 58cm height. A 9cm conical reduction zone connects the bed and freeboard section together. The whole reactor is constructed of stainless steel and heated through external electric oven with an operating temperature up to 1200°C. The temperatures in the reactor are measured by seven K-type thermocouples located at different heights. Three pressure drop transducers are installed to monitor the pressure drop fluctuation inside the reactor. A cyclone is positioned at the gas outlet and can separate particles with diameter larger than 5μm. Glass fiber thimbles are used to further remove particles bigger than diameter 0.8μm before the product gases will be fed into the gas analysis system.

The gases CO_2, CO, H_2, H_2O, CH_4, HCl, H_2S and NH_3 are continuously measured with FTIR and TCD gas analyzer. Tar content is measured in a tar analyzer by taking sample every 10 minutes during the steady state operation.

A post combustion unit (KNV) employing Pt/Al_2O_3 as catalyst is used to burn out the product gases before they will be emitted to atmosphere.

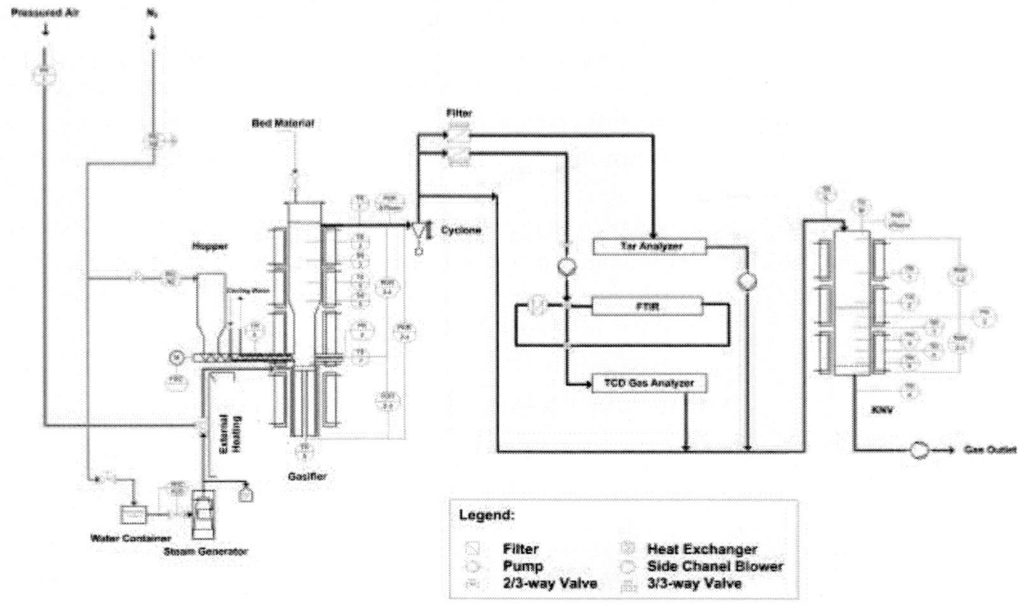

Figure 2. Flowchart of a lab-scale bubbling fluidized bed gasifier system.

2.1.2. Experiments and Results

Before each experiment the hopper is filled with the selected wood particles with diameter 1-2mm. The chemical properties of wood particles are listed in Table 1.

Table 1. Wood Particle Chemical Properties

Proximate Analysis	[wt%]
Fixed Carbon	14.95
Volatile Matter	76.97
Ash	0.23
Moisture	7.85
Ultimate Analysis	[wt%]
Carbon	46.22
Hydrogen	5.41
Sulphur	0.82
Oxygen	39.46
Lower Heating Value [MJ/kg]	16.89

Silica sand with diameter of 300 to 700 μm is chosen as the only bed material. The bed aspect ratio, defined as the ratio of static sand bed height to reactor diameter $H_{Sand\ Bed}/D_{Reactor}$, is kept between 2 to 4 in order to have a better fluidization performance. Sand minimum fluidization velocity u_{mf} at operating temperature is previously determined in the reactor.

Considering the bed as a mixture of sand and char particles under operation condition, the minimum fluidization velocity of the mixture is calculated as [Narváez]:

$$(u_{mf})_{\text{mixture of sand and char}} = 1.3(u_{mf})_{\text{sand}}$$

The fluidization number u_0/u_{mf} was selected between 2 to 5 to have a better mixing condition in the bed. On the basis of the minimum fluidization velocity u_{mf} and the fluidization number u_0/u_{mf}, the superficial gas velocity u_0 can be calculated.

Air is taken as gasification agent and is continuously introduced into the reactor. Equivalence ratio ER, defined as the air-to-fuel weight ratio used divided by the air-to-fuel weight ratio of stoichiometric combustion, varied from 0.20 to 0.50. As the temperature inside the reactor reaches operating temperature, biomass is fed to the gasifier and air flow is adjusted according to the previously selected ER. After about thirty minutes of gasification, the system reaches a steady state. Gas samples are taken to measure the tar content using the tar analyzer. Product gas compositions are measured continuously with FTIR and TCD gas analyzers and the data are recorded automatically in the control computer system.

Some experimental results of the product gas composition in the case of wood gasification with air and the corresponding operation conditions are listed in Table 2.

Table 2. Operation Parameters and Results for Some Gasification Experiments

Operation Parameters		
Fluidization Agent	Air	Air
Average Bed Temperature T_{Bed} [°C]	902	845
Freeboard Temperature $T_{Freeboard}$ [°C]	710-770	710-790
Bed Material	Silica Sand	Silica Sand
Fixed Bed Height H_{bed} [cm]	10	10
Mean Particle Size d_p (μm)	464	464
Minimum Fluidization Velocity u_{mf} (m/s)	0.078	0.078
Minimum Fluidization Velocity $u_{mf\ mixture}$ (m/s)	0.102	0.102
Air Volume Flow Rate [Nl/min]	15	15
Superficial Gas Velocity $u_{0,Bed}$ (m/s)	0.548	0.521
Fuel	Wood Particles	Wood Particles
Equivalence Ratio ER	0.3	0.3
Mass Flow [g/min]	13	13
Weight Hourly Space Velocity WHSV [h^{-1}]	2.82	2.82
Throughput [kg/h m^2]	398	398
Experimental Results		
Gas Composition [Vol%]		
CO	23.1	18.8
H_2	7.3	7.8
H_2O	26.9	27.3
CO_2	10.5	14.2
CH_4	3.8	4.7

As the bed temperature increases, CO content increases as well while CO_2 and CH_4 content decrease. This is because high temperature favors the Boudouard reaction and steam methane reforming reaction, which are endothermic reactions and CO is the end product.

2.1.3. Chemical Kinetics Simulation

Biomass gasification is a quite complex thermalchemical process. In general, the carbonaceous fuels mainly undergo four processes in the gasifier: drying, pyrolysis, oxidation and reduction. The pyrolysis process occurs at low temperature. Volatiles, including tars and gases (CO, CO_2, H_2, CH_4, H_2O and so forth) are released and char is produced. The volatiles partly react with oxygen, which provides heat for the subsequent reduction reactions.

The main reactions considered in the simulation study performed are listed below. The hypothetical empirical formulas are given as $C_1H_{2.43}O_1$ for tar and $C_1H_{0.62}O_{0.12}$ for char [Tepper].

1) Tar oxidation $C_1H_{2.43}O_1 + 0.5\ O_2 \Rightarrow CO + H_2O + 0.215\ H_2$
2) Methane oxidation $CH_4 + 2\ O_2 \Leftrightarrow CO_2 + 2\ H_2O$
3) Char combustion $C_1H_{0.62}O_{0.12} + 1.095\ O_2 \Rightarrow CO_2 + 0.31\ H_2O$
4) Heterogeneous water gas reaction $C_1H_{0.62}O_{0.12} + 0.88\ H_2O \Rightarrow CO + 1.19\ H_2$
5) Tar decomposition $C_1H_{2.43}O_1 \Rightarrow 0.66CO + 0.33CH_4 + 0.33H_2O + 0.215H_2$
6) Steam methane reforming $CH_4 + H_2O \Leftrightarrow CO + 3\ H_2$
7) Water gas shift reaction $CO + H_2O \Leftrightarrow CO_2 + H_2$
8) Boudouard reaction $2\ CO \Leftrightarrow C + CO_2$

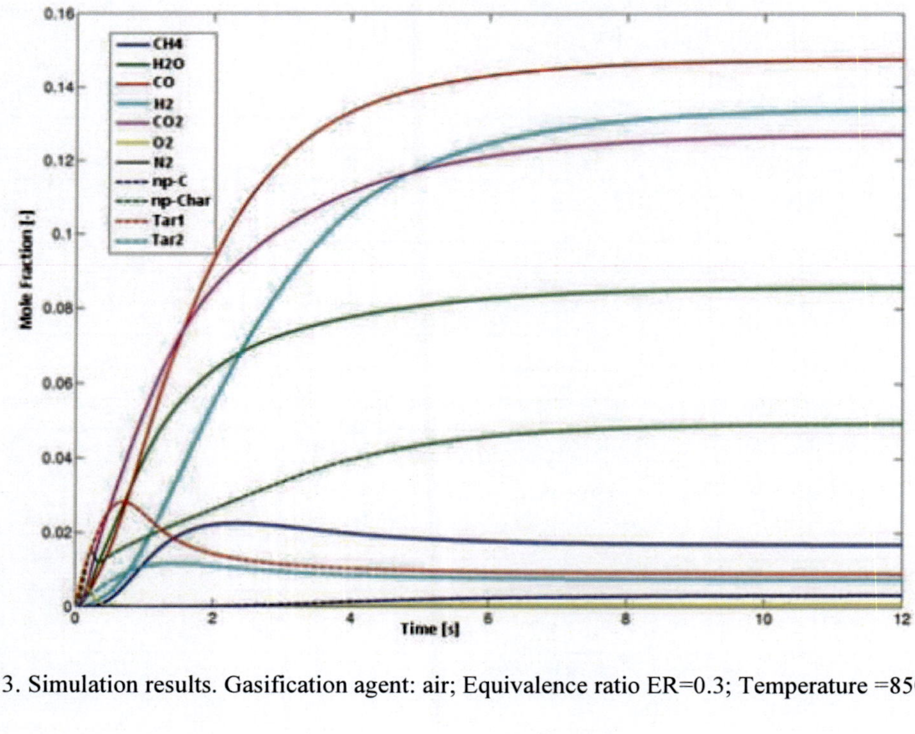

Figure 3. Simulation results. Gasification agent: air; Equivalence ratio ER=0.3; Temperature =850°C.

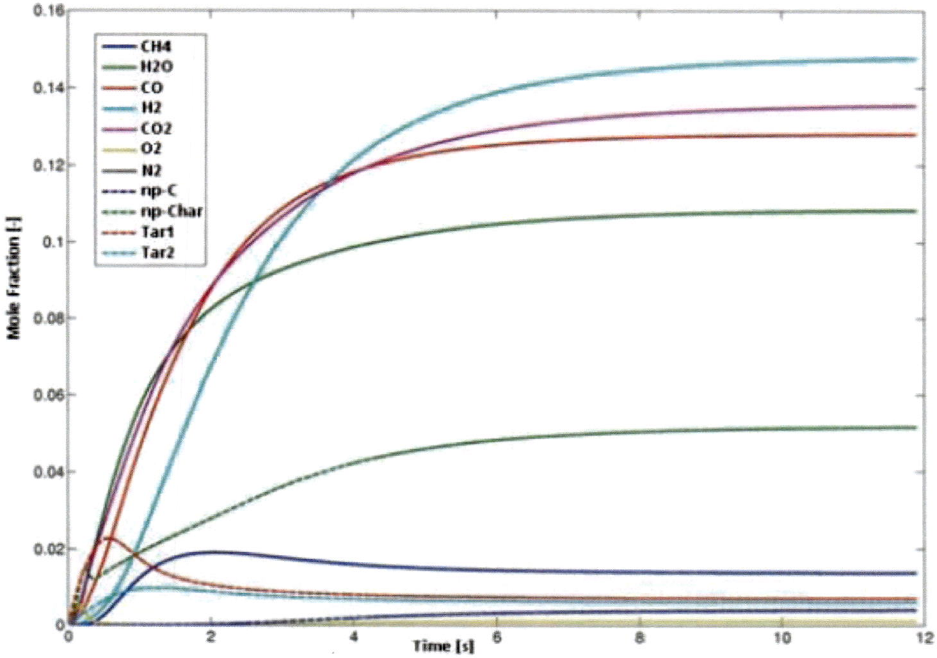

Figure 4. Simulation results. Gasification agent: mixture of air and steam; Equivalence ratio ER=0.3, Steam to biomass ratio S/B=0.3; Temperature =720°C.

Simulink/Matlab software is used for the simulation and it provides an interactive graphical environment and a customisable set of block libraries for modelling, simulating, analysing and testing of dynamical systems. Figure 3 and Figure 4 show the typical simulation results for biomass gasification process using air and air/steam mixture as gasification agent, respectively.

2.2. High Temperature Fuel Gas Cleaning

In the frame of the cooperation project ProBio the Fraunhofer Institute for Factory Operation and Automation IFF is developing as a first cleaning step a hot gas filtration unit to separate dust particles from the hot fuel gas stream at temperatures ranging between 700 and 800 °C. A granular bed filter with ceramic pellets is being employed. Interception, diffusion, inertial impaction, gravity or electrostatic forces activate the filtration process as the fuel gas flows through the filter material (Figure 5). To prevent collected dust particles from blocking the filter bed, the filter material has to be transported out of the apparatus together with the adhering dust particles. Once pneumatic transport and a subsequent cyclone have freed the pellets of dust particles, the refreshed filter material can be recharged into the granular bed filter. When the granular bed is operated in a steady-state, dust is separated from the fuel gas and continuously discharged from the separation unit.

The second function of the moving bed reactor beside the dust filtration is the reforming of tar components which are formed during the gasification process of biomass. For this, the bed material of the moving bed reactor must be catalytic active. As catalyst, Ni-alumina spheres are used for the filter bed operated at 800 to 900 °C. A fix bed reactor as a second purification step to clean the fuel gas is being developed to separate halogenides and sulphur components. To find the appropriate adsorbents, different bed materials, e.g. ZnO, CaO and dolomite, are being tested in a fixed bed reactor to determine their cleaning efficiency.

2.2.1. Experimental Setup

An experimental bench setup was constructed and installed to analyze the influence of operating parameters and to optimize the separation unit (Figure 6).

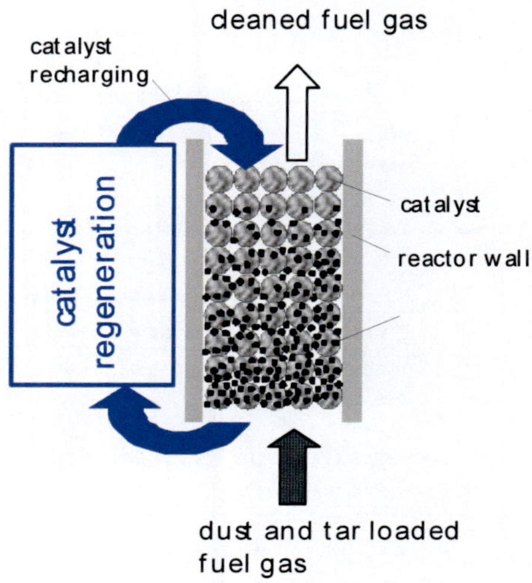

Figure 5. Granular bed filter schematic.

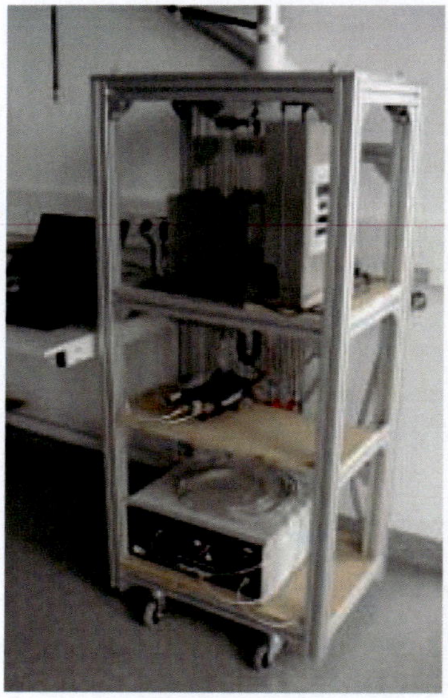

Figure 6. Experimental setup to test the filtration efficiency.

The test rig consists of a particle dispersing unit that regulates defined particle loading in a gas stream, a particle analyzer system (aerosol spectrometer) that measures filtration efficiency and a separation unit that contains the filter bed. The setup was used to investigate the influence of such parameters as filter material and dust particle sizes, dust loading of the gas, temperature and gas velocity.

A test bench to analyze the removal of halogenides and sulfur components is currently under construction. This setup will consist of a gas supply unit with mass flow controllers to dose test gases, a fix bed reactor containing the adsorption materials and a gas analyzing system consisting of FTIR and GC/MS to determine the gas composition.

2.2.2. Experiments and Results

Filtration experiments were conducted on the test bench setup. An air flow with defined particle loading was flowed through the filter bed and the particle content of the gas flow leaving the filter unit was analyzed. For a first characterization of the filtration process, experiments were carried out using a fix bed. Figure 7 presents typical results of two fix bed filtration measurements.

Figure 7a plots the particle collection efficiency for an inlet flow rate of 3 m^3/h at room temperature with an inlet particle loading of 3 g/m^3 and a filter bed height of 50 mm in dependence on the over dust particle diameter and time.

The number of dust particles in the gas flow was measured at the filtration unit outlet every ten minutes and particle collection efficiency was calculated in consideration of the inlet conditions. The increase of collection efficiency over time can be attributed to the fixed bed's porosity, which decreases over time because particles accumulate in the filter bed. Particle collection efficiency also increases because separated particles form filter cake on the surface of the filter bed. The occurrence of surface filtration in addition to depth filtration enhances particle separation.

Figure 7b plots the particle collection efficiency versus the dust particle diameter after ten minutes under different inlet conditions.

Parameters such as filter height, inlet flow dust loading, inlet gas flow rate and filter bed material (spheres, cylindrical pellets) were changed. The yellow line (state of reference) represents the results achieved with an air flow rate of 3 m^3/h, dust loading of 3 g/m^3 and a filter bed height of 50 mm (alumina spheres with diameters of 1.8 mm). Varying the filter parameters will enable determining their influence on particle collection efficiency with an eye toward optimizing the filtration process.

Experiments regarding tar reforming are currently in preparation. For the tar reforming of tar a nickel catalyst will be used. Main tar components are benzene, toluene, naphthalene and PAK (e.g. phenanthrene).

In Case of fluidized bed gasification of wood the tar components amounts to approx. 3 to 10 g/m^3. Regarding on the utilization of the produced fuel the tar components has to be decreased to a certain value. For the, use in a gas engine without any difficulties the tar concentration has not to exceed the limit of 50 mg/m^3. In contrast to this tar can be used as additional fuel in high serves partly temperature fuel cells as SOFC. But in this case exceed a limit defined by the coke formation process which deactivates the fuel cell.

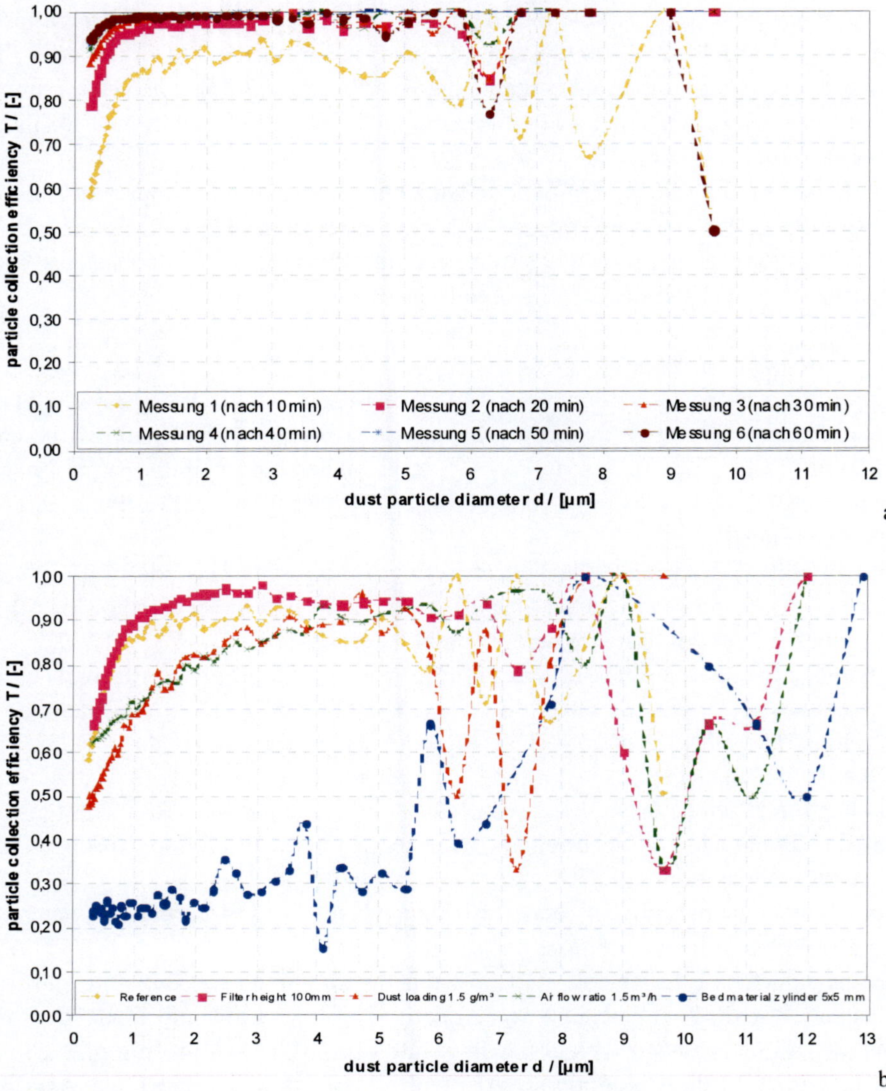

Figure 7. a) Particle collection efficiency versus dust particle diameter and time; b) Particle collection efficiency T as a function of different parameters (bed material height, dust loading, air flow ratio, and bed material).

2.2.3. Simulation

Dust Separation and Tar Reforming

A heterogeneous unsteady one-dimensional model was developed considering solid and gas phase to describe the dust filtration and the tar reforming process in a moving bed reactor. Depth and surface filtration effects are taken into account by the derived model. Parameters as bed porosity, velocity of the gas and of the solid material, bed particle diameter, diameter of the dust particles, dust loading of the gas flow, influence the filtration process and are considered by the model.

The filtration efficiency φ is calculated on the basis of the mentioned parameters according to Löffler [Löffler]. The tar reforming process is not completely implemented in the model at this time.

For the description of the filtration process, mass balances of the concentration of dust particles within the gas phase, c, and of the loading of dust particles on the bed material Θ were derived.

The dust concentration of the gas phase, c, refers the ratio of the dust particle mass to gas volume. The dust loading of the bed particles, Θ, represents the ratio of the dust particle mass and the bed material mass. The mass balances of the moving bed with particle filtration leads to the following partial differential equation system:

$$\frac{\partial c}{\partial t} = -\frac{\partial}{\partial z}(c \cdot u) - 1.5 \cdot \frac{1-\varepsilon}{\varepsilon} \cdot \frac{\varphi}{d_{cat}} \cdot c \cdot \varepsilon \cdot u \qquad \text{Gas phase}$$

$$\frac{\partial \Theta}{\partial t} = -\frac{\partial}{\partial z}(\Theta \cdot w) + 1.5 \cdot \frac{1-\varepsilon}{\varepsilon} \cdot \frac{\varphi}{d_{cat}} \cdot \frac{c \cdot \varepsilon \cdot u}{\rho_{cat}^{mean}} \qquad \text{Solid}$$

$$\varepsilon = 1 - (1-\varepsilon_0) \cdot \left(1 + \frac{\rho_{cat}}{\rho_{dust}} \cdot \Theta\right) \qquad \text{Bed porosity}$$

$$u = \frac{\dot{V}_{gas}}{\varepsilon \cdot A_0} \qquad \text{Gas velocity}$$

Figure 8 shows a typical simulation result of the dust loading of the solid Θ for a) a co-current filtration process, in which the gas and the solid possess the same flow direction, and b) a counter-current filtration, where the solid and the gas flows contrariwise. In both cases the gas enters the moving bed from the left and flows towards the reactors exit positioned on the right.

The gas flow rate amounts to 3 m^3(STP)/h with an inlet dust concentration of 1 g/m^3(STP), the diameter of the simulated moving bed is 40 mm and the moving bed height amounts to 50 mm.

It can be seen that for co-current operation the dust loading of the bed material increases till a maximum with increasing bed length but with low velocity the dust loaded bed material needs more time to leave the reactor.

Because of the low filtration time and bed velocity a Peak appears in the front reactor part. In case of the counter-current process the maximum dust loading of the bed material is under the filter bed surface and with increasing bed length decrease the dust loading of the bed material. Simply the front part of the filter-bed is used by the opposite movements of the bed and the gas stream to the filtration.

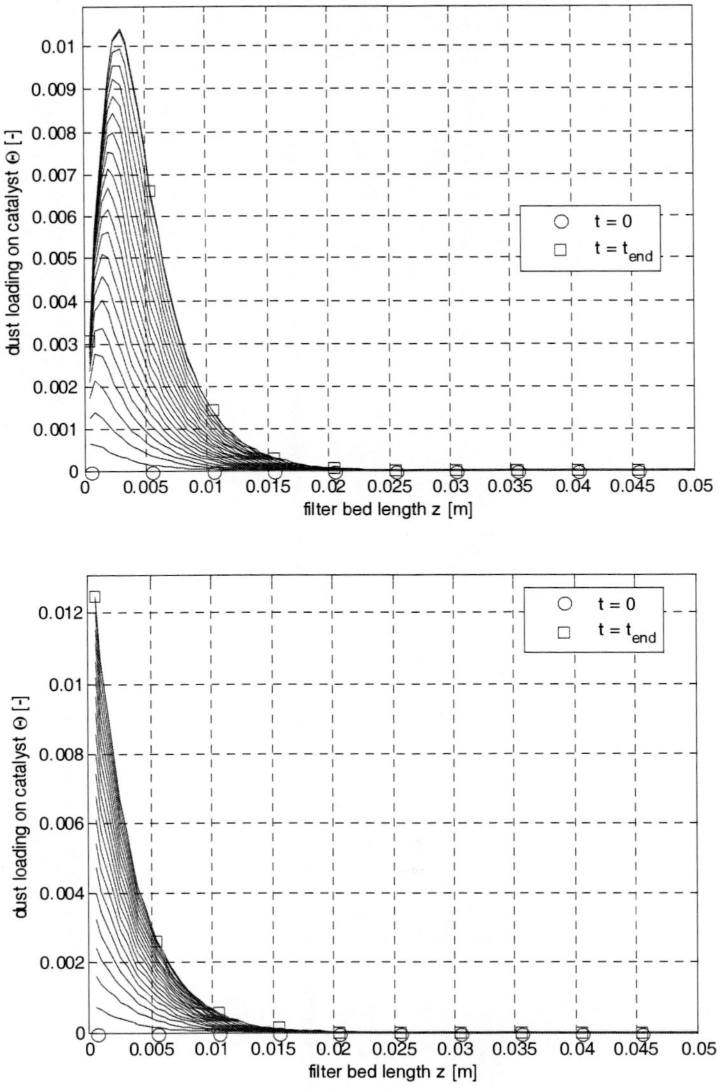

Figure 8. Simulated dust loading of the solid material during the filtration process in a moving bed for a) co-current and b) counter-current operation.

Halogenides Und Sulphur Removal

To select appropriate materials for the removal of hydrogen sulfide and halogen compounds as hydrogen chloride at higher temperatures, the partial pressures of H_2S and HCl at equilibrium were calculated for the following reaction with different metallic oxides (MeO):

$$MeO + H_2S <-> MeS + H2O$$

$$MeO + 2\ HCl <-> MeCl_2 + H2O$$

Figure 9 presents the calculated equilibrium partial pressure of hydrogen sulfide (Figure 9a) and hydrogen chloride (Figure 9b) with various adsorbents as a function of the temperature. As it can be seen, excepting CaCO3 and $Ca(OH)_3$ for H_2S the equilibrium partial pressure of H2S and HCl increases with increasing temperature. Thus, a separation of these components is promoted at lower temperature.

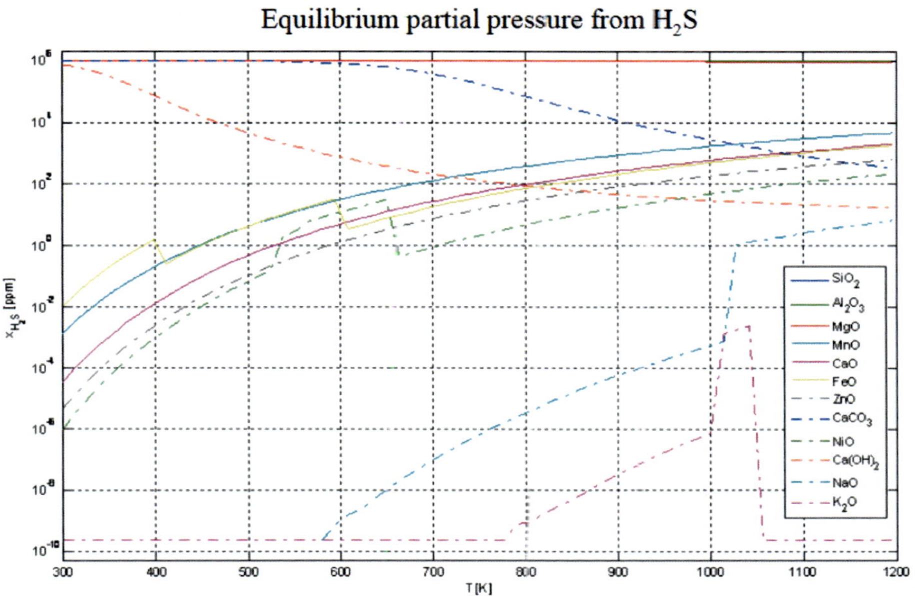

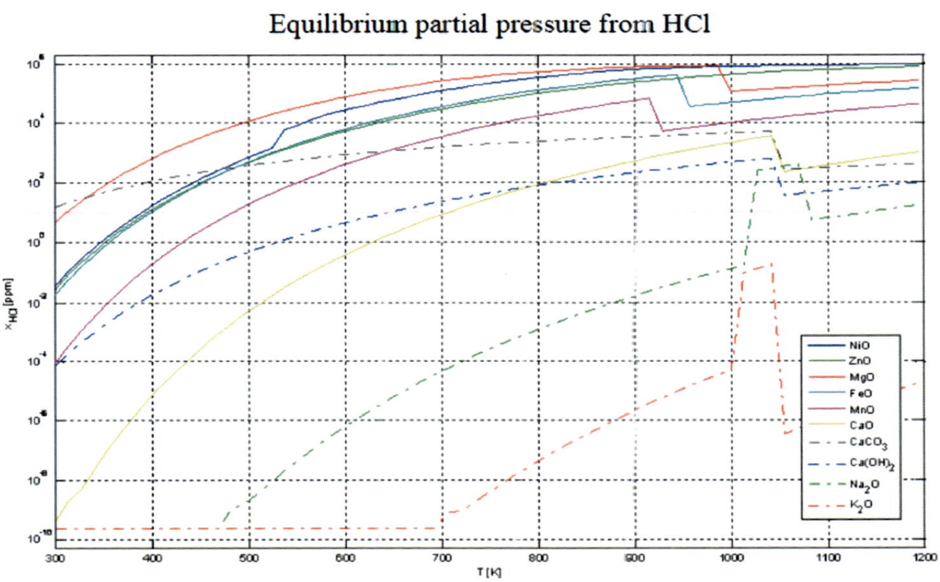

Figure 9. Equilibrium partial pressure of a) hydrogen sulfide and b) hydrogen chloride with various materials as a function of the temperature.

In preparation for the test and for the characterization of different metal oxide materials regarding the removal of hydrogen sulfur and hydrogen chloride in a fixed-bed reactor, a heterogeneous unsteady one-dimensional reactor model considering gas and solid phase under isothermal conditions was developed. The model allows the calculation of breakdown curves of fixed-bed reactors considering axial dispersion effects in the gas phase and competing reactions of the different gas phase components (H_2S, HCl) with the metallic oxide. Based on the mass balances, the following partial differential equation system results:

$$\frac{\partial x_\alpha}{\partial t} = -v\frac{\partial x_\alpha}{\partial z} + D_\alpha \frac{\partial^2 x_\alpha}{\partial z^2} + \frac{p_{ges}\Theta_B}{c_{total}}\left(\nu_{\alpha 1}k_1 x_A + \nu_{\alpha 2}k_2 x_E^2\right)$$

$$\alpha = A, D, E$$

$$\frac{\partial \Theta_\beta}{\partial t} = \frac{M_\beta p_{ges}\Theta_B}{\rho_{B,max}}\left(\nu_{\alpha 1}k_1 x_A + \nu_{\alpha 2}k_2 x_E^2\right)$$

$$\beta = B, C, F$$

x_α denotes the molar fraction of the gas component α, v refers to the gas velocity, D is the axial diffusion coefficient, k are the reaction rate constants for reaction 1 and 2 (with ν as the stoichiometric coefficient of component α in reaction 1 or 2), p denotes the total pressure and c is the total concentration. Θ means the coverage of the active sites, ρ is the density and M is the molar mass of the solid material.

The following reaction systems were considered in the simulation study done:

$$A_{gas} + B_{fest} \xrightarrow{r_1} C_{fest} + D_{gas} \qquad \text{e.g. } H_2S + ZnO \rightarrow ZnS + H_2O$$

$$2E_{gas} + B_{fest} \xrightarrow{r_2} F_{fest} + D_{gas} \qquad \text{e.g. } 2HCl + ZnO \rightarrow ZnCl_2 + H_2O$$

A typical simulation result is shown in Figure 4. In this case a mixture gas with 150 ppm HCl and 200 ppm H_2S diluted in N_2 enters with a flow rate of 0.125 l(STP)/min a fixed-bed reactor filled with ZnO-pellets (bed diameter 0.01 m and bed heights 0.1 m). At the time t=0, the ZnO bed material is fresh and the active sites for the reactions 1 and 2 are unconverted. After entering the reactor, H_2S and HCl reacts with the unconverted ZnO sites to ZnS or $ZnCl_2$ and H_2O. The concentration of H_2O in the product gas as a function of the time and the axial reactor coordinate is illustrated in Figure 10a. It can be observed that no reaction takes place after 60 min as all ZnO sites are consumed. In Figure 10b the breakdown curves of both reactants can be seen. Figure 10b shows analogue to Figure 10a that after approx. 60 min no reaction occurs and the concentrations of H_2S and HCl at the outlet correspond to the inlet concentrations.

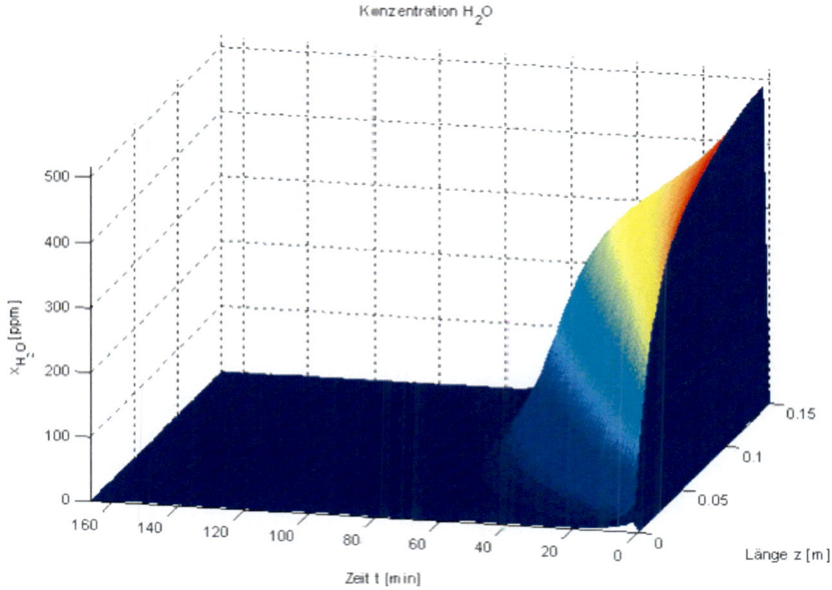

a)

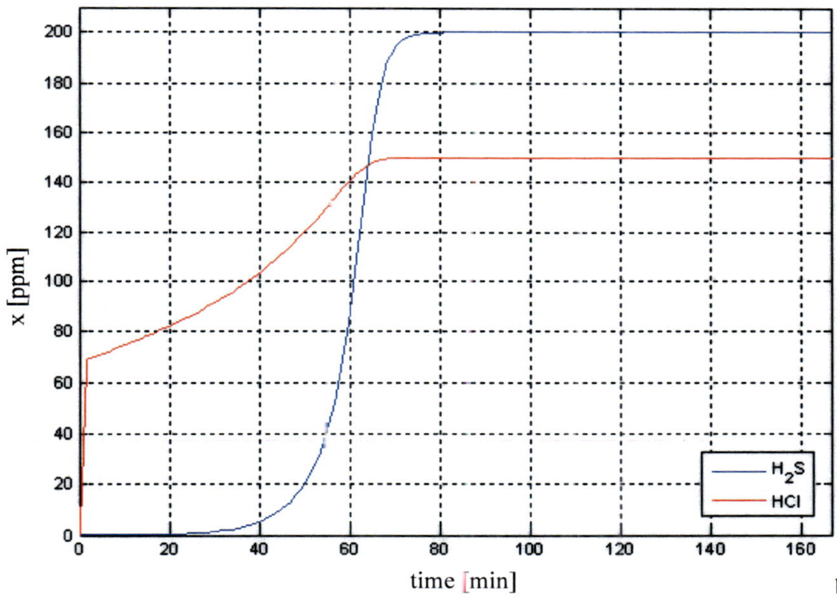

b)

Figure 10. a) H_2O concentration in the product gas versus time and axial reactor coordinate and b) breakdown curves of both reactants H_2S and HCl.

CONCLUSION

In order to develop an integrated biomass gasification and fuel cell system, theoretical and experimental investigation on high quality fuel gas production through biomass gasification with fluidized bed technology and the following high temperature gas cleaning processes are being intensively performed to meet the stringent specification of fuel gas for fuel cell. Results from experiment and simulation are discussed in this paper.

For biomass gasification with air, equivalence ratio ER is proved to be the most important factor. Higher ER will lead to a product gas with lower calorific value since part of the gases is burnt out with oxygen. Biomass gasification with lower ER, on the contrary, will produce a fuel gas with higher calorific value but also higher tar content. Further test runs are needed to find out the optimum values of operation parameters, e.g. equivalence ratio ER and reactor temperature.

In future work, gasification experiments with different gasification agents such as steam, oxygen and air/oxygen mixture will be carried out and additives, e.g. dolomite and olivine, will be used to reduce the tar content and increase the H_2 production.

The cleaning of fuel gas originating from biomass gasification by the presented concept is especially advantageous to the efficiency of the whole process, because the temperature level of the gas stream is held over the complete plant. The present state of technology requires a cool down of the fuel gas to approx. 250 °C for filter and wet scrubbing processes with a following heat up to approx. 800 °C for use in gas turbines or fuel cells (SOFC). This is completely avoided by the featured concept.

This concept offers additional benefits by saving of purchase costs and operating expenses because less apparatuses compared to established low temperature cleaning processes is needed. The simulation results as well as the experimental results have shown that the realization of this cleaning concept is very promising. With the dust separation and the tar reformation together in one apparatus the moving bed reactor is a matter of optimization.

A very important point in the fixed bed process for removal of halogenides and sulphur compounds is the bed material. From the huge number of suitable bed materials an optimum has to be found for the respective application. Depending on the limit values of the following prime mover the moving bed reactor for dust separation and tar reformation as well as the fixed bed reactor for halogenides and sulphur removal has to be adapted.

REFERENCES

[Löffler] Löffler, F. Staubabscheiden; Georg Thieme Verlag: Tübingen, 1988.

[Narváez] Narváez, I.; Orío, A.; Aznar, M.P.; Corella, J. Biomass gasification with air in an atmospheric bubbling fluidized bed. Effect of six operational variables on the quality of the produced raw gas. *Ind. Eng. Chem. Res.* 1996, 35, 2110-2120.

[Tepper] Tepper, H. Zur Vergasung von Rest- und Abfallholz in Wirbelschichtreaktoren für dezentrale Energieversorgungsanlagen. *Dissertation.* 2005.

ENVIRONMENTAL IMPACT

In: Renewable Resources: Obtaining, Processing… ISBN 978-1-6060692-534-8
Editors: R. Kozlowski, G. Zaikov and F. Pudel

Chapter 22

Environmental Problems in Production of High-Density Activated Carbon from Wood

Ausma Tardenaka**, Janis Rizhikovs, Baiba Spince and Janis Zandersons
Latvian State Institute of Wood Chemistry, 27 Dzerbenes Str., Riga, LV 1024, Latvia

Abstract

The production of high-density and mechanically strong granular activated carbon from deciduous wood is started with hydrothermal treatment of chips to obtain modified wood. The undesirable formation of effluents and exhaust vapours can be minimised by technological means. By switching over from a batchwise process in an autoclave type apparatus to a flow reactor and treatment of chips by saturated steam, the total consumption of steam, and the overall amount of effluents and exhaust vapour can be reduced from 5-6 kg to 2-3 kg to produce one kilogram of modified wood.

Introduction

A well-known fault of activated carbon of wood origin is its low density and high fragility. These poor mechanical properties are characteristic if the activated carbon is made from wood with a soft structure, e.g., from grey alder (*Alnus incana (L) Moench.*) charcoal. In general, the capacity of an adsorbent is expressed in terms of the surface area or the amount adsorbed per weight unit. High density adsorbents are required because only a certain volume of the adsorbent can be packed into adsorption facilities. In order to improve the adsorption capacity per unit volume of the carbonaceous adsorbent, an attempt was made to prepare granular activated carbon of increased density and mechanical strength from granulated wood sawdust. In our studies, an attempt was made to use the self-binding properties of the hydrothermally treated grey alder wood.

* e-mail: tpd@edi.lv

Hydrothermal treatment of grey alder chips in aqueous or saturated steam media leads to solubilisation of hemicelluloses, leaving a solid phase enriched in both cellulose and lignin. Hydrothermal treatment at 160-180°C causes autohydrolysis. The water-soluble destruction products located in the bulk of chips, hogged chips or sawdust are effective binders during the process of moulding or granulation at elevated temperatures. It is considered that, when autohydrolysis of xylan-containing lignocellulosic material is carried out under suitable conditions, the major reaction products are substituted xylooligomers, which can be further decomposed. The treatment of some agricultural residues (rice husks, corncobs) caused a substantial solubilisation of the raw material: 21.0–36.2 o. d. wt % of the raw materials, and the formation of acetic acid (0.48-0.64% o. d. mass basis) and furfural (0.26-0.47% o. d. mass basis) [1].

Grey alder is a widespread tree species in Northern Europe. Due to the fast growth rate and symbiosis with actinomyces of the genus Frankia, which develop nodules on its roots and fix atmospheric nitrogen, grey alder is not only an excellent pioneer species for re-establishing woodlands on even quite poor soils, but also is suggested for short rotation forestry stands. Our idea was to look for the feasibility to produce more value added products from grey alder wood than merely fuel chips. We chose carbon materials of specific use: high density and mechanical strength activated carbon and granules, slabs or boards of enhanced conductivity charcoal.

The basic processes of the technology are hydrothermal treatment, drying, granulation, carbonisation and activation with superheated steam. The main problems of technological by-products or emissions emerge in the wood modification process – during autohydrolysis and drying. On the basis of laboratory experimental data, the provisional amount of emissions during the production of activated carbon from grey alder wood and the feasibility of restriction are estimated.

EXPERIMENTAL

2.1. Starting Material

Debarked grey alder chips stored at room temperature were ground in a laboratory cutter mill using a 2-mm sieve. The obtained sawdust-like material was used for the experiments. The initial moisture content necessary for the experiments was achieved, mixing the chips with the equilibrium moisture content corresponding to room temperature (about 8% wet basis) with the calculated water amount and holding in an air-tight vessel at room temperature for 48-72 hours.

2.2. Hydrothermal Treatment

2.2.1. Treatment in an Autoclave

Hydrothermal treatment was carried out in a 1-L autoclave, which was placed in a thermostat and connected with a steam generator and a blow-off steam condenser-cooler. Saturated steam corresponding to the necessary temperature was used, which was periodically supplied in the autoclave with the chips. The process was controlled with steam and reaction mass temperature measurements using digital thermometers and a recording system.

The wood with the required moisture was placed in the autoclave and warmed-up to the designed temperature. Then steam was supplied into the autoclave from the steam boiler. The steam blow-off and regeneration were carried out every 15-20 min. The acids and furfural concentrations were determined in the condensate. For modified wood, moisture and hot water-soluble matter content [2] as well as acetic acid and furfural concentration in the extract were determined. The modified wood was dried at room temperature till the equilibrium moisture content.

2.2.2. Treatment in a Flow Reactor

Hydrothermal treatment was carried out in an 18-L cylindrical flow reactor with a heating jacket. Grey alder wood (particles measuring 0.63-2.00 mm) with appropriate moisture content in the amount approximately 2 kg (in terms of oven dry wood mass) was loaded in the reactor. The requisite temperature of the heating jacket was set, and the saturated or somewhat superheated steam flow was supplied at the rate 10-20 g/min. The condensate was collected and counted. The necessary temperature was maintained by adjusting the appropriate pressure in the reactor's chamber.

2.3. Determination of Acids Concentration

Reagents: Purified water; isopropyl alcohol; KOH (0.1 *M* solution in water).

Equipment: Analytical balance "Precisa" XB 220 A; "Radiometer Analytical" titrator, TIM 9000 and Tim Talk 9 computer program.

Analysis procedure: About 2 ml of the analysed solution was weighed on the analytical balance and placed in a 50-ml titration glass. Then 10 ml of purified water and 10 ml of isopropyl alcohol were added. Methods for acetic acid titration and titration parameters were set in the Tim Talk 9 program. Electrodes (glass and calomel) were immersed in the solution, and the solution was titrated with a 0.1 *M* KOH solution.

2.4. Determination of Monosaccharides (Sugars)

The α-diol groups reaction with periodate (Malaprade reaction) was used for the determination.

Reagents: Purified water; sodium periodate (0.2 water solution); ammonium molibdate, tetrahydrate (10% water solution); acetic acid 99.5%; potassium iodide (10% water solution); soluble starch (0.5% water solution); sodium thiosulphate (0.1 *M* water solution).

Equipment: Analytical balance "Precisa" XB 220 A; thermostated water bath "BioSan"WB 4; "Radiometer" automatic burette ABU 80.

Analysis procedure: About 0.2 ml of the analysed solution was weighed on an analytical balance and placed in a 50-ml measuring glass with a cup. Then 1 ml of 0.2 *M* Na periodate solution and 10 ml of purified water were added. The cup was put on the measuring glass, and the latter was placed in a thermostated water bath, with a temperature set to 35°C. A cover was placed on the bath, as, during the reaction with periodate, the samples should be in the dark. After 2 hours, the measuring glass was taken from the water bath, and 5 ml of 10% ammonium molibdate solution and 1 ml of acetic acid were added. The content of the glass was well mixed and left to stand at room temperature for 5 min. Then 1 ml of 10% Kl solution was added and, after 10 min, the released iodine was titrated with 0.1 m Na thiosulphate solution, using starch as the indicator.

Calculations: Hexoses react with 5 molecules periodate, forming 5 molecules iodate. In the reaction with iodide, each iodate forms 3 I_2. So, 15 I_2 molecules correspond to 1 hexose molecule, hence the hexose gram equivalent, which is used in the calculations according to this method, is 180:30 = 6. In the case of pentoses, the reaction proceeds with 4 molecules periodate, and the gram equivalent used in the calculations is 150:24 = 6.25.

2.5. Determination of Furfural

Reagents: Purified water; sulphuric acid 95-97%; anthrone (0.1% solution in sulphuric acid).

Equipment: Analytical balance "Precisa" XB 220 A; spectrophotometer Lambda 25 "Perkin Elmer".

Analysis procedure: About 0.15 ml of the analysed solution was weighed on an analytical balance and placed in a 50-ml volumetric flask. The volumetric flask was filled with purified water till the ringmark and mixed well. The aliquot 2.0 ml of the prepared solution was taken and transferred in a 50-ml beaker. 4 ml of anthrone solution in sulphuric acid was added to the solution, mixing carefully, and left to stay at room temperature for 45 min. Then light absorption measurements were carried out at the wavelength 600 nm using a spectrophotometer. From the calibration plot, the furfural amount was found, and the furfural content in the solution was calculated.

Results and Discussion

It is known that the solubility of wood in water grows with increasing temperature [3]. The reason for it is the increase of the medium's acidity as a result of acetyl groups' hydrolysis, when acetic acid is formed. The solution's pH can reach 4.5 to 3.5. So, heating the wood in the presence of water, the polysaccharide hydrolysis and lignin destruction products pass to the solution. The longer is pre-treatment time and the higher is temperature, the more is the water-soluble matter. At the temperature 150-175°C, the solubility of wood becomes essential, especially in the case of deciduous wood. Already within 1-2 hours, 20-30% of the wood bulk is dissolved. The content of hemicelluloses in wood decreases dramatically (even by 70%), while cellulose is susceptible to the hydrolytic action moderately. Sugars, oligomers, uronic acids, the product of partial dehydration of pentosanes – furfural, non-volatile organic acids, acetic acid as well as lignin destruction products pass to the solution. Soluble products of lignin and carbohydrates degradation, if they remain in wood, when pressing or granulating the modified wood at elevated temperature, can mutually react or partially condensate, forming substances with adhesive properties [4].

Table 1. Yield of modified wood and by-products in a blow-off autoclave at 170°C

Input wood		Modified wood		Consumption of steam, kg/kg M.W.	Blow-off vapour			Drying of modified wood		
Moisture, % wet basis	Measure of particles, mm	Yield, % o.d. wood basis	Water soluble matter, % o.d. M.W.*		Water, kg/kg M.W.	Furfural	Acetic acid	Water, kg/kg M.W.	Furfural	Acetic acid
						% condensate basis			% condensate basis	
30.0	2.00	98.7	22.5	2.2	1.33	0.50	0.38	1.29	0.03	0.03
	0.25	97.2	22.4	2.2	1.36	0.11	0.72	1.28	0.05	0.05
50.0	2.00	91.5	21.8	1.8	1.34	0.51	0.35	1.57	0.06	0.07
	0.25	92.3	24.1	2.6	1.66	0.25	0.37	2.04	0.04	0.04

* M.W. – modified wood (here and in the rest tables).

Our experiments demonstrated that, if grey alder wood is treated with water or steam at the temperature 170-172°C, more than 20% of wood are destructed and become hot water-soluble matter (see Table 1), from which 35 to 50% are monosaccharides. The volatile destruction products – acetic acid and furfural are less significant in terms of mass, while their presence in condensates, effluents and gas-vapour emissions would pose a certain range of technical problems and cause additional production costs due to wastewater treatment. Therefore, the main parameters of the choice of the optimum technological solutions are: a) the minimum quantity of furfural and acetic acid formation, minimum pollution degree of effluents and minimum energy consumption. For the same reason, the use of any hydrolysis catalysts was discharged, although a certain energetic advantage was therefore lost.

Three hydrothermal treatment versions were checked: hydrothermal treatment of water-saturated wood at 170-180°C in an autoclave, treatment of wood at the temperature 150-175°C in an autoclave with batch supply of saturated steam and blow-off, and treatment of wood with saturated or somewhat superheated steam in a flow reactor.

Experiments with water-saturated wood demonstrated that the heating-up of the reaction mixture from ambient temperature to 150-175°C as well as the drying of the resulting modified wood are time and energy consuming processes. While operating on a laboratory scale, drying of the wet modified wood mass is carried out in a thin layer at ambient temperature. On an industrial scale, drying of the highly wet mass will be a rather complicated and heat energy consuming process. To obtain 1 kg of dry modified wood, 5.0 to 5.6 kg of water should be evaporated together with furfural (0.14-0.55% o. d. m. w. mass basis) and acetic acid (0.10-0.23% o. d. m. w. mass basis).

More satisfactory results were achieved in a blow-off autoclave. The results are summarised in Table 1. The amount of the contaminated effluent and exhaust from the drying of modified wood is substantially less than in the case of the batch autoclave process.

The quality of the modified wood obtained in the batch process autoclave may be considered high if the water-soluble matter content is the accepted criterion of quality. Indeed, to a certain extent, the water-soluble matter content may be accepted as a criterion of the modified wood quality. Nevertheless, it must be pointed out that the mechanical properties of granulated modified wood, carbonised granules and granulated activated carbon are not in quite direct coherency with the hot water-soluble matter content of modified wood.

The amount of the emitted harmful substances per unit of modified wood produced is approximately only a half of that for the direct batchwise process. An especially positive feature of the blow-off process is the substantial decrease in the moisture content of modified wood that cuts the drying expenses.

Table 2. Yield of modified wood and by-products in a pilot scale flow reactor

Steam	Tempera-ture, oC	Treatment time, min	Modified wood		Consumption of steam, kg/kg M.W.	By-products, % o.d. M.W.	
			Yield, % o.d. wood basis	Water soluble matter, % o.d.M.W.		Acetic acid	Furfural
Saturated	170	90	80.3	15.0	6.6	0.6	1.2
Superheated	172	60	87.6	18.5	2.0	0.2	0.6
Superheated	172	30	92.4	18.0	1.0	0.32	0.3

Further improvement in the production of modified wood suggests the employment of a flow reactor of the prehydrolysis pilot installation of the Latvian State Institute of Wood Chemistry. The results of experiments on this apparatus are listed in Tables 2 and 3. A material contribution in the decrease of steam consumption and in the reduction of the treatment time is made by some superheating of steam (see Table 2). At the same time, the noxious by-products' yield is also reduced.

Table 3. The results of hydrothermal treatment of wood in a flow reactor at 180°C, pressure 0.9 MPa during 60 min

No	Moisture content		Condensate, kg/kg o.d.M.W.	Yield of modified wood, % o.d. wood basis	Consumption of steam, kg/kg o.d. M.W.
	Input wood, % wet basis	Modified wood, % wet basis			
1	40.0	56.3	1.28	86.7	1.82
2	41.7	49.6	1.87	82.8	1.99
3	39.3	59.2	1.59	82.4	2.26
4	41.0	56.9	1.80	85.4	2.30
5	43.1	57.3	1.96	88.9	2.45
Mean	41.0	55.9	1.70	85.2	2.16

To realise the practical approach, a pilot flow reactor was used in a series of experiments with damp wood at a slightly higher temperature of steam (see Table 3). The mean moisture of modified wood is only 55.9% on the wet basis, and that of the condensate yield 1.70 kg/kg of the modified wood demonstrates that the effluents and emissions are substantially reduced.

The environmental provisional situation is summarised in Table 4.

Table 4. Provisional situation of environmental problems in modification of wood by various technologies

Treatment type	Quantity of purifyable water, kg/kg o.d. M.W. mass basis	Acetic acid, % o.d. M.W. mass basis		Furfural, % o.d. M.W. mass basis	
		Blow-off condensate	Drying exhaust	Blow-off condensate	Drying exhaust
Autoclave	5.0 – 5.6	-	0.10 – 0.23	-	0.14 – 0.55
Blow-off autoclave	2.3 – 4.3	0.20 – 0.60	0.30 – 2.30	0.20 – 0.90	0.30 – 3.00
Flow reactor	1.8 – 2.4	0.60 – 2.30		1.91	

CONCLUSIONS

Hydrothermal treatment of water-saturated wood in an autoclave is simple, the quality of the modified wood is good, while the energy consumption is high and pollution with emissions, especially with furfural, is considerable. Treatment in a blow-off autoclave with batch supply of saturated steam makes it possible to diminish both energy consumption and the quantity of purifyable emissions to 2.3–4.3 kg/kg of modified wood.

A certain decrease in the emissions mass and energy resources will be reached with the help of pilot installations of the prehydrolysis flow reactor and some superheating of steam.

ACKNOWLEDGEMENT

The work has been financed by the Latvian State Research Program No. VP-4.

REFERENCES

[1] Garrote G., Falque E., Dominguez H., Parajo J.C. *Bioresour. Technol.* 2007, vol. 98, 1951-1957.

[2] Practical classes in wood and cellulose chemistry, ed. Nikitin V.M., *"Forest industry"* publishers, Moscow, 1963, pp. 50-95; 150-158 (in Russian).

[3] .Browning B.L. Wood chemistry, *"Forest industry"* publishers, Moscow, 1967, p. 16 (in Russian).

[4] Klauditz W., Stegmann G. Holz als Roh- und Werkstoff 1955, vol. 13, 434-440.

In: Renewable Resources: Obtaining, Processing... ISBN 978-1-6060692-534-8
Editors: R. Kozlowski, G. Zaikov and F. Pudel

Chapter 23

FIELDS OF OZONE APPLICATIONS

***S. Rakovsky*[1], *M. Anachkov*[1] *and G. E. Zaikov*[2]**
[1]Institute of Catalysis, Bulgarian Academy of Sciences, Sofia 1113, Bulgaria
[2]N.M.Emanuel Institute of Biochemical Physics Russian Academy of Sciences, Moscow 119334, Russia

INTRODUCTION

The ozone application is based on its powerful oxidative action. The ozonolysis of chemical and biochemical compounds, as a rule, takes place with high rates at low temperatures and activation energies [1]. This provides a basis for the development of novel and improved technologies, which find wide application in ecology [2], chemical, pharmaceutical and perfume industries, cosmetics [3, 4], cellulose, paper and sugar industries, flotation, microelectronics, veterinary and human medicine, agriculture, foodstuff industry and many others [5-8]. At present it is very difficult to imagine a number of high technologies without using ozone. Here should be mentioned the technologies for purification of waste gases, water and soils, manufacture of organic, polymer and inorganic materials, disinfecting and cleaning of drinking and process water, disinfecting of plant and animal products, sterilization of medical rooms and instruments, resolving cosmetics problems, sterilization and therapy in veterinary and human medicine, surface cleaning and functionalization in the manufacture of polymer and microelectronics articles, flotation, deodorization and decolourization of gases, liquids and solid substances, oxidation in the intermediate stages of various technologies, etc.

I.1. ECOLOGY

I. 1.1. Waste Gases

More than 100 contaminants of the atmospheric air have been identified. Among them SO_2, CO_2, nitrogen oxides, various hydrocarbons and dust constitute 85 %.

The main sources of harmful substances emissions, i.e. dust, SO_2 and CO_2 in the air, are the processing plants for coke, briquettes, coals, the thermal power stations, air, water, and road transport [9].The exhaust gases contain also, CO, organic and inorganic compounds, etc.[10]. Dust, sulfur dioxide, carbon dioxide, nitrogen oxides and organic compounds, etc. are the main pollutants released in the environment from metallurgy, the manufacture of fertilizers and petrochemistry. The most typical air pollutants emitted by some chemical productions are listed in Table 1.

Table 1. Main air contaminants emitted by some chemical productions

No.	Production	Pollutants	No.	Production	Pollutants
1.	Nitric acid	NO, NO_x, NH_3	12.	Ammoniumnitrate	CO NH_3, HNO_2, NH_4NO_3 - dust
2.	Sulfuric acid: a) Nitroso b) contact	NO, NO_x, SO_x, H_2SO_4, Fe_2O_3, dust	13	Superphosphate	H_2SO_4, HF, dust
3.	Hydrochloric acid	HCl, Cl_2	14	Ammonia	NH_3, CO
4.	Ocsalic acid	NO, NO_x, $C_2H_2O_2$, dust	15	Calcium chloride	HCl, H_2SO_4, dust
5.	Sulfamidic acid	NH_3, H_2SO_4, $NH(SO_3NH_4)_2$,	16	Chlor	HCl, Cl_2, Hg
6.	Phosphor	P_2O_3, H_3PO_4, HF, $Ca_5F_4(PO_4)_2$, dust	17	Caprolactam	NO, NO_2, SO_2, H_2S, CO
7.	Phosphoric acid	P_2O_3, H_3PO_4, HF, $Ca_5F_4(PO_4)_2$, dust	18	PVC	Hg, $HgCl_2$, NH_3
8.	Acetic acid	CH_3CHO, CH_3OH	19.	Artificial fibers	H_2S, CS_2
9.	Nitrogen fertilizers	NO_2, NO, NH_3, HF, H_2SO_4, HNO_3	20.	Mineral pigments	Fe_2O_3, $FeSO_4$
10.	Carbamide	NH_3, CO, $(NH_2)_2CO$, dust	21.	Electrolysis of NaCl	Cl_2, NaOH

The technologies with ozone participation are very promising for SO_2 utilization as sulfate, CO as carbonate, nitrogen oxides as nitrates.

Scrubbers used previously for removing only acids, for example HCl and HF, etc., now can be used for SO_2 separation via the addition of an oxidizing agent into the water intended for gas treatment.

Ozone as compared with the conventional oxidizers such as hydroperoxide, chlorine, sodium hypochloride, perchlorate , shows appreciably higher oxidizing efficiency [11].

The rate constant of ozone reaction with NO is about 1010 $cm^3/(mol.s)$, and with NO_2 it is ~ 107 $cm^3/(mol.s)$ whereby the oxidation of NO by ozone in the liquid phase is characterized by a higher absorption rate and rise in the concentration of the obtained HNO_3. Moreover the oxidation of NO can be accomplished in the exhaust gases or in the course of absorption [12].

The CO oxidation by ozone is carried out in the presence of Fe, Ni, Co and Mn oxides. In most cases ozonation is more appropriate than the conventional methods and sometimes it appears the only possible way for its preparation [13].

The purification of exhaust gases from burners working on liquid and gas fuel is accomplished by using oxidation catalysts of ABO3 perovskite-type oxides combined with ozonation. Ozone is injected prior the waste gases flow. A may be La, Pr or other alkali earth element; B may be Co, Mn or other transition metal. A may be also partially substituted by Sr, Ca or any other alkali earth element. A catalyst deposited on ceramic support of honeycomb type may contain $SrCo_{0.3}Mn_{0.7}O_3$. It is oxidized 24% CO in the absence of ozone and 80% in ozone atmosphere [14].

The removal of nitrogen oxides (NO_x) from waste gases is carried out through ozone oxidation in charged vertical column. The lower part of the column is loaded by 5-15% $KMnO_4$, and ozone is blown through the lower and upper part of the column and the waste gases are fed to the center of the column charge. The inlet concentration of 1000 ppm NO_x in the waste gas is reduced to 50 ppm in the outlet gas flow. Another method for cleaning of exhaust gases from NO_x involves the application of plasma generator and ozonator [15] which practically leads to the complete removal of nitrogen oxides.

Sulfur containing compounds are removed from gas mixtures by silicon oil scrubbing and ozone oxidation of absorbed pollutants. A model system providing that the gas mixture contains methandiol and deimethylsulphide is fed at a rate of 9-120 m^3/h into silicon oil charged scrubber followed by ozone oxidation of the absorbate. The oxidates are extracted with water and the silicon oil is regenerated. Thus the purified gas mixtures practically do not contain any sulfur [16].

A special apparatus is designed for decomposing acyclic halogenated hydrocarbons in gases. It includes a chamber for mixing of the waste gases with ozone coupled with UV-radiation, ozonator, and inlet and outlet units. This method is very appropriate for application in semiconducting industry whereby acyclic and halogenated hydrocarbons are used as cleaning agents [17].

The mechanisms of ozonolysis of volatile organic compounds such as alkenes and dienes are discussed and the products output is determined by matrix isolation FTIR spectroscopy [18].

The synthesis of a material from zeolite via pulverization, granulation and drying at 500-700°C, cooling to 100-200°C , electromagnetic radiation or ozone treatment is described in [19]. The material thus obtained is suitable for air deodorization, drying and sterilization.

The purification of gases containing condensable organic pollutants can be carried out by gas treatment with finely dispersed carbon, TiO_2, Al_2O_3, Fe_2O_3, SiO_2 and H_2O_2 and ozone as oxidizers [20].

A wet-scrubbing process for removing total reduced sulfur compounds such as H_2S and mercaptans, as well as the accompanying paramagnetic particles from industrial waste gases is proposed [21]. For this purpose a water-absorbing clay containing MnO_2 is used. The collected clay is regenerated by ozone oxidation.

The removal of mercury from waste gases is carried out by catalytic ozonation [22]. The used gases mixed with ozone are blown over a zeolite supported Ni/NiO catalyst. This procedure results in 87% conversion of mercury into mercuric oxide which is isolated by filtration. Pt and CuO_x/HgO system is proposed as another catalyst for this process [22].

I.1.2. Waste Water

In contrast to the cleaning of natural water by ozonation which is experimentally confirmed as the most appropriate, the purification of waste water by ozone is still an area of intensive future research. This could be explained by the great diversity of pollutants in the used water and the necessity of specific approach for each definite case.

The recycling of water from cyanide waste water (copper cyanide contaminant) is accomplished by ozone oxidation combined with UV radiation and ion exchange method [23]. Further, the oxidate is passed through two consecutive columns charged with cationite and anionite resins. The ozonolysis priority over the conventional chlorinating method is demonstrated by the fact that ozonation fails to yield chlorides whose removal requires additional treatment.

The ozonolysis of CN-ions leading to CNO^- -ion formation combined with UV-radiation proceeds at 3 fold higher rate as compared with that without radiation employment. Upon exchange in NaOCN, the Na^+ is substituted by H^+ followed by its decomposition to CO_2 and NH^{4+} via hydrolysis in acid medium. The ammonium ion is absorbed by the cation-exchange resin. The solution electroconductivity on the anion-exchange resin becomes lower that 10 muS/cm and cyanides, cyanates or copper ions have not been monitored. On the basis of these experimental results, a recycling method and apparatus for detoxification of solutions containing cyanides is proposed. The recycled water from the cyanide waste water may be re-used as deionization water in gold - plating. The process ion-exchange resins are regenerated after conventional methods. Actually, this method is not accompanied by the formation of any solid pollutants [23].

The waste water from the electrostatic and galvanic coatings containing CN^- -ions and heavy metals is treated consecutively by ozone and CO_2. The cyanides and carbonates being accumulated in the precipitate are already biodegradable [24].

Waste water containing Cr(III) is oxidized by ozone combined with UV-radiation [25]. The radiation reduces the oxidation time about three-fold. The oxidate is consecutively passed through columns charged with cationite and anionite resulting in the obtaining of waste water with electroconductivity below 20 mu.S/cm. This deionized water can be successfully used for washing of coated articles. The Cr(IV) concentrated solution from the anionite regeneration can be further treated by appropriate ion-exchange method yielding a high purity Cr(IV) solution [25].

Purification of waste water containing $ClCH_2COOH$ and phenol is carried out by ozone treatment and UV-radiation in the presence of immobilized photocatalyst. Thus the rate of decomposition is higher than the rate obtained only in the presence of ozone and UV-radiation [26].

Used water containing small amounts of propyleneglycol nitrate or nitrotoluenes is a subject to combined oxidation with ozone and H_2O_2, under pressure and heating to the supercritical point. Thus a nominal conversion of the contaminants higher than 96.7% is achived [27].

In laboratory experiments for reduction in residual COD in biologically treated paper mill effluents it is subject to ozonation or combined ozonation and UV radiation at various temperatures and pressures [28]. At a ratio of the absorbed ozone to COD less that 2.5:1 g/g, the elimination level with respect to COD and DOC is up to 82% and 64%, respectively. The ozone consumption is essentially higher in the case of the UV combined ozonation at pH>9 and elevated temperature to 40°C.

The decolourization and destruction of waste water containing surface active substances is performed by ozone treatment at ozone concentration of about 80 mg/l. It has been found that for waste water which has undergone partial biological treatment by fluent filters or anaerobe stages up to COD concentration > 500 mg/l and ratio 30D5/COD>0.2 ozonolysis with ozone concentrations up to 1.8 g/l does not change CBR, while after the complete biological treatment at COD<500 mg/l ozonation results in rise of the ratio 30D5/COD from <0.05 to 0.37. This evidences the increase of CBR [28].

The purification of waste water from paper manufacture characterized by enhanced biodegradability, as well as the removal of COD and halogenated compounds is carried out by a method based on ozonation and biofiltration [29].

Waste water from the paper pulp production is clean up by treatment with ozone and activated acid tar. The high values of pH favor the lignin decomposition and carboxylic acids formation since in this case the $^{\bullet}OH$ - radicals and not ozone are the oxidizing agent. The application of activated acid tar has a positive effect on the dynamics of the microbiological growth, substrate consumption and CBR of organic acids. For example, maleic and oxalic acids are decomposed completely at ozonation of waste water in the presence of activated acid tar. For immobilization of biological culture the use of polyurethane foams appear to be very suitable [30].

The used water in collectors of a plant for paper manufacture are ozonized in a foamy barbotating contactor. The results indicate that ozone is very efficient in the oxidation of coloring and halogenated compounds. Its activity is proportional to the amount of absorbed ozone regardless of the variations in the gas flow rate, the inlet ozone concentration and contactor design. The amount of consumed ozone depends on the operating conditions and waste water characteristics. The absorption rate rises in the presence of stages including ozone decomposition, particularly at higher rates [31].

In biological granular activated carbon (GAC) columns the effect of pretreated ozonation on the biodegradability of atrazine is investigated. The metabolism of isopropyl-^{14}C-atrazine gives higher amount of $^{14}CO_2$ than the ring UL-^{14}C - atrazine which shows higher rate of dealkylation that the process of ring cleavage. The pretreatment with ozone increases the mineralization of the ring UL-^{14}C - atrazine and consequently raises the GAC-columns capacity. 62% of the inlet atrazine is transformed into $^{14}CO_2$ in columns charged by ozonized atrazine and water. However, in columns with nonozonized atrazine and water only 50% of the inlet atrazine is converted into $^{14}CO_2$ and in columns supplied only with nonozonized atrazine only 38% of it are converted to $^{14}CO_2$ [32].

The waste water from petrochemical industry contains substantial amounts of phenol. The phenols mixtures appear to be more toxic that phenol itself and possess a synergetic effect. Their decomposition to nontoxic products is important and modern problem. The oxidation of phenols such as *p*-cresol, pyrocatechin, rezorzine and hydroquinone by O_3 in aqueous medium appears to be a promising method for their degradation. At concentrations of 100 mg/l and pH = 11.5 their complete oxidation is performed for 30 min.

The phenol content in used water is reduced from 145-706 mg/l to 2.5 mg/l at ozone consumption of 1.1 - 2.6 g/l. In real conditions the ozone consumption varies from 5 to 10 g/l per g phenol (Table 2).

The treatment of concentrated aqueous solutions of phenol (1.0g/l) with ozone causes the appearance of a yellowish coloring after absorption of 1 mol O_3 per 1 mol phenol which gradually fades away. In this case the higher phenols undergo slight oxidation and H_2O_2 is identified in the water after the treatment.

Table 2. Results from the ozonation of waste water from coke processing $[O_3]$ = 5 mg/l, flow rate 6 l/min and treatment time - 4 h

Pollutant, mg/l	Before ozonation	After ozonation
Monophenols	710	0.8
Polyphenols	380	198
Cyanides	3	29
Thiocyanides	384	0
Thiosulfates	538	0
Sulfides	43	2
H_2O_2	0	12
Bases	114	27

The tests demonstrate that upon varying content of phenols in the used water the ozonation is one of the most promising method for their removal. The ozonation was found to be applicable for the decomposition of rhodanide in neutral and weak acid medium in the temperature range of 9-25°C. The ozone consumption in this case is 2 mg/mg. The ozone uptake for the oxidation of the cyanide ion (CN^-) is 1.8 mg/mg.

The complex Zn cyanides are oxidized similarly to the simple soluble cyanides. Regardless of the higher resistance of Cu cyanide complexes, the ozone treatment reduces the CN^- content substantially (traces). Copper carbonate is precipitated as a reaction product.

The ozonation can be successfully employed as a method for cleaning the waste water from the manufacture of ammonium nitrate containing CN^-, S^{-2}, CSN^- ions. The purification level is 83, 98 and 95% respectively at O_3 consumption of 0.7 g/l/g water for 10 min contact.

The ozonation of waste water from sulfate-cellulose production results in 60% purified used water which is returned in the process cycle.

The cleaning of waste water from the petrochemical plants by ozone is efficient under wide range of pH (5.8 - 8.70 and temperatures (5 - 50°C). Thus the level of petrochemicals in the waste water at pH 5.8, ozone consumption of 0.52 mg/mg and 10 min contact is reduced from 19-33 mg/l to 2 mg/l. In Table 3 is shown the efficiency of various treatment procedures for deodorization of petrochemicals contaminated water.

Ozonation is very economical and highly efficient method for decomposing carcinogenic substances such as 3,4-benzopyrene, etc., particularly after the biological purification of waste water.

Table 3. Result from deodorization of petrochemicals contaminated water (grades/dilution)

Petroleum products	Before treatment	Coagulation and filtration	Charcoal and filtration	Chlorination	Ozonoation	Coagulation, filtration and ozonation
Petrol	5/80	5/60	5/50	5/80	2÷3/10	1/0
Kerosin	5/100	5/80	5/70	5/100	4÷5/25	1/0
Petroleum	5/100	5/60	5/60	5/100	5/20	2/5
Oil	5/10	1/0	1/0	5/10	1÷2/2	1÷0/0

The application of ozone is also advisable for mercury oxidation in chlorine producing, for decontamination of the used water from the poisonous tetraethyllead, for destroying pesticides traces, etc.

The highest efficiency of cleaning of petrochemicals contaminated water is achieved at the following sequence: primary cleaning (mechanical), secondary (biological) and complete purification (ozonation).

Ozone is also successfully applied for the complete purification of waste water form dyes manufacture. Their color decreases about 10 000 times at exposure to 0.22 g/l ozone and the biological activity of the reaction products is reduced almost to zero. The semi-industrial experiments in chemical plant "Kostenetz" (Bulgaria) were carried out on the dyes manufactured in the plant (Table 4):

Table 4. Ozonated days in the plant Kostenetz

Order number	Day's name	Empiric formula	Molecular weight	Structural formula
1.	Direct Congo red	$C_{32}H_{26}O_6N_6S_2Na_2$	698	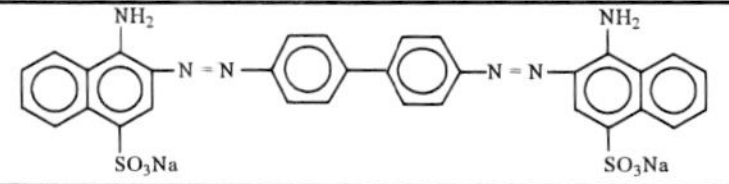
2.	Direct blue KM	$C_{34}H_{30}O_5N_5S_1Na_1$	643	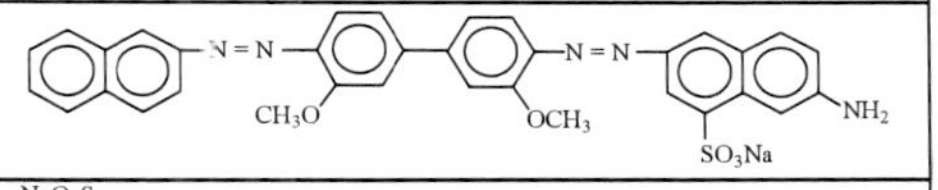
3.	Acid blue ZK	$C_{26}H_{20}O_{10}N_3S_3Na_3$	699	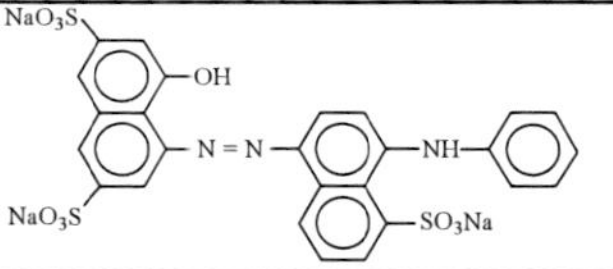
4.	Acid chromium green	$C_{16}H_{11}O_{10}N_4S_2Na_2$	529	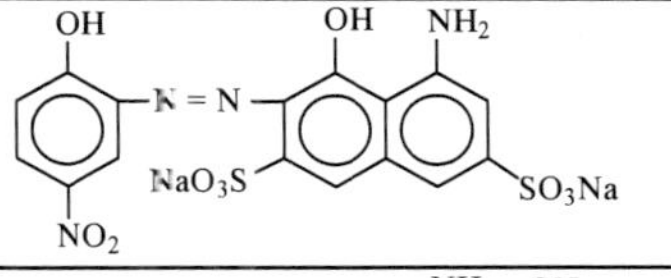
5.	Acid black ATT	$C_{22}H_{16}O_9N_6S_2Na_2$	618	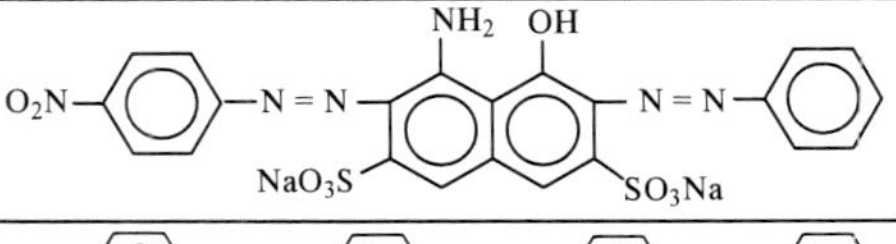
6.	Acid sulfonic blue	$C_{32}H_{25}O_6N_5S_2Na_2$	685	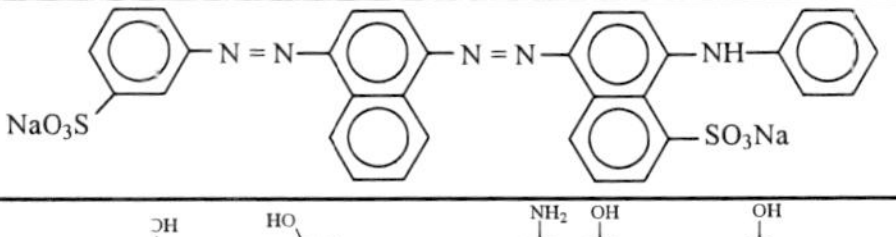
7.	Acid chromium black C	$C_{22}H_{19}O_5N_6S_1Na_1$ + $C_{16}H_{13}O_7N_4S_1Na_1$,	502 428	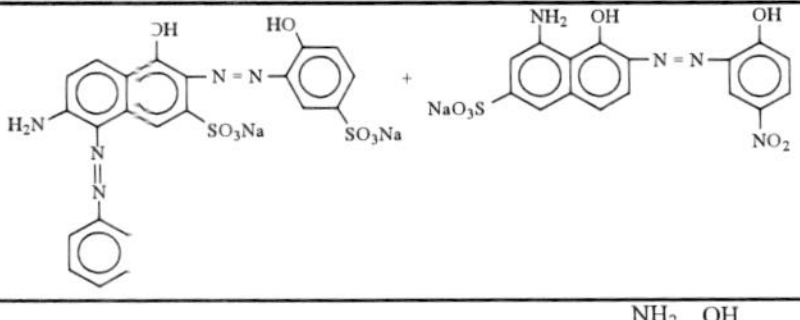
8.	Direct black Z	$C_{34}H_{26}O_7N_8S_2Na_2$	768	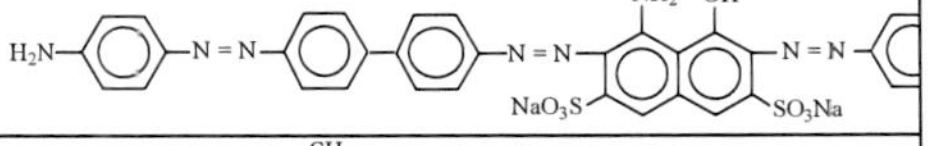
9.	Acid chromium yellow R	$C_{17}H_{14}O_3N_3S_1Na_1$	367	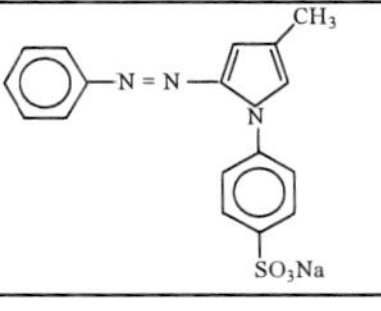

In the process of developing a technology for water purification from various organic contaminants, we have proposed a method and apparatus for quantitative determination of chemical compounds, separated by means of liquid chromatography. Special attention is paid to those compounds susceptible to ozone attack such as olefin hydrocarbons and their derivatives, phenols, amines, thiocarbamates, inorganic compounds, etc. The characterization of the separated compounds is carried out by spectral, mass spectral, refractometer, electrochemical analyses, but the detectors in most of these equipments are rather expensive and complex units without sufficient identification capabilities, particularly for the compounds pointed above. In this sense the apparatus designed by us allows to overcome the disantvantages of the known methods and equipments. The scheme of the ozone detector applicable for identification and liquid chromatography separation of chemical compounds is presented in Figure 1.

The ocean power stations operate on the basis of the temperature difference between the upper and lower water layers. Taking advantage of the one-step solution of ozone in depth, the COD is decreased which prevents the befouling in the piping systems found in the upper layers. The ozone-enriched water from the lower layers is supplied to a cyclic filter and is directly injected into upper seawater for control of the sea microorganisms' growth [33].

The removal of benzofurenes from ash, clays, soils, water and oils by combustion, ozonolysis of supercritical water, cracking-processes of petroleum, as well as thermal, photochemical and biological decomposition, has been the subject of many investigations [34]. The direct blowing of ozone through water pipers results in slimes removal from the inner pipers walls [35].

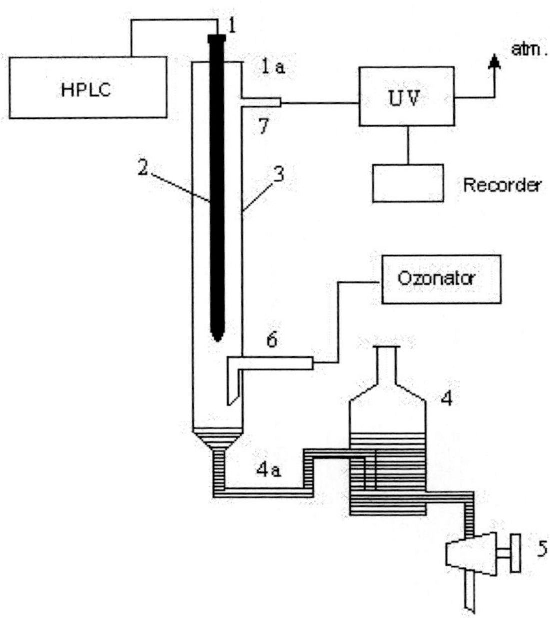

Figure 1. Set-up of ozone detector for identification of separated by liquid chromatography chemical compounds. 1 - eluent access; 1a - entry for pouring out the elate as thin film on 2 - solid glass stick; 3 - outer glass jacket; 4 - vessel for eluent accumulation; 4a - hydraulic détente; 5 - tap for pouring out the eluate; 6 - ozone inlet; 7 - ozone outlet.

The principal mechanism of advanced oxidation processes (AOPs) function is the generation of highly reactive hydroxy-radicals. Consequently, combination of two or more AOPs expectedly enhances free radical generation, which leads to higher oxidation rates.[36] Among various wastewater treatment options ozone-based AOPs such as ozone/H_2O_2, ozone/UV and ozone/H_2O_2/UV are key technologies for degrading and detoxifying of all major groups of pesticides, namely carbamates, chlorophenoxy compounds, organochlorines, organophosphates, aniline-based compounds, pyridines and pyrimidines, triazines, and substituted ureas [37, 38]. AOPs and ozonation are also promising for efficient degradation of pharmaceuticals in wastewater [39].

Photocatalytic oxidation and ozonation appear to be among the most popular treatment technologies for the wastewater compared with other (AOPs) as shown by the large amount of information available in the literature [40]. Photocatalytic oxidation and ozonation is a promising way to perform the mineralization of the wastewater substances like organohalogens, nitrocompounds, organic pesticides and insecticides, surfactants, and colouring matters [41-43].

The ozone blowing at a rate of 2.87 mg/min through aqueous solutions of 15 pesticides with 10 μg/l concentration leads to their transformation and their concentrations are reduced by about 20% for 60 min. The identified intermediates from the rearrangement of EPN, phenitrothione, malathion, diazinon, izoxathion and chloropiriphos are the corresponding oxones, while for phenthione they are the corresponding sulfoxide, sulfone and sulfooxide oxone analogues, and for disulfothione the intermediate is its sulfo analogue. 2-Amino-4-ethylamino-6-chloro-1,3,5-thriazine is identified as symazin and its 4-isopropylammonium derivative is atrazine intermediate. The isopropylthiolane results in obtaining of its 1-oxo-derivative. Upon replacement of ozone by air changes are actually not registered [44].

At the purification of process water in a soil decontamination plant containing pesticides the removal of organophosphorous compounds such as Thiometon and Disulphoton is of particular importance. In this connection four methods have been applied: 1) ozonolysis at pH = 2.5; 2) ozonolysis at pH = 8; 3) UV/H_2O_2 treatment and 4) oxidation by Fenton reagent [45]. The laboratory experiments were carried out with pure compounds soluted in buffered deoionized water and in process water which is extracted by solid-liquid extraction of the contaminated soil. The use of ozone in an acid medium turns to be ineffective since the Thimeton reaction practically stops after the formation of PO-derivatives. In all cases when the oxidation is carried out by HO - radicals, a sufficient removal of the pesticides and their metabolites is achieved. Oxadixyl, a cyclic nitrogen containing compound which is present in high concentrations in the process water is the most stable one, thus being the main soil contaminant. The investigated experimental conditions and the results obtained show that these methods are quite acceptable for universal application [45].

The water-soluble agrochemicals ASULAM and MECOPROP are decomposed for 30 min at ozone blowing (flow rate - 5.1 mg/min and concentration of 10 ppm)) through their aqueous solution (200 ml). The decrease in the flow rate of ozone retards the decomposition process. Similar treatments for 5h give in ACEPHATE and DICAMBA decomposition to 20%, the rate of decomposition being promoted with H_2O_2 addition [46].

The primary ozonation products of organophosphorous pesticides in water such as diazinone, phenthione (MPP) and ediphenphos (EDDP) [47], are identified by means of GC-MS analysis. The massspectra of the ozonation products of 17 organophosphorous pesticides point oxones as the primary reaction products. This fact is also confirmed by the SO_4^{2-}

generation resulting from the ozonation of their thiophosphoryl bonds. Oxones are relatively stable towards ozone attack but they are further hydrolyzed to trialkylphosphate and other hydrolysis products. However, with MPP the thiomethyl radicals are first oxidized to thiophosphoryl bonds giving MPP-sulfoxide, MPP-sulfone, MPP-sulfoxide-oxone are also obtained. Two main oxidation products have been identified at the oxidation of bis-dithio-type ethiones. Phosphate type EDDP is stable towards ozonolysis but its oxidation products are identified after hydrolysis [47].

Pesticides and their degradation products which are not mineralized to CO_2, NH_3, H_2O and inorganic salts can damage and contaminate the water piping. The treatment by photolysis or ozonation may substantially increase the mineralization rate. The photodestructive products of S-triazoles, chloroacetanylide and parakaute are declorinated and/or oxidized. The ozonation of these herbicides results in formation of products whose side alkyl chains are oxidized or removed, and the aromatic ring is oxidized or soluted but dechlorination does not occur. In some cases the medium may cause microbiological effect [48].

A deodorization method for treatment of ill- smelling air containing ammonia, hydrogen sulfide or amines from refrigerator chambers, lavatories, cattle-sheds, etc. involves the ozone contact with air in the presence of porous catalyst of 120 m^2/g specific surface. The used gas passes through a layer of carbon black for adsorption of the residual ozone and the nondegradated smelling substances. The active component of the catalyst may be: transition metal or its oxides supported on porous honeycomb type supports [49].

The contaminated smelling gases in drain and fecal water are dried to 60-30 % humidity, followed by ozone treatment on an oxidizing catalyst. For example, contaminated gas containing 30 ppm methylsulfide is dried to 30% humidity and then is treated with ozone in the presence of honeycomb type catalyst containing 83:12.5 = TiO_2:SiO_2:MnO_2 with a volume rate of 50 000 h^{-1}. In this case the deodorization efficiency is 99 % while at 100 % humidity it drops to 80% [50].

I.1.3. Natural (Drinking) Water

The basic sources of drinking water usually contain various organic admixtures in mg/l: carbon - 30, nitrogen - 0.8, fatty acids - 30, highmolecular organic acids - 0.2. naphthenolic acids - 1.5, phenols - 1.2, luminescent substances: neutral resins - 17%, oils, humus - 55%, acid resins - 19%, hydrocarbons, naphthenolic acids - 4-7% [51].

The coloring of natural water varying from pale yellow to brown is due to the presence of humus substances, Their decolorization is carried out by adsorption on coagulated $Al(OH)_3$ or $Fe(OH)_3$. The increasing demand for fresh water requires the use of more colored water which however is pre-chlorinated and then subject to adsorption up to their complete decolorization [52].

The ozone treatment removes the bad taste and odor of water, the resistant phytoplankton being oxidized to 20%. The ozone consumption for decolorization of natural water varies in broad limits - from 1 to 18 mg/l and depends on the humus composition [51]. The process of humus compounds oxidation by ozone is presented in Scheme 1.

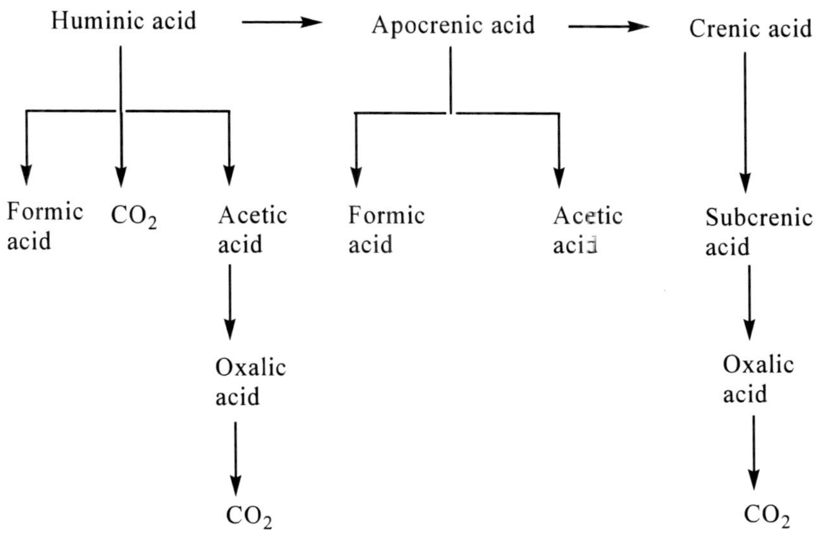

Scheme 1.

The various biological processes lead to formation of substances causing the bad taste and smell of water - mercaptanes, sulfides, alcohols, carbonyl compounds, acids, terpenes, amines, etc. The deodorization effect of ozone is preserved in wide range of pH values, temperatures and ion composition. Due to its high oxidation potential, ozone can also degrade the resistant toxic substances. It is known that during the blooming in the water reservoirs are accumulated biologically active substances among which algotoxine is particularly harmful for the human health. Its complete degradation is achieved by using ozone concentration of 14 mg/l and 50 min contact time [53].

At ozone concentration of 8 mg/l in water, only 6% of *Scendezmus* algae remain alive after 10-15 min. However is to be noted that the ozone action on the various algae is quite different and specific. Thus *Astetionella* type algae are very ozone resistant [52].

The oxidative, disinfective, decolorizative and deodorizative properties of ozone have found wide application for preparation of drinking water [54, 55]. Many world corporations such as the French Taileygas, Degramont, Ozonia, the German Fisher, AEG, Siemens, the Swiss Braun and Bovery, the American General Electrics, General Motors etc., are among the biggest producers of drinking water [56]. Twenty of the biggest world ozonator stations for drinking water treatment are listed below (Table 5).

Ozonation stations for drinking water preparation work in Portugal, Greece, Bulgaria, USA, etc. all over the world.

The technology of drinking water ozonolysis leading to its deodorization, disinfecting and decolorization includes gas units for dry air producing, ozonator, barbators or other mixing units, apparatus for HO•-radicals generation, promoting the ozone formation and filter-catalyst for decomposition of the residual ozone [57]. The preferable materials for barbators are zeolites or ceramics on the base of aluminum and/or silicon with pores diameter of 5-100Å, and the catalysts should contain one or more of the following components: Pt, MnO_2, CuO, Ni_2O_3, Zn, YiO_2 and SiO_2 [58].

Table 5. Some of the world biggest ozonation stations for drinking water treatment

City	State	Water, m^3/day	Ozone, kg/h
Moscow	Russia	1 200 000	200
Montreal	Canada	1 200 000	150
Ashford Common	GB	-	146
Coppermills	GB	-	144
Neuilly-sur-Marne	France	600 000	120
Choisy-le-Roi	France	900 000	120
Masan	Korea	-	112.5
Kiev	Ukrain	400 000	80
Helsinki	Finland	495 000	63
Manchester	GB	480 000	50
Orly	France	320 000	50
Singapore	Singapore	230 000	48
Minsk	Belaruss	200 000	47
Mery-sur-Oise	France	300 000	45
Lodz	Poland	190 000 290 000	40
Chiba	Japan	270 000	38
Nizhni Novgord	Russia	350 000	30
Bodenzee	Germany	480 000	30
Nant	France	380 000	30
Amsterdam	Holland	125 000	25
Brussels	Belgium	250 000	24
Quebec	Canada	220 000	19
Wrotzlaw	Poland	180 000	17

For disinfection of drinking water, the ozone-containing air is mixed with water in the water pipe, after which the bubbled water is supplied into a special designed chamber filled by porous ceramics charge for intensification of ozone dissolution into water. Some problems concerning the optimization of the UV stimulated oxidation of organic contaminants in soil water with ozone and hydroperoxide is discussed [59].

The disinfection of drinking water by ozonation can be also carried out at the outlet of various water supplying units before and after consummation, such as taps, fountains, water heaters, dental units [60], etc. In this connection are used small dimensional and low capacity personal ozonators coupled or installed additionally to these equipments.

The attempts for treatment of drinking water with ionization radiation appear to be ineffective mainly because of nitrates and H_2O_2 formation. These problems, however, do not occur when the radiation is performed in the presence of sufficient ozone amount [61]. The addition of ozone before and in the course of radiation converts the radiation into a purely oxidation process, a modification of the so called - Advanced Oxidation Process (AOP). The combined ozone-ionization radiation method is unique since two processes for HO-radicals generation are simultaneously induced. This method yields higher concentrations of HO-radicals as compared with other AOP. This in its turn results in precipitation of smaller amounts at equal ozone concentrations or reduces the ozone consumption at one and the same amount of precipitates. This makes the process particularly suitable for remediation of weakly contaminated soil water. The ozone concentrations at this process are higher than those used for drinking water disinfection. Thus, the drinking water subject to this treatment is disinfected more rapidly. The purification of waste water using this method is more efficient than the UV-radiation treatment only. This combination reduces COD without causing the rise of BOD. In this sense this method appears to be more attractive than the conventional two-stage processes of ozonation and biological treatment. The injection of gaseous ozone into the radiation chamber improves the water flow turbulence and substantially increases the efficiency of contaminants degradation.

The European organization "European regulation" has allowed the use of ozone for iron and manganese removal from mineral water, these initiating studies on secondary products formation, amount and composition, particularly during ozonation of bromine water. This water, usually contain BrO_3^- ions and various halogenated organic compounds (HOC). Some projects regulate a standard for admissible concentration limit (ACL) of HOC of 25 μg/l BrO_3^--ions, while after other it is from 10 to 0.5 μg/l. The present standard for LD of HOC is 100 μg/l [62]. Upon ozonation of bicarbonate mineral water - one sodium-rich water, and the other, calcium rich, it has been established that: 1) the process depends on the initial Fe and Mn concentrations and the Na and Ca concentrations; 2) HOC and bromate ions are secondary products and 3) the specific components characterizing a definite mineral water affect the ozonolysis. This requires definition of the conditions for Fe and Mn removal for each concrete type of mineral water. The removal of iron and manganese is efficient when the amount of ozone is three times greater than the stoichiometric one (0.43 mg O_3/mg Fe and 0.87 mg O_3/mg mn).

The disinfection of swimming pool water is carried out by controlled addition of compounds- bromine and iodine donors (NaBr or NaI) and water oxidation by hypochloride, ozone or potassium peroxymonosulfate [63]. We have designed a compact module system (stationary and mobile) which includes module for air preparation, ozonator, absorption column and unit for the complete decomposition of the residual ozone. Its capacity is 12-15 g/h and it is very suitable for sanitation of drinking water and used water from swimming pool (3.5 l/s capacity).

I.1.4. Soils

The decontamination of dredging wastes containing spilled petroleum is carried out by using a system for the direct injection of ozone (100-1000 ppm) containing air into pipes at the bottom; means for pumping out the spilled petroleum to a tank for its storage; units for its mixing with ozone for decomposition of organic matter in the tank; units for returning of the processed petroleum for the complete degradation of the residual organic matter by means of aerobic bacteria. The efficiency of organic matter decomposition is by about 20% higher than that by using the conventional methods including only treatment by ozone [64].

Another technology for treatment of contaminated soils [65] involves the ozonolysis of the water slurry of the contaminated soil layer, in the presence of H_2O_2 or in its absence and consecutive biological decomposition. The priority of the proposed technology over the conventional ones is that the removal of organic contaminants is by *on-site* treatment of the soils.

The *on-site* recyclation of petroleum contaminated soils by ozonation requires the good knowledge of the geological setting and the recycling system. Usually the recycling stages include: 1) system for air evacuation from the soil; 2) means for injection of ozone combined with air evacuation from the soil; 3) development of means for ozone generation; 4) screening of soils contaminants; 5) use of systems for safety control; 6) organization of the cleaning units - pumping and injecting drills; 7) decision for air injection into the soil; 8) methodology for measurement and regulation of ozone concentration; 9) hydraulic purification of the highly contaminated soil areas, etc. [66].

The treatment by ozone can be successfully applied for decomposing mineral oils, polycyclic aromatic compounds (PACs), phenols and some pesticides to biodegradable, nontoxic compounds. It should be noted that although ozone decomposes a great number of microorganisms, the soil microflora can be easily remediated after this treatment [67]. The latter can be carried out on- and out of the contaminated area. A number of problems related to the ozonolysis of organic compounds, the PACs degradation products, the effect of ozone on the soil microflora, the laboratory and pilot results, etc. are discussed.

A method and apparatus for decontamination of wastes such as soil, porous granulated sludge, are suggested. The polluted materials are placed in a special vessel and ozone is bubbled through. The waste gases obtained from the ozone oxidation are further treated with ozone [68].

A method for ozone treatment both *in-situ* and *on-site* has appeared to be applicable for the degradation of organic substances adsorbed on the solid surface [69].

Wet-scrubber method (based on MnO_2) for removing total reduced S-containing compounds such as H_2S, mercaptanes and non-magnetic particles from industrial waste gases using ozone, is suggested [21].

I.2. INDUSTRY

I.2.1. Manufacture of Organic Compounds

We will focus our attention to some of the numerous ozone applications in organic polymer and inorganic productions, passivation, cleanup and preparation of surfaces for electronics, superconductors, etc.

The reaction of ozone with olefins is the main reaction for ozonides preparation (1, 2, 4-trioxalanes) which can be converted into a mixture of carbonyl compounds, the exact composition of which depends on the olefin structure. This reaction is used for synthetic purposes as with small exceptions the C=C bond is cleaved quantitatively under very mild conditions. Usually the ozonolysis is carried out at low temperatures (as a rule at $-70 \div -30^{o}C$) as demonstrated in [3] Chapters 2 and 3. Mostly, ozonolysis of olefins, except in solid phase, is carried out in various solvents such as paraffins, halogenated hydrocarbons, oxygen, sulfur and nitrogen containing hydrocarbons like ahcohols, glycols, aldehydes, ketones, acids, ethers, esters, amines, amides, nitriles, sulfoxides, mineral acids, bases and mixtures from them, etc. In most cases the reaction products are not separated from the reaction mixture except for the purposes of ozonide obtaining, and the oxidates are subject to hydrolysis, reduction or oxidation [1,3].

The hydrolysis is applied for preparation of aldehydes [70] and ketones [71]. The reductive decomposition is carried out in the presence of Zn/AcOH, SO_3^{-2}, HSO_3^{-1}, $LiAlH_4$, $SnCl_2$, ArP_3, Ph_3P, SO_2, $(CH_3)_2S$, H_2/catalyst. The reduction by metal hydrides leads to alcohol formation in the decomposition products mixture while the use of other reductors yields carbonyl compounds. Actually there are no clear requirements for selection of the most appropriate reductor, but it turned out that dimethylsulfoxide is probably [72] the most preferable reductor in methanol solution. The ozonolysis in the presence of tetracyanoethylene yields directly carbonyl moieties omitting the reduction step [73]. A novel modification of the conventional methods of ozonolysis is exemplified by the ozonolysis of silica gel adsorbed alkenes (acetylenes) [74] and selective ozonolysis of polyalkenes, controlled with the help of dyes introduced in the reaction mixture [73]. Some examples of ozonolysis using various reductors for ozonides decomposition are given in Table 6 [75-80].

Ozonolysis is a powerful chemical method for economical and ecologically pure preparation of various carbonyl compounds [81].

Table 6. Preparation of aldehydes, ketones and alcohols during alkenes ozonolyis with consecutive reduction

Nî.	Substrate	Conditions	Product	Yield, %
1.	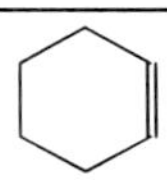	EtOAc, -20^0- -30^0C 2. $Pd/CaCO_3/H_2$	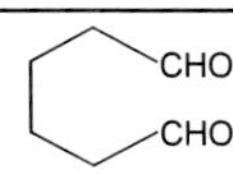	61
2.	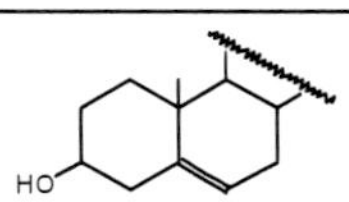	$n-C_6H_{14}$ Zn/HOAc	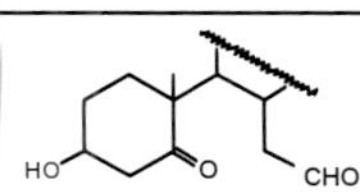	59
3.	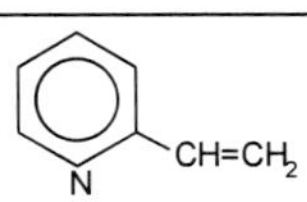	1. MeOH/-40^0C 2. Na_2SO_3/H_2O	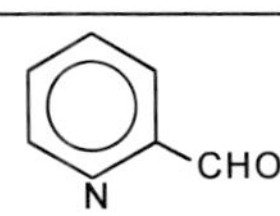	65
4.	$Me(CH_2)_5CH=CH_2$	1. MeOH/$-30^0\div -60^0C$ 2. Me_2S	$Me(CH_2)_5CHO$	75
5.	n-Bu-OOCCH=CHCOO-Bu-n	1. MeOH/$-65^0\div -70^0C$ 2. $(MeO)_3P$	n-Bu-OOCCHO	78
6.	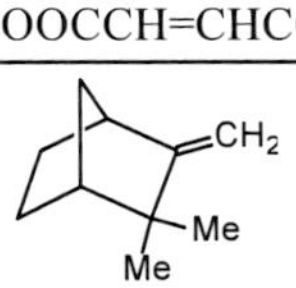	1. MeOH/$-35^0\div -75^0C$ 2. NaI/MeOH/HOAc	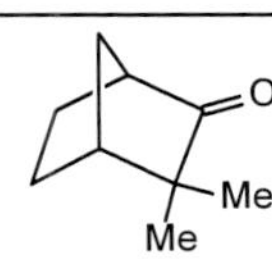	75
7.	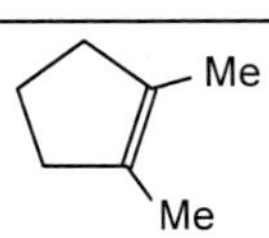	EtOAc/$(CN)_2C=C(CN)_2$, -70^0C	$MeCO(CH_2)_3COMe$ + 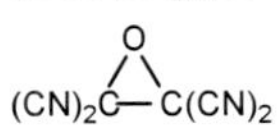	61
8.	$Me(CH_2)_5CH=CH_2$	1. n-Pentane/-38^0 - -42^0C 2. $LiAlH_4/Et_2O$	$Me(CH_2)_5CH_2OH$	93
9.	$H_2C=CH(CH_2)_8COOH$	1. MeOH/0^0C 2. $NaBH_4$/NaOH/ $EtOH/H_2O$	$HOCH_2(CH_2)_8COOH$	91

The ozonolysis of cyclic alkenes in protic solvents followed by reductive decomposition of the hydroperoxides formed is a classical method for dialdehydes synthesis [82]. This method is applied for the preparation of 3-ethoxycarbonylglutaric dialdehyde by ozonolysis of ethyl 3-cyclopentenecarboxylate. The dialdehyde is an intermediate in the synthesis of dolazetrone mesitilate which is an active medical substance in ANZEMET anti-emetik. Zn/AcOH, phosphines, amines and sulfides such as 3, 3'-thiodipropionic acid and its salts are the most suitable reductors for this system. Their efficiency is comparable to that of methyl sulphide but without its shortcomings. The polymer immobilized 3,3'-thiodipropionic acid is also very active in this reduction reaction [82].

The efficient conversion of oximes into the corresponding carbonyl compounds can be also carried out through ozone oxidation [83].

The application of ozone is a new and convenient way for the preparation of cyanoacetylaldehyde (3-oxopropylonitrile) (1) and its stable dimethyl acetal (3,3-dimethoxypropylonitrile) (2) used as valuable intermediates for organic synthesis [84]. For this purpose the ozonolysis of (E)-1,4-dicyano-2-butene or 3-butylonitrile is carried out. Then the oxidates are treated by DMS yielding 1. Further, compound 1 can be used either directly in the next reactions or is transformed into 2. The 2 output amounts to from 67 to 71%. The acetal can be again hydrolyzed to 1 by treatment with ion-exchange resin Amberlyst-15 [84].

The easily available allylphenyl ethers are ozonized at -40°C, treated with DMS giving solutions of the corresponding phenoxyacetaldehydes which are purified by column chromatography. Their reaction with 1-methyl-1-phenylhydrazine yields the corresponding hydrazones [85].

Quinolinealdehyde derivatives can be synthesized by ozonolysis of the respective quinolineolefins at -60 - -72°C in methanol or ethanol solution with consecutive reduction with DMS at the same temperatures. Then the reaction mixture is heated to room temperature for 1 h. The yield of the products amounts to 29%. [86].

6-Chloro-2-hexanone is prepared from 1-methylcyclopentane via three-step scheme: a) ozonolysis of 1-methylcyclopentane in carboyilic acid solution to 1-methylcyclopentanol; b) conversion of 1-methylcyclopentanol into 1-methylcyclopentyl hypochloride using NaClO; c) cyclization of 1-methylcyclopentyl hypochloride in 6-chloro-2-hexanone [87].

Ozonation is a stage in the synthesis of optically pure (R)-(+)-4-methyl-2 cyclohexen-1-one from (R)-(+)-pulegone and hydroxy ketone from (-)-*cis*-pulegol [88].

Polycarbonyl compounds and aldehyde-acids are the ozonolysis products of (+)-4a-[1-(triethylsiloxy)-ethenyl]-2-carene [89].

Carbonyl oxides obtained from vinyl ethers ozonolysis undergo rapid [3+2] cycloaddition with imines giving the corresponding 1,2,4-dioxazolidines in yields of 14-97% [90].

Carbonyl oxides can be also used for the synthesis of 3-vinyl-1,2,4-trioxalanes (α-vinylozonides) by a [3+2] cycloaddition with α,β-unsaturated aldehydes. However, α, β-unsaturated ketones are practically inactive in this reaction. The reaction of 3-vinyl-1, 2, 4-trioxalanes with ozone leads to the formation of the corresponding dizonides [91].

The carbonyl oxides prepared from the ozonolysis of enol ethers (for example, 1-metoxy-4,8-dimethylnone-1,7 diene) undergo stereoselective intramolecular cycloaddition with inactivated alkenes yielding bicyclic dioxalanes. The latter are easily converted into β-hydroxycarbonyl species and 1, 3 - diols by catalytic hydrogenation thus providing a new approach to the synthesis of 1,3-oxygenated products [92].

The ozonolysis of vinyl ethers is discussed in detail in Ref. 93.

The synthesis of 1,2-dioxalanes is carried out by ozonolysis of 1,1-disubstituted nonactivated olefins. Thus the ozonolysis of cyclopropyl-1,1-disubsituted olefins does not produce carbonyl oxides, but formaldehyde oxide. The latter can react with the initial olefin leading to the formation of 2-dioxalanes with 10% yield. At the addition of "foreign" olefins or aldehydes other dioxalanes and normal ozonides can be obtained [94].

The cycloalkenes ozonolysis in the presence of methyl pyruvate results in tri-substituted ozonides formation. The latter contain three reaction centers (peroxides, proton on the ozonides cycle and methoxy-carbonyl group) accessible for various functionalization. The cyclohexene ozonolysis in the presence of methylpyrovate gives ozonide whose treatment with PPh_3 or Et_3N yields $CHO(CH_3)_4CHO$ and $CHO(CH_3)_4COOMe$ (after esterification), respectively. This method proved to be very convenient and practical way for the synthesis of linear compounds containing various terminal groups from symmetrical cycloalkenes, in two steps and with good yields [95].

The direct conversion of olefins into esters is accomplished during mono-, di- and thi-substituted olefins ozonolysis in 2.5 M methanol sodium hydroxide-dichloromethane solution. The methyl esters are obtained in high yields. Thus, 3-benzyloxy-1-nonene (5b) is transformed into 2-benzyloxyoctanoate (7b) in 78% yield [96].

The ozonolysis of acyclic alkenes including terpenes is reviewed in Ref. 97.

The role of ozonolysis as ecological process for the selective and specialized oxidation of petrochemical olefins and cyclic alkenes, for the manufacture of biologically active substances and normal organic compounds is thoroughly discussed in Ref. 98.

In Ref. 99 are summarized the data on ozonolysis of acyclic and cyclic mono-, di- and trienes in the various stages of the synthesis of insect pheromones and juvenoids.

N-acylated esters of (cyclohexa-1,4-dienyl)-L-alanine are ozonized aiming at the synthesis of novel unnatural amino acids. The combined reduction and ozonolysis followed by condensation with a suitable nucleophile results in transformation of the aromatic ring of L-Phe to isooxasolyl, N-phenylpyrazolyl and to bicyclic pyrazolo[1,5-a]pyrimidine groups. The preparation of heterocyclic alanine derivatives is reported [100].

Thymidine diphospho-6-deoxy-a-D-ribo-3-hexulose synthesized by D-glucose ozonolysis of methyl-glucophosphate. is used as a central intermediate in the biosynthesis of di- and tridioxy sugars [101].

The ozonolysis of vinyl halides followed by reductive regio- and stereo-controlled intramolecular cyclocondensation is a key stage in the synthesis of amino sugars [102].

The ozonolysis of pyrols, oxazoles, imidazoles, and isooxazoles demonstrates another application of this reaction to the organic synthesis. Pyrols are efficient protective groups of the amino functions in the synthesis of α-aminoalcohols, α-aminoketones, α-aminoaldehydes and some peptides. The oxazole ring is also known as a protective group in the peptide synthesis [103].

The 1-substituted imidazoles ozonolysis leads to the formation of the corresponding N-acylamides, which are important amine or acyl derivatives [104].

Some novel tetraacetal oxa-cages and complex tetraquinone oxa-cages are synthesized from alkylfuranes by three stage reaction. Oxo rings in the tetraacetals are obtained by the ozonolysis of *cis*-endo-1,4-diones (norbornene derivatives) in dichloromethane solution at -78°C and consecutive reduction with DMS. The tetraquinone oxa-cages are obtained in the cis-endo-1,4-diones ozonolysis in dichloromethane solution at -78°C and TEA treatment [105].

A method for cleavage and oxidation of C_{8-30} olefins to compounds containing terminal carboxylic acid groups is reported in Ref. 106. The process is promoted by oxidation catalysts, such as Cr, Mn, Fe, Co, etc.

The conversion of ethane to methanol and ethanol by ozone sensitized partial oxidation at near atmospheric pressure has been quantitatively studied. The effect of temperature, oxygen concentration in the inlet gas, contact time in the reactor and ozone concentration in oxygen has been evaluated. The selectivity in regard to ethanol, methanol, as well as the combined selectivity towards formaldehyde-acetaldehyde-methanol-ethanol is also discussed [107].

Upon studying the ozone-induced oxidative conversion of methane to methanol and ethane to ethanol it has been established that these reactions do not occur in the absence of ozone which clearly suggests that the partial oxidation is initiated by the oxygen atoms generated from the ozone decomposition [87, 108].

Ortho-selective nitration of acetanilides with nitrogen dioxide in the presence of ozone, at low temperatures results in ortho-nitro derivatives formation in high yields [109].

The ozone mediated reaction of nonactive arenes with nitrogen oxides in the presence of suitable catalysts is reported as a new method for the synthesis of the corresponding nitro derivatives with high yields [110].

The ozone induced reaction of polychloro benzenes and some related halogeno compounds with nitrogen dioxide is a novel non-acid methodology for the selective mono nitration of moderately deactivated aromatic systems [111]. In the presence of ozone and preferably with methanesulfonic acid as a catalyst, the polychloro benzenes undergo mononitration with nitrogen dioxide at low temperatures giving polychloronitrobenzenes in nearly quantitative yields [111].

The ozone mediated reaction of aromatic acetals and acylal with nitrogen dioxide is suggested as a novel methodology for the nuclear nitration of acid sensitive aromatic compounds under neutral conditions [112].

Mineral and acid-free nitro compounds are prepared from CH_2Cl_2 soluted pyridine or its derivatives treated with $NO_2/O_2/O_3$ - mixture for 8 h [113].

The nitration of aromatic compounds with nitrogen oxides in the presence of ozone is a catalytic process [114].

The stereoselective synthesis of vinyl ethers is accomplished by N - (arylidene (or alkylidene) amino) - 2-azetidinones reaction with ozone and $NaBH_4$ treatment resulting in di- and trisubstituted olefins derivatives [90].

The steroeselective synthesis of (2s, 3s) norstatine derivatives is carried out through aldehydes ozonolysis in the presence of lithium methoxyallene [94].

An interesting method of succinic acid preparation from butadiene rubber ozonolysis is suggested [115].

The preparation of α-phenyl ketone, ω-carboxylate-ended telehelic methyl methacrylate oligomers by the ozonolysis of regioregular methyl methacrylate-phenylacetylene copolymers is described in Ref. 116. The oligomers have a number molecular mass varying from 1600 to 4500 (with respect to the initial copolymers) and polydespersity less than 2.

Ozone is also very convenient agent in the manufacture of organic ceramics [117].

The ability of ozone to destroy the double C=C bonds in organic compounds is the reason for its wide application in the preparation of bifunctional compounds. This is the principle that lies in the organization of the manufacture of dodecanedicarboxylic acid (1) and azelaic acid (2). The initial substrate for the preparation of (1) is cyclododecane which is obtained from the trimerization of butadiene and partial hydrogenation and (2) is prepared on the base of oleic acid:

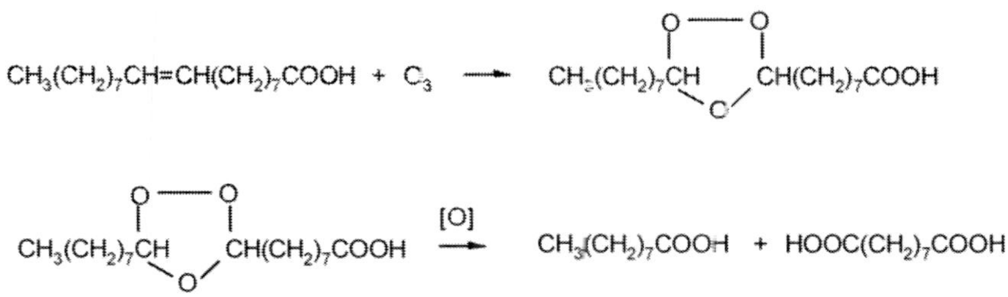

The main part of dicarboxylic acids is used in the manufacture of polyester fibers, and the azelaic acid esters (*n*-hexyl, cyclohexyl-, *iso*-octyl- and 2-ethylhexyl ester) are excellent plastisizers and synthetic lubricants.

The action of ozone is also used in pharmaceuticals in the preparation of valuable hormone products. Thus the C=C bond at C_{17} in the side chain of stigmasterol is destroyed by ozone yielding progesterone - an initial source for many hormones such as cortisone, male and female sex hormones, etc.

The method of selective decolorization of fabrics containing cellulose materials, such as cotton and oxidizing dyes [118], includes the following steps: application of oxidation blocking agent to the fibers; contact of the fibers with the oxidizing reagent in gas or evaporated state in the presence of moisture till the oxidation and decolorization of the dyer is carried out; interruption of this contact before the beginning of a substantial destruction of the fiber. The oxidizing agents are selected among ozone, chlorine and nitrogen oxides and sulfur dioxide flow. The application of ozone for decolorizing indigo-painted cotton jeans after this procedures results in jeans material which does not turn yellow for 6 months while the untreated goes yellow much more rapidly [118].

The changes in the composition and chemistry of UV/ozone modified wool fiber surfaces [119] are investigated by means of photoelectronic spectroscopy (XPS). The oxidation of the disulfide sulfur to sulfone groups ($^{-}SO_3H$) containing S^{6+} approaches almost 90% conversion. This is much more that the conversion levels by using oxygen plasma. Ozonolysis results in rise of C-O - groups content, particularly of the carbonyl ones [119].

The process for producing cellulose fibers and moldings such as fibers, filaments, threads (yarn), films, membranes in the form of flat, pipe and empty fibrous membranes, etc. is carried out using ozone [120]. They are produced by extrusion of cellulose solutions in tertiary amine aminooxides and in some cases in water (particularly N-methylformaline N-oxide and water); regenerating bath and water for washing by treatment with hydrogen peroxide, peracetic acid, ozone or chlorine dioxide for the regeneration of tertiary amines oxides. The introduced compounds can be enzymatically or catalytically destroyed before solvent and water regeneration.

Pitch-based carbon or graphite fibers of high tensile strength are manufactured by primary treatment of the pitch-based fibers with high concentrated ozone for a short period; then the treatment by ozone-free gases follows; in the next step they are subject to carbonization of graphitization to give elliptical fibers of long, wool similar structure. Upon treatment in the absence of ozone the fibers stitched during graphitization [121].

The EPR analysis of thermal decomposition of peroxides in ozonized polypropylene fiber for grafting shows that the decomposition of the peroxide groups begins at about 70°C. The generation of several peroxides radicals is registered; the access of the fiber to the spin sample is enhanced through oxidation. Small amounts of RO_2 - radicals with lifetime of couple of weeks have been identified. The integral intensity of the EPR-signal rises with time and ozonation temperature [122].

Ozonolysis modifies the diffusion pattern of liquid monomers in polypropylene matrix as their addition occurs in the amorphous phase. The appearance of intra-morphological structure and peroxides localization upon ozonolysis of polypropylene (granules and fibers) is monitored using electron spin resonance and transmission electron microscopy [123].

Upon ozonolysis, UV-radiation and plasma treatment polymer peroxides are generated on the surface of films from polypropylene, polyurethanes and polyester fibers [124]. Their thermal and reductive-oxidative decomposition have been studied by means of functional analysis using peroxidase and iodide. However, their disposition in the polymer specimen is quite various thus impeding their analysis and depends on the treatment agent. For example, the treatment with plasma generates easily accessible peroxides in polyurethane films while the UV-radiation and ozonation leads to formation of peroxides incapable of reacting with aqueous solution of peroxidase. The redox decomposition of the peroxide groups by ferro-ions at 25°C has shown that less than 50% of the peroxides may react with ferro-ions at rate constants similar to those of hydrogen peroxide in aqueous solution. The thermal degradation of peroxides does not follow first order kinetics, most probably because of the generation of various peroxide species, characterized by different rates of decomposition. The lowest rate constant of decomposition observed at 62^0C is 3.10^{-3} min^{-1} which does not depend on the polymer nature and the method of peroxides generation [124].

Ozone-induced graft polymerization onto polymer surface is an important and convenient method for polymer modification [125].

Ozone-induced graft copolymerization of polyethylene glycol monomethyl ether methacrylate onto poly(etherurethane) improves the hydrophility and water absorption. The autoaccelerated effect in ozone-induced polymerization has been also discussed [126].

The modification of the surface properties of polypropylene and block copolymer is carried out by ozone treatment. Thus, the wet ability of the polymers is improved - the contact angle of water becomes $\leq 67^0$. The break of the polymer chain and carboxyl groups formation are accelerated by using high ozone concentrations. The cut-off fragments resulting from the ozone treatment are removed from the surface by ultrasound and organic solvents [127].

The low temperature nonelectrolytic nickel plating onto three types of polypropylene is carried out by substrate pretreatment with ozone. The latter modifies the polymer surface for galvanization while the combination of polar with anchor effect as a result of the ozone etching enhances the adhesive properties of the polymer surface. The washing of the material after ozonolysis is obligatory for ensuring a good adhesion of the material [128].

The surface properties of polypropylene (I), 100 μm films from (I), ethylene-propylene block polymer (II), or ethylene-propylene polymer prepared by the random method (III) can be improved by ozonation at concentrations of 1.38, 0.64 and 0.41 mol %, respectively. The adhesion of dyes, coloring agents, dye layers, on (I) is substantially improved and depends on the substrates in the following order: (II) > (III) > (I) It has been found that for each sample there are optimum conditions of ozonation. Reactive dyes such as epoxy and acrylo urethane resins impart a better adhesion force that the conventional nonreactive acrylic or vinylchloride resin [129].

Freshly extruded, 50 μm C_2H_4-ethyl acrylate-maleic anhydride copolymer film is treated by 500 ml/m^2 oxygen containing 10 g/m^2 ozone and is calendered with 200 μm monolayer C_2ClF_3 - polymer film at 15°C to produce moisture-proof packaging material with intra-layer adhesion of 800 g/15 mm and thermosealing force of 3.6 kg/15 mm against 250 and 1.3, respectively in the absence of ozone treatment [130].

Methylmethacrylate polymers with good thermal degradation resistance are prepared by ozonolysis [131] whereby the end unsaturated groups are converted into carbonyl or carboxyl ones. Ozone/air mixture is blown through a 10 g Acrypet VH solution in CH_2Cl_2 at -78°C followed by exposure only to air for 60 min and after solvent removal the solid residue is dried in vacuum at 297°C for 8 h. The polymer thus obtained is characterized by an initial temperature of thermal degradation of 315°C against 297°C for the polymer without ozone treatment [131].

The surfaces of propene polymer moldings are treated with ozone for improving their hydrophility. The ozone treatment of the surfaces for 8 h reduces the contact angle of water from 100 to 81^0 [132].

Ozonation of PVC latex is also carried out for removal of vinyl chloride residues. The aqueous dispersions of saturated polymers are treated with ozone and the vinyl monomers are removed [133].

Graft polymerization of vinyl monomers onto Nylon 6 fiber is carried out after ozone oxidation of the fibers or films from Nylon 6 with vinyl monomers such as acrylamide, methylmethacrylate and vinyl acetate. The molecular mass of Nylon 6 decreases slightly upon ozonation. For acrylamide graft polymerization system the preliminary treatment in air or vacuum by γ-rays radiation prior ozonation results in higher graft percent. For methyl methacrylate the apparent graft percent does not rise with the ozonation time. However the apparent graft percent for vinyl acetate is increased with the ozone time treatment [134].

The adhesion of PVC-, fluropolymer, or polyester-coated steel panels is improved by treatment with 5-50% solution of H_2O_2 or ozone [135].

Synthesis of water-soluble telehelic methyl-ketone-ended oligo- N,N-dimethylacrylamides by the ozonolysis of poly(N,N-dimethylacrylamide-stat-2,3-dimethylbutadiene)s is reported in Ref. 136.

The synthesis of telehelic methylmethacrylate and styrene oligomers with fruorophenyl ketone end groups is accomplished by the ozonolysis of copolymers containing 4-flurophenyl butadiene units [137].

Graft polymerization of acrylic acid is carried out into preliminary ozonized siloxane matrixes [138].

The manufacture of base discs for laser recording material is realized by covering of the plastic substrates surface with solid polymer layer possessing directing channels and/or signal holes. The surface of the plastic substrate is preliminary cleaned by UV/ozone exposure before the formation of the hardened polymer layer [139].

The controlled degradation of polymers containing ozonides in the main chain takes place in the presence of: 1) periodate suported on Amberlyst A26 (SPIR); 2) diphenylphosphine deposited on polystyrene (SPR); boron hydride supported on Amberlyst A26 (BER) [129]. Poly(butyl)methacrylate-copolymers are prepared by emulsion polymerization. These materials and the homopolymer poly (butyl) methacrylate are ozonized at various temperatures and treated by any of the reagents described above, thus giving telechelic oligomers in 99% yield. Molecular mass varies when the ozonation temperature is changed. The end aldehyde groups are registered using ^1H and ^{13}C -NMR spectroscopy; the end hydroxyl groups are observed by means of FTIR and ^{13}C-NMR; the presence of hydroxyl groups is confirmed by tosylate formation; the carboxyl groups are identified by FTIR and ^{13}C-NMR and quantitatively determined by titration; in the SPR-generated oligomers the content of aldehyde groups constitutes about 80% of the end functional groups. The oligomers obtained in BER and SPIR contain 99% hydroxyl and carboxyl groups [140].

Polyethylene fibers are subject to ozone treatment for modifying their surface [141]. The analysis of the surface is carried out by means of X-ray photoelectron (ESCA) and IR (FTIR) spectroscopy. Carbon (C) and oxygen (O) were the main atoms monitored with ESCA (C-1s, O-1s areas) on the treated fibers. The analysis of C-1s peaks (C_1, C_2 and C_3) reveals that the oxidation level depends of the ozonation time. The components of 1s peak (O_o, O_1, O_2) are very useful for carrying out the surface analysis. They demonstrate the presence of carbonyl groups (1740-1700 cm^{-1}) even onto untreated fibers whose intensity rises with treatment time. Ozonolysis is directed from the surface to the fiber bulk. This is confirmed by the great enhancement of the carbonyl bond band after 3 h ozonation. The thermal analysis suggests structural and morphological changes of the fiber when ozonation time exceeds 2 h [141].

The processing of fiber-reinforced plastics is performed by blowing an ozone-oxygen mixture (flow rate of 0.4 l/min) for 5 h through a CH_2CL_2 solution containing glass reinforced fiber particles (1mm diameter) filled with $CaCO_3$ and unsaturated polyester. The fibers emerge on the surface till a fine powder is precipitated [142].

The manufacture of laminates through heat-sealing method involves the application of electric crown, UV-radiation or ozone exposure [143].

The manufacture of pour point depressants for oils , particularly useful for diesel oils, is carried out by oxidative degradation of waste polyethylene and/or polypropylene (I) with ozone at 30-150°C. 1000 g waste (I) is exposed to ozone action at 150°C for 5 h giving the pour point depressant. The addition of 1% depressant to diesel oil reduces the solidification of the oil from -20°C to -35°C and the temperature of filter plugging from -9 to -20°C [144].

I.2.2. Inorganic Productions

Stainless steel parts are treated by ozone for surface passivation. The parts are heated in oxidative or inert atmosphere at a temperature of condensation lower or equal to -10°C. The unreacted ozone is re-used [145].

The main passivating agent is oxygen in combination at least with ozone. The system is particularly appropriate for passivation of metal (for example, stainless steel, Ti, etc.) equipment, used in chemical plants and exposed to strong corrosion action at high temperatures and pressures [146].

For increasing the corrosion resistance of metals and alloys they were exposed from 1 s to 10 min in cold plasma under pressure 1-103 Pa and 100-5000 V in atmosphere containing O_2, O_3, N_2, H_2, air, CO_2, CO, N-oxides, H_2O (gas), combustible gas and/or inert gas. Thus,

17% C0-ferrite stainless steel is subject to plasma treatment for 4 min at 103 Pa, 100 mA and 250 V in nitrogen-oxygen mixture with 20% oxygen. The corrosion resistance was evaluated by treatment of the sample with a solution containing 17 ml 28% $FeCl_3$, 2.5 ml HCl, 188.5 ml H_2O and 5 g NaCl. The sample shows relatively good resistance as compared with the untreated one [147].

Electrochemical tests reveal the influence of dissolved ozone on the corrosion behavior of Cu-30 Ni and 304L stainless steel in 0.5N sodium chloride solutions [148]. These experiments include: measurements of the corrosion potential as a function of the time and ozone concentration; cyclic polarization experiments; isopotential measurements of the current density and study of the film components. The results of these experiments show that for Cu-30 Ni and 304L -stainless steel the corrosion potential is shifted to the more noble values (300 mV) at $[O_3] < 0.2$-0.3 mg/l. At higher concentrations it remains unchanged. The dissolved ozone reduces the corrosion level for Cu-30 Ni - alloys which is evaluated by the substantial decrease in the current density at constant applied potential. The improvement of the corrosion resistance should be related to the decrease of the thickness of the corrosion products film and to the higher oxygen fraction as compared with the chloride in the same film. For stainless steel the differences in the passivating films in ozonized and nonozonised solutions are negligible as it is shown by spectroscopy [148].

Laboratory experiments have been carried out to study the ozone application for acid oxidizing leaching of chalcopyrite with 0.5 M H_2SO_4. For evaluating the reaction mechanism the effect of particles size distribution, stirring and acid concentration, the dissolution reaction and reaction kinetics on the leaching have been investigated. The reaction rate is governed by ozone diffusion to the reaction mixture. Ozone is an efficient oxidizer and the process is most effective at 20°C [149].

The rate of acid leaching of chalcopyrite depends on the use of ozone as an oxidizing agent in sulfuric acid solutions. The leaching of chalcolyrite follows a parabolic law [150]. The use of ozone as an oxidizer provides conditions for the formation of elemental sulfur on the leached surfaces. The rate of leaching is reduced with temperature as the ozone solubility decreases with temperature. The results show that ozone is the best oxidizing agent for acid oxidizing leaching of chalcopyrite and may be applied in pilot plants for regular manufacture.

We have studied the possibility of using ozone for improvement Ag extraction form polymers deposits in a flotation plant in the town of Rudozem (Bulgaria). It has been found that upon bubbling of ozone (1% vol. concentration, flow rate - 300 l/h) through flotation machine the degree of Ag extraction is increased by 1-2% [150].

The redox leaching of precious metals from manganese-containing ores carried out by other authors show also positive results [151].

Molybdenite flotation from copper/molybdenum concentrates by ozone conditioning results in relatively pure copper-free molybdenum. The process including multi-step ozone flotation proves to be a technical and economical profitable method [152].

The manufacture of potassium permanganate is accomplished by melting of Mn-containing compounds with KOH, dissolution of the melt and the solution oxidation. For reducing the energy consumption $Mn(NO_3)_2$ is used as a Mn-containing compound. It can be easily alloyed with KOH in a 1:5 - 1: 10 ratio at 250-300°C. Then the product is dissolved in 20-25% KOH solution, the solid residues is removed, saluted in 3-5% KOH and the solution is oxidized by ozone-air mixture [153].

The manufacture of silicon carbide ceramics is performed as melted organosilicon polymer is oxidized with 0.001% vol. ozone. Polycarboxylstyrene is the preferred polymer. The ceramics obtained is characterized by high thermal resistance and acceptable physical properties [117].

Mixtures containing In- and Sn-compounds are molded and sintered in a furnace in air atmosphere containing ≥ 1000 ppm ozone. The ITO ceramics thus obtained are characterized by high density at low temperature sintering for a short time [154, 155].

A method for oxidizing carbonaceous material and especially for bleaching gray kaolin for subsequent use as coating or filler for paper in the presence of ozone is reported in [156].

The manufacture of mercury (I) chloride includes the reaction of Hg with hydrochloric acid in the presence of water, subsequent removal and drying at 95-105°C. Ozone (0.1-0.1 g/g product) is bubbled through the reacting mass to increase the product yield and quality [157].

Arsenic acid is prepared from $(As_2Cl_6)^{2-}$ -ions by ozonolysis [158].

The removal of color and organic matter in industrial phosphoric acid by ozone and the influence on activated carbon treatment is described in Ref.148. Industrial phosphonic acid containing 42-45% P_2O_5 and 220-300 mg/l organic matter (OM) is subject to combined treatment with ozone and activated carbon. The independent ozonation results in removal of the initial dark color of the acid and the organic matter. It is only through absorption on activated carbon that the level of OM could be reduced to 80% per 25 g/kg P_2O_5. The ozonation prior adsorption enhances the efficiency of the activated carbon effect and decreases its specific consumption [159].

The fabrication of high-Tc superconductors is carried out using ozone-assisted molecular beam epitaxy (MBE). It includes the simultaneous evaporation of the elementary components and application of ozone as a reactive oxygen source. The ozone priority over the other oxygen forms is that it is rather stable and could be produced and supplied to the substrate in a very pure state using simplified apparatus ensuring a well defined flow of oxidizing gas. In order to prepare films with high temperatures of superconductivity the growth should be carried out at relatively low stresses in the system. In addition, the surface during film growth can be analyzed by reflectance high energy electron diffraction. The most recent improvements in ozone-assisted MBE for family of $YBa_2Cu_3O_7$- delta films are described. The results show that this technique is very appropriate for growth of high quality superconductive films and could be ideal for the manufacture of such structures [160].

The growth of superconductive oxides under vacuum conditions compatible with MBE requires the use of activated oxygen. The atomic oxygen or ozone appears to be such species [150]. The characteristics of a radio-frequency plasma source for molecular beam epitaxial growth of high-Tc superconductor films (200 Å) from $DyBa_2Co_3O_7$ on $SrTiO_3$ are described and discussed in Ref. 161.

An apparatus for the preparation of pure ozone vapor for use *in-situ* growth of superconducting oxide thin films is designed [151]. Pure condensed ozone is produced from distillation of diluted ozone-oxygen mixture at 77K. The condensed ozone is heated until the pressure of its vapors approaches the necessary for an adequate flow of ozone- gas to the chamber of thin films growth. The thin films from $YBa_2Cu_3O_7$ with zero resistance at temperatures of about 85K grow at ozone pressure in the chamber of 2.10^{-3} Torr even without subsequent untempering. It should be noted that in contrast to other highly reactive oxygen species ozone can be prepared and stored in very pure form which makes it very convenient for studies on the kinetics of growth and oxidation of films with a well defined gas [162].

A patent for preparation of oxides superconductors M-M[1]-M[2]-M[3] includes ozone oxidation where: M = elements of III B groups such as: Y, Sc, La, Yb, Er, Ho or Dy; M[1] = elements from II A group, such as: B, Sr, Ba or Ca; M[2] - Cu and one or more elements from I B group such as Ag or Au; M[3] = O and on or more elements belonging to VI A group like S or Se and/or elements from VII A group as F, Cl or Br. The superconductors thus prepared have high critical temperature and high critical current densities [163].

The manufacture of bismuth-, copper- alkaline earth oxide high-temperature superconductors is carried out through calcination and/or sintering in ozone-containing atmosphere with $Bi(OH)_2$, $Ca(NO_3)_2$ and $Sr(NO_3)_2$ and subsequent treatment by $CuCO_3(OH)_2$; the components are dried, calcinated in ozone containing air, molded and dried in air atmosphere. Their temperature of superconductivity is 107K and the critical current density is 405 A/cm^2 [164].

A low temperature method for preparation of superconductive ceramic oxides is described in Ref.165. It includes the treatment of the substrate heated surface by ozone to the complete evaporation of the other components thus forming a superconductive ceramic oxide.

The manufacture of rare earth barium copper oxide high-temperature superconductor cermics is carried out by sintering at 930 - 1000°C in ozone atmosphere containing oxygen. Y_2O_3, $BaCO_3$ and CuO are mixed, calcinated, pulverized, pressed and sintered at 950°C in the presence of ozone-oxygen mixture after which they are gradually cooled to the critical superconductive temperature 94K [166].

A review devoted to the growth of co-evaporated superconducting yttrium barium copper oxide (YBa_2CuO_7) thin films oxidized by pure ozone is presented in [167].

The removal of organic pollutants from the surface of supports for microelectronics purposes is conducted by UV/ozone treatment [168].

Low temperature silicon surface cleaning is carried out by fluoric acid (HF) etching, washing, cleaning with deionized water, N--gas blowing, /UV-ozone treatment. This is a treatment preceding the process of silicon epitaxy [169-171].

The method for total room temperature wet cleaning of silicon surface comprises the use of HF, H_2O_2 and ozonized water and is with 5% more economical than that the standard procedure of wet cleaning [172].

The removal of resist films supported on semiconductive substances is accomplished by using ozone [173].

The application of ozone for reducing the temperature and energy in the process of very large scale integration (VLSI) is described in [174, 175]. The main problems under discussion are as follows: cleaning of Si-support with ozone by two methods: dry process combined with UV and wet one - with ozonized water; stimulation of Si thermal oxidation with ozone; deposition of SiO_2 under atmospheric pressure and low temperature via chemicals evaporation using tetraethylorthosilicate and ozone; 4) deposition of Ta_2O_3 at atmospheric pressure and low temperature applying $TaCl_5$ and ozone; 5) tempering with ozone and UV-radiation of Ta_2O_5 films used for dielectric for memory units , this treatment reduces the current permeability into the film; 6) etching of photoresistant materials by $O_2/O_3/CF_4$ and the effect of ozonator charge and injection of CF_4 [174, 175].

The cleaning of synchrotonic radiation optics with photogenerated reactants has a number of priorities over the methods of discharge cleaning [176]. Upon discharge cleaning, the discharge particles should react with the surface contaminants until its sheilding by the rough discharge elements which may pollute or destroy the surface. Contrary, if the particles can be photon generated near the surface, the problem with the protection drops off and in some cases the cleaning can be more efficient. An estimation of the various methods for cleaning was made comparing the rates of polymethylmethacrylate films removal. A number of various light and geometry sources have been tested. The highest rate of cleaning was achieved upon using UV/O_3 cleaning method at atmospheric pressure. This method has been widely applied for cleaning semiconductive surfaces from hydrocarbon contaminants. It is noted that it is also effective in removing graphite-like pollutants from synchrotonic radiation optics. It proves to be more simple, economical and selective method as compared with other cleaning method. Although it requires the drilling of a hole in the vacuum chamber, the cleaning of the optics can be carried out without dissemble which saves much time. This method is successfully applied for cleaning grates and reflectors in several beams [176].

Native oxide growth and organic impurity removal on silicon surface is carried out by ozone-injected ultrapure water [166]. In order to manufacture high-efficient and reliable ULSI-units the further integration and minimization is in progress. The cleaning with $H_2SO_4/H_2O_2/H_2O$ is accompanied by serious problems: a great amount of chemical wastes is obtained which must be suitably treated. The cleaning technology with ultrapure water includes ozone injection in concentrations of 1-2 ppm. This method is very efficient in removing the organic impurities from the surface for a short time and at room temperature. The process wastes can be treated and recycled. The ozonized water components can be easily controlled [177].

The reaction mechanism of chemical vapor deposition using tetraethylorthosilicate and ozone at atmospheric pressure is reported and discussed in Ref.178.

Covering of semiconductor devices with silica films is carried out via CVD using $Si(OEt)_4$ and ozone. The deposition process is followed by heating in oxygen atmosphere with simultaneous UV radiation. The SiO_2-film thus formed can improve the water resistance of semiconductive devices [179].

The fabrication of nondoped silicate glass film with flat structure by O_3-TEOS deposition includes three stages: 1) substrate treatment; 2) formation of Al conducting layer on it using oxygen plasma through heating; 3) formation of SiO_2-film on the substrate by plasma using ozone and TEOS [180].

In Ref. 170 are reviewed the future trends for interlayer dielectric films production and their formation technologies in ULSI multilevel interconnections. The properties of the interlayer dielectric films and their preparation technologies should satisfy the following three requirements: 1) available for disposition (setting) on large surface; 2) low dielectric constant; 3) low deposition temperature. Two techniques have been developed for the selective deposition of SiO_2-films which is the best way for achieving a complete planirization of the interlayer surface of dielectric films: low temperature liquid phase deposition using a saturated aqueous solution of hydrofruorosilicon acid H_2SiF_6; 2) half-selective SiO_2-films deposition at 390°C with $Si(OC_2H_5)_3$ and ozone and preliminary treatment with tetraflurocarbonic (CF_4) plasma on TiW or TiN surfaces [181].

Lead zirconate and titanate thin films are successfully prepared by reactive evaporation. The elements Pb, Zr and Ti are evaporated in ozone-oxygen mixture. The films obtained have equal thickness and composition per a large surface (in the range of ±2% from the 4-inch support) [182].

Ceramic coatings on substrates are formed in the presence of ozone [183]; the substrates (silicon checks) are covered (dipped) in a solution of one or more (partially) hydrolyzed pre-ceramic silicon alholates with general forrmula $R_xSi(OR)_{4-x}$ ($R = C_{1-20}$ alkyl, aryl, alkenyl or alkinyl; x = 0-2). Further, the solvent is evaporated to form a coating which is subsequently heated in the presence of ozone up to 40-100°C thus converting into a ceramic coating. It in its turn may be covered also by a protective layer containing Si, or Si and C, or Si and N, or Si, C and N, or SO_2 and oxide. These coatings which are abrasive-, corrosive and thermo-resistant have also a small number of defects and suppress the diffusion of ionic contaminants such as Na and Cl ions and are particularly convenient for electronic units [183].

Method of forming zinc oxide light-shielding film for liquid crystals shields includes the injection of vapors of alkyl zinc compound with ozone or atomic oxygen in the activation chamber after which they pass through a chamber for ZnO film deposition at low temperature heating at about 200oC. The method is applicable for large scale production of these films [184].

An evidence for a new passivating indium rich phosphate prepared by UV/O_3 oxidation of indium phosphide, InP, is provided in Ref.185. The phosphate does not exist as crystal compound and its composition is $InP_{0.5}O_{2.75}$. The passivating ability of the latter with respect to InP surface is discussed.

For improving the light-absorption characteristics of oxide optical crystals they are heated in to ozone-oxygen atmosphere [186, 187]. Thus the light absorption from the optical crystal in a wave range different from that with which the crystal affects its own absorption is reduced to the most possible level. Devices using such oxide optical crystals with improved absorption characteristics work with high efficiency as optical amplifiers, optical isolators, optical recording medium, optical generators.

The manufacture of solar cell modules with transparent conducting film covered by amorphous Si layer and electrode on the backside linked to the transparent isolator layer is described in Ref.188. The preparation includes the application of a laser beam for electrode molding and exposure in oxidizing atmosphere containing 0.5-5% O_3 or ≥10% O [188].

A method for strong oxidation in ozone atmosphere is proposed for surface activation of photoconductive PbS films [189].

Adhesion-producing materials for electroless plating and printing circuits contain particles of thermostable material slightly soluble in the oxidizing agent; the particles are dispersed in the thermostable resin which becomes almost insoluble in the oxidizing agent at hardening. These materials find application for printing circuits and are treated by an oxidizing agents (for example, chromic acid, chromate, permanganate or ozone) to create a concave material surface [190].

REFERENCES

[1] S.K. Rakovsky, G.E. Zaikov, *Kinetics and Mechanism of Ozone. Reactions with Organic and Polymeric Compounds in Liquid Phase,* NOVA Sci. Publ. Inc., *Commack,* N.Y. (second edition), p. 340 (2007).

[2] L.M. da Silva abd W.F. Jardim, *Qim. Nova*, 29, 310 (2006).

[3] P.S. Bailey, *Ozonattion in Organic Chemistry*, Academic Press, New York, v.I, 1978, v.II, 1982, *Adv. Chem. Ser. - A,* (Ed. HH Wasserman), 39 -I, 39 - II, 1978, 1982.

[4] R.G. Rice, *Ozone: Science and Engineering*, 24, 1 (2002).

[5] F. C. Pedroza, M.S. Aguliar, A.M. Luevanos, and J.G. Anaya, *Ozone: Science and Engineering*, 29, 307 (2007).
[6] F. De Smedt, S. De Gendt, M. Claes, M.M. Heyns, H. Vankerckhoven, and C. Vinickier, *Ozone: Science and Engineering*, 24, 379 (2002).
[7] M.A. Khadre, A.E. Yousef, and J.-G. Kim, *Journal of Food Science*, 9, 1242 (2001).
[8] A.A. Shatalov, and H. Pereia, *Carbohydrate Polymers*, 3, (275) 2007.
[9] E.A. Pergut, , and D.O. Gorelik. *Instrumental Methods to Control Air Pollution*. Leningrad, Khimiya, 1981.
[10] E.T. Chanlett,. *Environmental Protection*., McGraw-Hill, 1979.
[11] A.B. Flaekt, *Res. Discl.,* 326, 453 (1991).
[12] W. Ruck, *Vom. Wasser,* 80, 253 (1993).
[13] M.S. Mccoustra, and A.B. Horn, *Chemical Society Reviews,* 23, 195 (1994).
[14] JP 01,270,928 (Oct. 30, 1989), K. Tabata, I. Matsumoto, and Y. Fukuda (Jpn. Kokai Tokkyo Koho).
[15] . Pol. PL 137,426 (Jul. 31, 1987), I. Pollo, J. Jarosszynska- Wolinska, J. Malicki, J. Ozonek, and W. Wojcik.
[16] De Guardia A., A.Bouzaza, G.Martin, and A. Laplanche, *Pollut.Atmos*., 152, 82 (1996).
[17] JP 01,236,925 (Sep. 21, 1989), K. Takeyama, and K. Nitta (Jpn. Kokai Tokkyo Koho).
[18] G.K. Moortgat,. G. Horie, and B.C. Zahn, *Pollut.Atmos.,* 29, 44 (1991).
[19] JP 02,152,547 (Jun. 12, 1990), K. Tomita, (Jpn.Kokai Tokkyo Koho).
[20] WO 92 19,364 (Nov. 12, 1992), J.F. Vicard, G.Vicard. (PCT Int. Appl).
[21] U.S. US 4, 923,688 (May 08, 1990), J. Iannicelli.
[22] DE 9,931,891 (Apr. 04, 1991), F. Buettner, and P. Koch.(Ger. offen).
[23] H. Wada, T. Naoi, Y. Kuroda, *Nippon Kagaku Kaishi*, 9, 834 (1994).
[24] EP 330,028 (Aug. 30, 1989), K. Matt, and H.G. Guttenberger, (Eur. Pat. Appl).
[25] Waga, H., T.Naoi, Y.Kuroda, *Nippon Kagaku Kaishi,* 4, 306 (1995).
[26] K. Tanaka, *PPM*, 27, 10 (1996).
[27] U.S. US 4,861,497 (Aug. 30 1989), J..F. Welch, and J.D .Siehwarth..
[28] H.J. Oeller, I. Daniel, and G. Weinberger, *Water Sci.Technol., Forest Industry Wastewaters V,* 35, 269 (1997).
[29] C.H. Moebius., and M. Cordes-Tolle, *Water Sci.Technol., Forest Industry Wastewaters V,* 35, 245 (1997).
[30] Y. Nakamura, T. Sawada, and F. Kobayashi, *Water Sci.Technol., Forest Industry Wastewaters V,* 35, 277 (1997).
[31] H. Zhou, and D.W. Smith, *Water Sci.Technol., Forest Industry Wastewaters V,* 35, 251 (1997).
[32] C.M. Huang, and M.K. Banks, *J.Environ.Sci.Health, Part B*, 31, 1253 (1996).
[33] JP 03,111,670 (May 13, 1991), H. Sumimoto, and T. Yamazaki.(Jpn.Kokai Tokkyo Koho).
[34] K. Kawamoto, *Kogaito Taisaku,* 27, 617 (1991).
[35] JP 03 80,996 (Apr. 05, 1991), S. Ukita, (Jpn.Kokai Tokkyo Koho).
[36] R.B.Gunukula, and M.E. Tittlebaum, *Journal of Environmental Science and Health*, A, 36, (307)2001.
[37] K. Ikehata, and M.G. El-Din, , *Ozone: Science and Engineering*, 27, 1(2005).
[38] K. Ikehata, and M.G. El-Din, , *Ozone: Science and Engineering*, 27, 173(2005).

[39] K. Ikehata, N.J. Naghashkar, and M.G. El-Din, , *Ozone: Science and Engineering*, 28, 353(2005.

[40] T.E. Agustina, H.M.Ang, and V.K. Vareek, *Journal of Photochemistry and Photobiology C: Photochemistry Reviews*, 6, 264(2005).

[41] E. Gilbert, *Ozone: Science and Engineering*, 24, 75(2002).

[42] Y. Yanga, J. Min, Q. Qina, and X. Zhaia, *Journal of Mclecular Catalysis A: Chemical*, 267, 41(2007).

[43] U. Cernigoja, U.L. Stangara, and P. Trebse, *Applied Catalysis B: Enviromental*, 75, 229(2007).

[44] Y. Miriguchi, H.Hayashi, and T.Umetani, *Osaka-shi Suidokyoku Suishtsu Shikensho Chosa Kenkyu narabini Shiken Seiseki*, 46, 10 (1994), /Pub. 1995/.

[45] C. Munz, R.Galli, and R. Egli, *Chem.Oxid.*, 2, 247, (1992), /Pub. 1994/.

[46] H. Kojima, E. Katsura, H.Ogawa, and H.Kaneshima, *Hokkaidoritsu Eisei Kenkyushoho,* 41, 71 (1991).

[47] N. Ohashi, Y. Tsuchiya, H. Sasano, and A. Hamada, *Jpn .J. Toxicol.Environ.Health,* 40, 185 (1994).

[48] C.J. Hapeman-Somish, *ASC Symp.Ser., (Pestic. Transform. Prod.: Fate Signif. Environ.,* 459, 133 (1991).

[49] JP 02 139,017 (May 29, 1990) S. Terui, K. Sano, K. Nishikawa, and A. Inone (Jpn.Kokai Tokkyo Koho).

[50] *JP 01 56,124 (Mar 03, 1989), S* Terui, K. Sano, T. Kanazaki, K. Mitsui, and A. Inoue (*Jpn.Kokai Tokkyo Koho*).

[51] E.K. Golubovskya,. *Biological Bases of Water Purification,* Visshaya Shkola, Moscow, 1978.

[52] Ch. Gottschalk, J. Libra, and A. Saupe, *Ozonation of Drinking Water and of Wastewater*, Willey-VCH, 2000, p.189.

[53] F.J. Beltran, *Ozone Reaction Kinetics for Water and Wastewater Systems*, Taylor and Frances, 2007.

[54] U. von Gunten, *Water Research*, 37, 1443(2003).

[55] U. von Gunten, *Water Research*, 37, 1469(2003.

[56] F. Geering, *Ozone: Science and Engineering*, 21, 187(1999).

[57] K.L. Rakness, *OZONE in Drinking Water Treatment: Process Design, Operation, And Optimization*, American Water Works Association, 2005, p. 302.

[58] JP 04,243,597 (Aug. 31, 1992), K. Ogawa, N. Seki (Jpn.Kokai Tokkyo Koho).

[59] JP 02 40,289 (Feb. 09, 1990), Y. Hoshima (Jpn.Kokai Tokkyo Koho).

[60] A. Filippi, F. Tilkes, E.G. Beck, and H. Kirschner, *Dtsch.Zahnaerztl.Z.*, 46, 485 (1991).

[61] Gehringer, P. Radiation- induced oxidation for water remediation. *Oesterr. Forschungszent. Zeitbersdorf, [Ber.] OEFZS 1995, (OEFZS - 4738), 21pp.*

[62] L. Charles, D. Pepin, Ph. Puig, *J.Eur.Hydrol.*, 27, 175 (1996).

[63] UK Pat. GB 2,219,790 (Dec. 20, 1989), R. W. Thornhill.

[64] JP 04,367,799 (Dec. 21, 1992), Y. Nitsuta, and Y. Nitta (Jpn. Kokai Tokkyo Koho).

[65] G. Gordon, and G.E. Pacey, *Chem.Oxid.*, 2, 230 (1992), /Pub. 1994/.

[66] F. Bruckner, *In situ* remediation of petroleum contaminated soils by ozone treatment. *ECOINFORMA 92, Int.Tag.Ausstellung Umweltinf.Umweltkommun.*, 2, *211 (2nd 1992), /Pub. 1993/,*

[67] M. Preusser, H. Ruholl, H. Schneidler, E. Wessling, and C. Wortmann, *Wiss.Umwelt*, 3, 135, (1990).
[68] DE 19,512,448 (Oct. 10, 1996), M. Rimpler (Ger. Offen.).
[69] E. Wessling, *Erzmetall*, 44, 196 (1991).
[70] M. Mori, H. Yamakoshi, M. Nojima, S. Kusabayashi, K.J. Mccullough, K. Griesbaum, P. Kriegerbeck, and I.C. Jung, *Journal of the Chemical Society - Perkin Transactions* I, 12, 1335 (1993).
[71] K.J. Mccullough, K. Teshima, and M. Nojima, *J.Chem.Soc. - Chem.Commun,.*: 11, 931 (1993).
[72] T. Sugiyama, H. Yamakoshi, and M. Nojima, *J.Org.Chem,* 58, 4212 (1993).
[73] K. Ishiguro, T. Nojima, and Y. Sawaki, *J.Phys.Org.Chem.,* 10, 787 (1997).
[74] T. Nojima, K. Ishiguro, and Y. Sawaki, *J Org Chem*, 62, 6911 (1997).
[75] R. Fukagawa, and M. Nojima, *Journal of the Chemical Society - Perkin Transactions 1*: 17, 2449 (1994).
[76] K.J. Mccullough, K. Teshima, and M. Nojima, *J.Chem.Soc, - Chem.Commun.,* 11, 931 (1993).
[77] K.J. Mccullough, T. Sugimoto, S. Tanaka, S. Kusabayashi, and M. Nojima, *Journal of the Chemical Society - Perkin Transactions,* 6, 643 (1994).
[78] S. Satake, Y. Ushigoe, M. Nojima, and K.J. Mccullough, *Journal of the Chemical Society - Chemical Communications,* 14, 1469 (1995).
[79] K. Teshima, S.I. Kawamura, Y. Ushigoe, M. Nojima, and K.J. Mccullough, *Journal of Organic Chemistry,* 60, 4755 (1995).
[80] K.J. Mccullough, S. Tanaka, K. Teshima, and M. Nojima, *Tetrahedron,* 50, 7625 (1994).
[81] B.D. Schobert, *Chim.Oggi,* 13, 21 (1995), /Pub.1995/.
[82] R.B. Appell, I.A. Tomlinson, and I. Hill, *Syn. Comm.,* 25, 3589 (1995).
[83] Y.T. Yang, T.S. Li, and Y.L. Li, *Synthetic Comm.,* 23, 1121 (1993).
[84] M. Jachak, M. Mittelbach, U. Kriessmann, and H. Junek, *Synthesis- Stuthgart*, 3, 275 (1992).
[85] W. Jellen, M. Mittelbach, and H. Junek, *Monatsh.Chem.,* 127, 167 (1996).
[86] JP 08 03,138 (Jan. 09, 1996), H. Matsumoto, H. Kanda, Y. Obada, H. Ikeda, and T. Murakami (Jpn.Kokai Tokkyo Koho).
[87] U.S. Pat., US 5,491,265 (Feb. 13, 1996), D.E. Allen, Ch.E. Ticker, Ch.C. Hobbs, and R. Chidambaram.
[88] H.W. Lee, S.K .Ji, I.Ch. Lee, and J.H. Lee, *J.Org.Chem.,* 61, 2542, (1996).
[89] F.Z. Mikaev, and F.Z. Galin, *Izv. Akad. Nauk SSSR, ser. khim.*, 10, 1984 (1995).
[90] K.J. Mccullough., M. Mori, T. Tabuchi, H. Yamakoshi, S. Kusabayashi, and M. Nojima, *J.Chem.Soc.-Perkin Transactions* 1, 14, 41 (1995).
[91] M. Mori, T. Tabuchi, M. Nojima, and S. Kusabayashi, *J.Org.Chem.,* 57, 1649 (1992).
[92] M. Casey, A.J. Ulshaw, *SYNLETT*, 3, 214 (1992).
[93] R.L. Kuczkowski, *Ozonolysis of vinyl ethers,* Series: Advances in Oxygenated Process: A Research Annual 3, (1991).
[94] R. Reiser, R. Seeboth, C. .Suling, G. Wagner, J. Wang, and G. Schroder, *Chem.Ber.,* 125, 191 (1992).
[95] Y.-S. Hon, and J.-L. Yan, *Tetrahedron,* 53, 5217 (1997).
[96] J.A. Marshall, A.W. Garofalo, and R.C. Sedrani, *SYNLETT.,5,* 643 (1992).

[97] VN. Odinokov, *Bashk. Khim. Zhurn.*, 1, 29 (1994).

[98] VN. Odinokov, *Bashk. Khim. Zhurn.*, 1, 11 (1994).

[99] G.Yu. Ishmuratov, R.Ya. Kkharisov, V.N. Odinokov, and G.A. Tolstikov, *Uspehi Khimii,* 64, 580 (1995).

[100] G. Zwilichovsky, and V.Gurvvich, *J.Chem.Soc , Perkin Trans.1.*, 19, 2509 (1995).

[101] T. Mueller, and R.R. Schmidt, *Angew.Chem., Int.Ed.Engl.,* 34, 1328 (1995).

[102] K. Pitzer,. and T..Hudlicky, *Synlet.,* 8, 803 (1995).

[103] Ch. Kashima, T. Maruyama, and H. Arao, *Yuki Gosei Kagaku Kyokaishi,* 54, 132 (1996).

[104] C. Kashima, K. Harada, and A. Hosomi, *Heterocycles*, 33, 385 (1992).

[105] H.-J. Wu, and Ch.-Ch. Lin, *J.Org.Chem.*, 60, 7558 (1995).

[106] U.S. Pat. US 5,543,565 (Aug. 06, 1996), K.R McVay, D.G. Gaige, and W.S. Kain.

[107] H.D Gesser, G. Zhu, and N.R.Hunter, *Catalysis Today*, 24, 321 (1995).

[108] H.D. Gesser, N.R. Hunter, and P.A. Das, *Ñatalysis Letters*, 16, 217 (1992).

[109] S. Hitomi, T. Ishibashi, M. Takashi, and T. Kenkichi. *Tetrahedron Letters*, 32, 6591, (1991).

[110] H. Suzuki, T. Murashima, K. Shimizu, and K. Tsukamoto, *Chemistry Letters,* 817 (1991).

[111] H. Suzuki, T. Mori, and K. Maeda, *SYNTHESIS- Stutgart*, 8, 841 (1994).

[112] H. Suzuki, S. Yonezawa, T .Mori, and K. Maeda. *J.Chem.Soc.-Perkin Trans l.,* 11, 1367 (1994).

[113] JP 04,282,368 (Oct. 07, 1992), H. Suzuki, T. Murashima, and K. Tsukamoto (Jpn.Kokai Tokkyo Koho).

[114] H. Suzuki, and T. Murashima, *Kagaku to Kagyo (Tokyo),* 46, 43 (1993).

[115] Russ. Pat. RU 2,036,895 (Jun. 09, 1995), A.G. Lyakumovich, Y.D. Samuilov, Ya.M. Goldberg, and M.N. Kondrashova.

[116] J.R. Ebdon, and N.J. Flint, *Macromolecules,* 27, 6704 (1994).

[117] J P 02 26,868 (Jan. 29, 1990), T. Sawaki, S. Watanabe, and K. Shimada (Jpn.Kokai Tokkyo Koho).

[118] U.S. Pat. US 5 261,925 (Nov.16, 1993), E. Wasinger, and D. Hall.

[119] R.H. Bradley, I.L. Clackson, and D.E. Sykes, *Applied Surface,* 72, 143 (1993).

[120] WO 96 18,761 (Jun. 20, 1996), H.G. Conner, and D. Budgell (PCT Int. Appl).

[121] JP 01,250,415 (Oct. 05, 1989), M. Miura, T. Naito, T. Hino, and H. Kuroda (Jpn. Kokai Tokkyo Koho).

[122] R. Queslati, S. Roudesli, *European Polymer Journal*, 27, 1383 (1991).

[123] B. Catoire, V. Verney, R. Hagege, and A. Michel, *Polymer,* 33, 2307 (1992).

[124] E.A. Kulik, M. Ivanchenko, K. Kato, S. Sano, and Y. Ikada, *J. Polym. Sci., Part A- Polymer Chemistry,* 33, 323 (1995).

[125] K. Fujimoto, Y. Takebayashi, H. Inoue, and Y. Ikada, *J. Polym. Sci., Part A- Polymer Chemistry,* 31, 1035 (1993).

[126] Ch. Wang, A. Wang, B. Che, C. Zhou, I. Su, S. Lin, and B. Wang, *Gaofenzi Xuebao,* 1, 114 (1997).

[127] Y. Jian, and Sh. Shiraishi, *Hyomen Gijutsu,* 40, 1251 (1989).

[128] Y. Jian, and S.H. Shiraishi, *Hyomen Gijutsu,* 40, 1256 (1989).

[129] Y. Jian, and S.H. Shiraishi, *Hyomen Gijutsu,* 41, 273 (1990).

[130] JP 02,198,819 (Aug. 07, 1990), H. Yamamoto (Jpn. Kokai Tokkyo Koho).

[131] JP 03,45,604 (Feb. 27, 1991), M. Yoshito (Jpn. Kokai Tokkyo Koho).
[132] JP 03,103,448 (Apr. 30, 1991), S.H. Shiraishi, U. Ken, K. Suchiro, and K. Nitsuta (Jpn. Kokai Tokkyo Koho).
[133] U.S. Pat. US 5,498,693 (Mar. 12, 1996), R.A. Marshall, D.K. Parker, and J.W. Hershberger.
[134] T. Iwasaki, *Kobunshi Ronbunshu,* 50, 115 (1993).
[135] JP 04,197,472 (Jul. 17, 1992), Y. Hosoda, K. Ikishima, and A. Yanai (Jpn. Kokai Tokkyo Koho).
[136] J.R. Ebdon, and N.J. Flint, *J. Polym. Sci., Part A- Polymer Chemistry,* 33, 593, (1995).
[137] J.R. Ebdon, and P. Hodge, *Polymer*, 34, 406 (1993).
[138] A.M. Eftushenko, I.P Chikhachova, G.V. timofeeva, S.T. Stavrova, and VP Zubov, *Visoko molek. soed., ser. A,* 35, 312 (1993).
[139] JP 01,248,339 (Oct. 03, 1989), S.H. Tashibana, and R. Yokoyama (Jpn. Kokai Tokkyo Koho).
[140] S. Rimmer, J.R. Ebdon, and J.M, *Reactive and Functional Polymers,* 26, 145 (1995).
[141] H. Chtourou, B. Riedl, B.V. Kokta, A. Adnot, and S. Kaliguine, *J. Appl. Polym. Sci.,* 49, 361 (1993).
[142] WO 96 18,484 (Jun. 20, 1996), A. Vallet, M. Delmas, T.H. Fargere, and G. Sacher (PCT Int.Appl).
[143] JP 09 01,665 (07 Jan. 07, 1997), K. Kawabe, and Y. Hatakeyama (.Jpn. Kokai Tokkyo Koho).
[144] Eur. Pat. EP 383,969 (Aug. 29, 1990), W. Czerwiec, and A. Kwaitek.
[145] JP 08 337,867 (Dec. 24, 1996), S.H. Shimizu, and N. Iwata (Jpn. Kokai Tokkyo Koho).
[146] EUR Pat. EP 504,621 (Dec. 23, 1992), V. Lagana.
[147] EUR Pat. EP 340,077 (Nov. 02, 1989), R. Berneron, and P de Gelis.
[148] Lu, HH, DJ Diguette. Effect of dissolved ozone on the corrosion behaviour of Cu-30 Ni and 304L stainless steel in 0.5N sodium chloride solutions. *Report 1989, Order No.AD-A213804, 37pp.*
[149] T. Havlik, and M. Skrobian, *Rudy,* 37, 295 (1989).
[150] T. Havlik, and M Skrobian, *Can. Metall.Q.,* 29, 133, (1990).
[151] Aust. Pat. AU 607,523 (7Mar. 07, 1991), T. Clough, J.W. Siebert, and A.C. Riese.
[152] Y . Ye, W.H. Jang, M.R. Yalamanchili, and J.D. Miller, *Trans.Soc.Min., Metall., Explor.,* 288, 173 (1990), /Pub. 1991/.
[153] U.S.S.R. Pat. SU 1,541,190 (Feb. 07, 1990), T.K. Chkoniya.
[154] JP 08 169,769 (Jul. 02, 1996), T. Saski, H. Sasaki, and K. Iida (Jpn. Kokai Tokkyo Koho).
[155] JP 07 277,821 (Oct. 24, 1995), T. Saski, and H. .Sasaki (Jpn. Kokai Tokkyo Koho).
[156] U.S. Pat. US 4,935,391 (Jun. 19, 1990), F.E. Caropreso, H.M. Castrantas, and J.M. Byne.
[157] U.S.S.R. Pat. SU 1,567,520 (May 30, 1990), O.V. Emel'yanova, and B.V. Emel'yanov.
[158] A.D. Ruhlandtsende, A.D. Bacher, and U. Muller, *Zeitschrift fur Naturforschung, Section B,* 47, 1677 (1993).
[159] H. Belfadhel, A. Ratel, A. Ouederni, R.S. Bes, and J.C. Mora, *Ozone: Sci.Eng.,* 17, 637, (1995).
[160] V.S. Achutharaman, K.M. Beauchamp, N. Chandrasekhar, G.C. Spalding, B.R. Johnson, and A.M. Goldman, *Thin Solid Films,* 216, 14 (1992).

[161] J.P. Locquet, *J.Vacuum Sci .and Tech., A.Vacuum surfaces and films,* 10, 3100 (1992).
[162] D.D. Berkley, A.M. Goldman, B.R. Johnson, J. Morton, and T. Wang, *Rev. Sci .Instrum.,* 60, 3769 (1989).
[163] JP 01 122,921 (May 16, 1989), T. Usui, and Y. Osanai (Jpn. Kokai Tokkyo Koho).
[164] JP 01 224,258 (Sep. 07, 1989), K. Odan, H. Miura, and Y. Bando (Jpn. Kokai Tokkyo Koho).
[165] WO 90 02,215 (Mar. 08, 1990), A.M. Goldman, D.D. Berkley, and B.R. Johnson (PCT Int. Appl).
[166] JP 01 179,752 (Jul. 17, 1989), T. Nakamori, (Jpn. Kokai Tokkyo Koho).
[167] B.R. Johnson, K.M. Beauchamp, D.D. Berkley, J.X. Liu, T. Wang, and A.M. Goldman,*Proc. SPIE-Int.Soc.Opt.Eng., (Process.Films High Tc Supercond.Electron),* 1187, 27 (1990).
[168] J.R. Vig, *Proc. Electrochem. Soc (Semicond.Clean.Tachnol.)* , .90, 105 (1990).
[169] M. Suemitsu, T. Kaneko, and N. Nobuo, *Jpn. J .Appl. Phys., Part 1,* 28, 2421 (1989).
[170] T. Kaneko, M. Suemitsu, and N. Miyamoto, *Jpn. J .Appl. Phys., Part 1,* 28, 2425 (1989).
[171] JP 01 135,023 (May 26, 1989), A. Inada, and K Kawasumi (Jpn. Kokai Tokkyo Koho).
[172] T. Ohmi,. *Semicond. Int.,* 19, 323 (1996).
[173] JP 02 106,040 (Apr. 18, 1990), Y. Kamikawa, and K. Matsumura (Jpn. Kokai Tokkyo Koho.
[174] K. Maeda, *Kurin Tekunoroji,* 4, 62 (1994).
[175] H. Choi, K. Kyoo, S.D. Jung, and H. Jeon, *Han'guk Chaehyo Hakhoechi,* 6, 395 (1996).
[176] R.W.C. Hansen, J. Welske, D. Wallace, and M. Bissen, *Nucl.Instrum.Methods Phys.Res., Sect.A,* 347, 249 (1994).
[177] T. Ohmi, T. Sagawa, M. Kogure, T. Imaoka, *J. Electrochem. Soc.,* 140, 804 (1993).
[178] K. Takaaki, Y. Akimasa, and M. Yasuji, *Jpn. J .Appl. Phys.,* 31, 2925 (1992).
[179] JP 01 82,634 (Mar.28, 1989), Y. Hishamune (Jpn. Kokai Tokkyo Koho).
[180] JP 09 08,026 (Jan. 10, 1997), H. Kurihara, M. Saito, and T. Kozuki (Jpn. Kokai Tokkyo Koho).
[181] T. Homma, *Mater. Chem. Phys.,* 41, 234 (1995).
[182] K. Toris, S. Saiton, and Y. Ohji, *Jap .J .Appl .Physics, Part,* 133, 5287 (1994).
[183] FR 2,697,852 (May 13, 1994), I.A. Haluska (Fr. Demande).
[184] DE 3,916,983 (Nov. 30, 1989), S.H. Nishida (Ger. offen).
[185] G. Hollinger, D. Gallet, M. Gendry, M.P. Besland, and J. Joseph, *Appl. Phys .Lett.,* 59, 1617 (1991).
[186] Eur Pat. EP 407,945 (Jan. 16, 1991), H. Tawada, M. Saitoh, and C.H. Isobe.
[187] H. Tamada, A. Yamada, and M. Saitoh, *J. Appl. Phys.,* 70, 2536 (1991).
[188] JP 08 148,707 (Jun. 07, 1996), M. Sadamoto, Y. Ashida, and N. Fukuda (Jpn. Kokai Tokkyo Koho).
[189] M.F. Bianchetti, H.R. Canepa, S. Larrondo, and N.E. Walsel de Reca, *An Asoc.Quim.Argent.,* 84, 73 (1996).
[190] DE 3,913,966 (Nov. 09, 1989), R. Enomoto, M. Asai (Ger. offen).

CONTRIBUTORS

M. Anachkov
A. Andruszewska
V. Antonov
M. I. Artsis
P. Baraniecki
Wolfgang Behns
Carmen Behns
Wolfgang Behns
A. P. Bonartsev
G. A. Bonartseva
Matthias Börner
K. Bujnowicz
H. Burczyk
M. Byczyńska
V. V. Chernova
H. Fischer
Elmarie Fischer
Kristina Florschütz
K. Z. Gumargalieva
Hartmut Haida
Katja Heide
Stefan Heinrich
K. Heller
André Herrmann
A. L. Iordanskii
Kozłowska Jadwiga
S. O. Jekayinfa
Christian Kaiser
I. G. Kalinina
Połtowicz Katarzyna
S. V. Kolesov
Paul P. Kolodziejczyk
W. Koncewicz
Martina Körner
Mirjana Kostova
Jakub Kowaliński
R. Kozlowski
E. I. Kulish
Kirstin Kunath
Gotthard Kunze
S. K. Laga
N. Lukaszewska
A. Luwanska
G. Mankowska
Biskupski Marcin
J. Marek
Gromek Marta
Vera Miteva
Lothar Mörl
Vicky Mosch
G. Oleszak
G. Perez Sanchez
S. Rakovsky
Janis Rizhikovs
S. A. Semenov
M. Sc. Ling He
Sabine Schlüsselburg
Katarzyna Schmidt-Przewoźna
V. Scholz
Eyck Schotte
Andreas Schuster
Kirsten Simon
R. Słomski
Dieter Sonnenholzner
Baiba Spince
Wężyk Stanisław
M. Strybe
Kristin Suckau
M. Szalata
A. Talarek
Ausma Tardenaka
Sascha Thomas
Steffen Uhlig
S. V. Kolesov
M. J. A. van den Oever
V. P. Volodina
A. I. Wasif
Christoph Wieland
K. Wielgus
G. Yilmaz
G. E. Zaikov
Janis Zandersons

INDEX

A

B

D

E

F

G

H

I

J

K

L

M

P

Q

R

S